COUNTERINSURGENCY LAW

Counterinsurgency Law

NEW DIRECTIONS IN ASYMMETRIC WARFARE

William Banks

OXFORD
UNIVERSITY PRESS

OXFORD
UNIVERSITY PRESS

Oxford University Press is a department of the University of Oxford. It furthers the University's objective of excellence in research, scholarship, and education by publishing worldwide.

Oxford New York
Auckland Cape Town Dar es Salaam Hong Kong Karachi Kuala Lumpur Madrid Melbourne Mexico City Nairobi New Delhi Shanghai Taipei Toronto

With offices in
Argentina Austria Brazil Chile Czech Republic France Greece Guatemala Hungary Italy Japan Poland Portugal Singapore South Korea Switzerland Thailand Turkey Ukraine Vietnam

Published in the United States of America by
Oxford University Press
198 Madison Avenue, New York, NY 10016

Library of Congress Cataloging-in-Publication Data
Counterinsurgency law : new directions in asymmetric warfare/edited by William Banks.
p. cm.—(Terrorism and global justice series)
Includes bibliographical references and index.
ISBN 978-0-19-994144-5 (hardback)
1. War (International law) 2. Counterinsurgency 3. Asymmetric warfare
4. War—Protection of civilians 5. Guerrillas (International law) I. Banks, William C.
KZ6398.A87C68 2013
341.6'3—dc23

2012026924

Note to Readers
This publication is designed to provide accurate and authoritative information in regard to the subject matter covered. It is based upon sources believed to be accurate and reliable and is intended to be current as of the time it was written. It is sold with the understanding that the publisher is not engaged in rendering legal, accounting, or other professional services. If legal advice or other expert assistance is required, the services of a competent professional person should be sought. Also, to confirm that the information has not been affected or changed by recent developments, traditional legal research techniques should be used, including checking primary sources where appropriate.

(Based on the Declaration of Principles jointly adopted by a Committee of the American Bar Association and a Committee of Publishers and Associations.)

Counterinsurgency Law

NEW DIRECTIONS IN ASYMMETRIC WARFARE

William Banks

OXFORD
UNIVERSITY PRESS

OXFORD
UNIVERSITY PRESS

Oxford University Press is a department of the University of Oxford. It furthers the University's objective of excellence in research, scholarship, and education by publishing worldwide.

Oxford New York
Auckland Cape Town Dar es Salaam Hong Kong Karachi Kuala Lumpur Madrid
Melbourne Mexico City Nairobi New Delhi Shanghai Taipei Toronto

With offices in
Argentina Austria Brazil Chile Czech Republic France Greece Guatemala Hungary
Italy Japan Poland Portugal Singapore South Korea Switzerland Thailand
Turkey Ukraine Vietnam

Oxford is a registered trade mark of Oxford University Press in the UK and certain other countries.

Published in the United States of America by
Oxford University Press
198 Madison Avenue, New York, NY 10016

Library of Congress Cataloging-in-Publication Data

Counterinsurgency law : new directions in asymmetric warfare/edited by William Banks.
p. cm.—(Terrorism and global justice series)
Includes bibliographical references and index.
ISBN 978-0-19-994144-5 (hardback)
1. War (International law) 2. Counterinsurgency 3. Asymmetric warfare
4. War—Protection of civilians 5. Guerrillas (International law) I. Banks, William C.
KZ6398.A87C68 2013
341.6'3—dc23
2012026924

Note to Readers
This publication is designed to provide accurate and authoritative information in regard to the subject matter covered. It is based upon sources believed to be accurate and reliable and is intended to be current as of the time it was written. It is sold with the understanding that the publisher is not engaged in rendering legal, accounting, or other professional services. If legal advice or other expert assistance is required, the services of a competent professional person should be sought. Also, to confirm that the information has not been affected or changed by recent developments, traditional legal research techniques should be used, including checking primary sources where appropriate.

(Based on the Declaration of Principles jointly adopted by a Committee of the American Bar Association and a Committee of Publishers and Associations.)

Contents

Introduction
Shaping a Global Legal Framework for Counterinsurgency: Placing Postmodern War in Context

William Banks*

CICERO IS CREDITED with saying "*silent enim leges inter arma*"—in times of war the laws fall silent.[1] Cicero was correct, literally, but even in biblical times, when war was a quotidian occurrence, informal guidelines limited armed combat. Early Roman military codes recognized criminal offenses. Egypt agreed with neighboring states on standards for treating prisoners, and the Hindu Code from around 200 B.C. forbade some kinds of weapons.[2] Throughout human history, however, the law has been reactive, catching up episodically to provide legal guidelines based on lessons learned from the last war. Admittedly, the processes and institutions of lawmaking, whether international or national, are deliberative, prone to political sideshows that often derail the substantive agenda, and beset by procedural roadblocks. Yet law can and should embody our noblest objectives and serve to inspire the people to further its normative aims—in this instance, serving humanitarian goals of protecting civilians from the ravages of war.

There remains great skepticism that law matters in regulating warfare. Clausewitz opined that the laws of war are "almost imperceptible and hardly worth mentioning,"[3] and his dismissive attitude continues to have considerable resonance with those who use force with impunity around the globe. Still, states and their armies developed battlefield codes over time, and a customary law of war based on state practice emerged alongside the codes. By the mid-nineteenth century, the first significant U.S. code was written by

Francis Lieber for the eventually victorious Union forces. Lieber's *Instructions for the Government of Armies of the United States in the Field* (1863) essentially codified what was then the customary law of war for soldiers fighting on the battlefield.[4] The success of the Lieber Code is demonstrated by its adoption in similar forms in several countries in Europe, Russia, and Argentina. Around the same time, a first Geneva Convention was adopted in Geneva in 1864, providing ten articles of protection concerning care of the wounded. Twelve nations signed the Convention, and the United States joined the Convention in 1882. Meanwhile, the 1868 St. Petersburg Declaration Renouncing the Use in War of Certain Explosive Projectiles became the first multilateral treaty banning a particular weapon.

World wars and myriad other military conflicts large and small occurred throughout the twentieth century. In keeping with the pattern, however, the legal regimes for limiting the use of force on the battlefields—the Hague and Geneva Conventions and their Protocols, some specific multilateral treaties, and emerging customary law—continued to lag behind changes in armed conflict. To be sure, many significant advances in the laws of war, or what has come to be called international humanitarian law (IHL), were achieved over the last century, based on the understanding that wars are fought by states against other states, or in a minority of circumstances, by a revolutionary or insurgent group against a state.

By the turn of this century, some observers began to recognize, as did General Sir Rupert Smith, that instead of a "linear process" where "peace is understood to be an absence of war... we are in a world of constant confrontation." In today's wars, Smith opined, civilians "are part of the terrain of your battlefield... [and] war is directed against non-combatants."[5] Speaking in 2006, General Smith took note of the most recent half-decade, and he presciently predicted the nature of future wars.

I. MODERNIZING IHL FOR POSTMODERN WAR

The September 11 attacks spurred the United States and its allies toward an extraordinary decade of military actions against al Qaeda, the Taliban, and then Saddam Hussein and his Iraqi regime. What the George W. Bush administration called the "Global War on Terror" routed the Taliban in Afghanistan and then attacked al Qaeda and Taliban terrorists in the Afghan/Pakistani border areas. The Iraq war was launched by the United States in 2003, despite the unwillingness of the U.N. Security Council to authorize military action. Before long, what began as a counterterrorist operation in Afghanistan and a conventional invasion in Iraq changed, dramatically.

Over the next few years, insurgencies flared in both countries, in different ways and from uniquely local circumstances, but both enmeshed in urban areas and local populations. Similarly, nonstate terrorist groups were carrying out attacks in the Middle East. In 2006, Hezbollah launched cross-border raids from Lebanon and fired rockets into Israel, killed civilians, and kidnapped Israeli soldiers. When the Israel Defense Forces (IDF) countered and attempted to root out and neutralize the Hezbollah fighters, attacks against them in civilian neighborhoods and in urban areas brought widespread criticism from human rights groups and the International Committee of the Red Cross (ICRC). These conflicts brought renewed attention to asymmetric warfare—where conventionally weaker nonstate enemies fight the stronger foe in ways that neutralize the conventional strengths and exploit apparent weaknesses—such as the willingness to engage in unlawful attacks with prohibited means, knowing that the dominant state will abide by the laws of war and not reciprocate.

This volume presents a series of insights from an interdisciplinary initiative at Syracuse University's Institute for National Security and Counterterrorism (INSCT), a research institute jointly sponsored by the Maxwell School of Citizenship and Public Affairs and the College of Law at Syracuse University, in partnership with the International Institute for Counter-Terrorism (ICT) at the Interdisciplinary Center (IDC) in Herzliya, Israel. Beginning in 2007, scholars, government experts, and practitioners from humanitarian, military, and other nongovernmental organizations (NGOs) have been working collaboratively to debate whether the post–9/11 security environment and the challenges posed by asymmetric conflicts require adaptations to the traditional laws and customs of war. The fruits of this dialogue have so far been encapsulated in our first book: *New Battlefields/Old Laws: Critical Debates on Asymmetric Warfare*, published in 2011 by Columbia University Press. This book, our second, addresses future warfare, particular counterinsurgency (COIN) and its variants, and considers whether a more forward-looking legal framework might provide useful normative guidance for this era of postmodern warfare.

In using the term "postmodern," I embrace John Kiszely's characterization of postmodern as warfare that is not new, but different—involving "challenges that are not primarily overcome with the tools of modernity: more advanced technology, firepower, lethality, speed, stealth, digitization, logistics, network-centric warfare or hi-tech 'shock and awe.'"[6] Postmodern warfare is anything but linear, and the distinction between war and peace is not clear. Enemies are not easily identifiable, and their identity may change frequently. Victory is achieved not only on the battlefield but also by winning public support, and the battlefield itself is not so much a military one as political, based on a battle of ideas and the support of the local population. At the same time, how the postmodern war is fought—the humanitarian quality of its campaigns—is crucially important in winning the local population. Modern labels are insufficient and sometimes plainly wrong—peacekeeping, peace enforcement, counterrevolutionary operations, and humanitarian operations. Even COIN has come to represent "a kaleidoscope of different types of operation, remarkably resistant to neatness in delineation."[7]

Modern IHL does little to accommodate the asymmetric form. After the horrors of World War II, the nations of the world agreed that the Common Articles codified in the revised 1949 Geneva Conventions constituted the exclusive threshold criteria for triggering IHL. Under Geneva, there are two kinds of wars: interstate (or international) armed conflicts and intrastate (or internal, noninternational) armed conflicts. The former invoke the full panoply of the laws of war, which in turn regulate the conduct of war (through the principles of distinction, proportionality, and military necessity); the latter do not trigger all the regulations for the conduct of war but provide limited humanitarian protections for civilians and those captured or detained. Over time, state practices and customary law have extended some of the IHL principles to irregular warfare.

The Geneva provisions mostly fail to anticipate wars involving nonstate actors, first, by providing a regulatory scheme designed for wars between states, except for minimal protections for those involved in noninternational armed conflicts that may include nonstate belligerents. Second, the Geneva criteria for earning the status of a lawful combatant are defined on the model of the state soldier. Requirements include a responsible command structure, wearing a fixed insignia recognizable at a distance, carrying arms openly, and conducting operations in accord with international humanitarian law. Only persons meeting these criteria gain full prisoner of war (POW) protections under the Conventions. It is virtually impossible for nonstate actors to meet these criteria and, thus, to become lawful combatants under humanitarian law. Because humanitarian law includes a "combatant privilege," a form of legal immunity for acts that would be criminal if performed during peacetime, nonstate actors may neither engage in lawful combat nor be its deliberate target. Although there are clearly downside risks to conferring combatant status on nonstate insurgents, the either/or orientation of the Geneva Conventions leaves little room for consideration of the dynamic qualities of asymmetric war.

As a further illustration of the state-centric nature of IHL, to invoke IHL and its humanitarian provisions, there must be an armed conflict within the meaning of the Geneva Conventions—that is, either an international (between two or more states) or noninternational (civil war or insurgency) armed conflict. Although the Conventions do not define armed conflict, knowing whether the threshold requirement has been met has not been especially difficult to apply in state versus state conflicts. What constitutes armed conflict—the use of armed forces by opposing regular armed forces—is fairly clear. If the armed conflict threshold is not met, then the humanitarian bulwark of the Geneva Conventions may not apply.

Determining the existence of a noninternational (or internal) armed conflict has, by contrast, not been easy. Many violent internal conflicts are civil disturbances that are amenable to and may be managed locally by domestic law and its arrangements. The ICRC maintains that only if an internal conflict involves several criteria—identification of parties, protracted hostile acts, minimum level of organization of armed forces, for instance—do the laws of war clearly apply.[8] Yet the U.S. Supreme Court decided in 2006 that the U.S. conflict with al Qaeda is a noninternational armed conflict, at least for the purposes of affording fair trial procedures to detainees accused of war crimes. In a similar vein, the U.S. government treats its ongoing conflict with al Qaeda, the Taliban, and associated forces as an armed conflict subject to IHL principles.[9] However, IHL is not generally applicable to a war against "terrorism," since there is no state enemy; nor is it necessarily applicable to a conflict that involves a state and a transnational terrorist network, if interventions do not meet the initial armed conflict threshold requirements. At the same time, many asymmetric conflicts are not "wars" either as defined under IHL—there is no state versus state or internal conflict. In these cases, the IHL threshold criteria present definitional conundrums in addition to practical limitations.

Facing a postmodern set of insurgencies, states have been compromised in setting strategies and choosing tactics in part because of the shortcomings in the legal structure to govern these conflicts. While the legal regime for countering terrorists has its own set of limitations and unanswered questions, countering insurgencies reminded the United States and its allies that kinetic warfare has limited utility in counterinsurgency. COIN is one variety of asymmetric warfare; the insurgent is taking advantage of the weaknesses of the stronger foe. Based on so-called "soft power" and a win-the-population objective, COIN includes armed conflict and warfare, but its elements require careful calibration and refinement of issues not traditionally at the center of IHL, including modifications of the principles of distinction and proportionality.

In addition to its blending and mixing of kinetic warfare and soft-power operations, COIN takes place in a variety of conflict situations. Depending on the setting, COIN occurs:

- where there is/is not an armed conflict of an international/noninternational nature;
- where human rights law (HRL) applies in the absence of or as an addition to IHL;
- in the face of customary or treaty-based principles of international law; and
- alongside domestic laws of the nation managing the COIN operation, which may or may not have incorporated international laws as part of the domestic law.

Although actual operations continue to evolve in COIN as it is practiced around the globe, scholars and practitioners have not critically and carefully examined the legal bases for COIN. In COIN operations, IHL, HRL, international law, and domestic laws do not provide adequate legal and policy guidance for a variety of reasons, some of them to be explored in this book. Meanwhile, in the absence of a legal framework that effectively regulates COIN, commanders and legal advisers have promulgated various operational rules that supply rough guidance for soldiers.

As a result, in the context of COIN strategy, operational rules have become more protective of civilians than the law. The authors of the chapters that follow will explore these overarching questions:

1) Is it prudent to reshape the overlapping and converging legal paradigms that conceivably apply in COIN and related forms of postmodern warfare into a discrete legal framework? Are the convergences damaging and, if so, how may the damage be minimized, avoided, or repaired? Do the overlaps threaten

the integrity of any one legal framework, and, if so, how may the threat be managed?

2) If operational rules exceed legal requirements in favor of a quest for legitimacy in a particular theater of operations, what are the implications to these operations of the loss of legal control over postmodern war? To the extent that the rules for COIN or other forms of postmodern war are fashioned by nonstate entities (such as nongovernmental organizations or private contractors) do states risk losing sovereign control over the postmodern battlefields for which they are held responsible? If so, what are the implications?

Reflecting the realities of postmodern warfare, COIN operations have been characterized as a subset of what the U.S. military refers to as stability operations. In addition to COIN, stability operations include providing humanitarian disaster relief, peacetime support to other nations, peace building and peace enforcement, counterterrorism operations, counterdrug operations, noncombatant evacuation operations, mass atrocity responses, and postconflict reconstruction operations. Despite recently promulgated and revised military doctrine in the United States, the law of COIN remains undefined.

Just as the use of the term "war" became problematic when wars were basically outlawed after World War II, the use of "armed conflict" to modernize the legal language has not helped set the parameters for COIN or other forms of postmodern war. Sometimes COIN goes on where there is "armed conflict" ("armed force between states or protracted armed violence between governmental authorities and organized armed groups") and sometimes where there is not a conflict as IHL defines it.[10] Where there is not an armed conflict, does human rights law control the battle space instead? Are the operations a part of policing, using law enforcement methods only? If law enforcement and military operations against insurgents go on at the same time, what is their interface? In 2009, in one of his occasional guidance memoranda to the NATO International Security Assistance Force (ISAF) and U.S. forces in Afghanistan, General David Petraeus admonished the force simultaneously to "pursue the enemy relentlessly," while living "among the people," to "fight hard" and "be a good guest," and to "hold what we secure" while we "live our values."[11] Imagine the brigade Judge Advocate General (JAG) officers fashioning their operational orders based on that guidance memorandum. What legal doctrine applies? Imagine the enlisted infantry private reading the Petraeus memo. What is his takeaway?

Throughout modern history it has been understood that insurgencies are not defeated on the battlefield, not exclusively anyway. COIN works only when the local population supports it. As practiced now in Afghanistan, COIN imposes operational restrictions on what military personnel may do in their COIN operations that go well beyond what IHL permits or requires. Instead of the traditional doctrine of proportionality—where only "excessive" harm to civilians is forbidden in relation to the military gain—COIN doctrine disfavors any harm to civilians. Similarly, the principle of distinction—requiring military forces to target only those who can be identified as combatants or civilians taking an active part in hostilities—is ratcheted up in the Afghan COIN operation to the extent that no forces were "declared hostile" in the operational environment, requiring fresh determination of combatant status at each confrontation.[12]

As the evolution of the operational controls in U.S. COIN operation in Afghanistan demonstrates, COIN is not a strategy, but a set of tactics that operate as means to some strategic end, undefined in the case of Afghanistan. President Obama stressed early on that the aim of the war in Afghanistan was counterterrorism, "to disrupt, dismantle and defeat al Qaeda." General Stanley A. McChrystal echoed the counterterrorism objectives in the summer of 2009, just before his removal. The political context in which this COIN operation goes on is complex, dynamic, and not always clear. In Afghanistan, the political context for COIN is perhaps the most important part—it would be possible to defeat the Taliban and other insurgents there and still lose the war.

II. OUR COLLECTIVE THESIS

A basic premise of our book is that we live in an era of postmodern warfare. If modern war was interstate and typically large scale between professional state armies, postmodern warfare discounts the value of overwhelming force, traditional battlefield tactics, and even technological advances, and substitutes equally lethal, smaller scale, urban-based "war among the people."[13] Postmodern war can be just as significant strategically as modern war, and its effects on a political order just as profound. Insurgents, after all, reject the existing political order.[14]

Law and strategy, like law and tactics, are necessarily connected. Law imposes limits on operations, but legal authority is also necessary to enable the use of force in any kind of war. Positive legal authority is precisely what gives war its legitimacy. Legal legitimacy is as central to postmodern war as any form of war, and we argue that the legal framework that has evolved for modern warfare is not sufficient for legitimizing postmodern war.

The postmodern iterations of COIN have many of the same characteristics as Western responses generally to asymmetric warfare challenges since 9/11—operations are sometimes effective, sometimes not; they evolve and improve, but they lack a strategic purpose and normative legitimacy. The outdated legal framework for conventional armed conflicts has compromised our settling on strategy and its core normative elements. As a result, for example, at the same time that the U.S. military attempts to negotiate with the Taliban, the State Department shows its human rights orientation and looks askance at negotiating with any group that disfavors education for girls and widened roles in society for women.[15]

There are gaps, lacunae, and needs for clarification in deciding what law applies in postmodern warfare.[16] Postmodern warfare will necessarily include components of policing and law enforcement, tasks and methods not conventionally part of military training or orientation. Even if these obstacles are overcome, performing military operations against insurgents at the same time as policing the population involve complex judgments and actions and careful attention to the legal parameters of the actions.

The tendencies to improvise and do what works practically permeate COIN, for good reason. The *U.S. Army Field Manual*, FM 3-24 (*Counterinsurgency Manual*), published in 2006 and updated regularly since then, became an especially high-profile document, and its "win-the-population" mantra dominated COIN in Afghanistan, after cutting its operational teeth in Iraq.[17] Taking lessons from the history of COIN successes and failures, including the U.S. adaptations in the Philippines, British success in Malaya, French failures and successes in Algeria, U.S. failures in Vietnam, and Sri Lanka's eventual defeat of the Tamil Tiger insurgency, modern U.S. COIN doctrine owes its doctrinal heritage to the cumulative lessons learned and from several classical and contemporary theorists, including David Galula, whose *Counterinsurgency Warfare*, based on his experiences as a French Army officer who fought in Algeria, has had considerable influence.[18]

Now, more than a half-decade since the publication of the *Counterinsurgency Manual* 4, the popularity of COIN ascended after apparent successes in Iraq and then descended after confronting the challenges of the Afghan/Pakistani insurgency. COIN arguably has failed in Afghanistan for lack of a strategy. The objectives of taking down al Qaeda and preventing their return is a counterterrorism plan based on building a stable state in Afghanistan. But the COIN precepts of securing the population, building governance institutions, and developing legitimacy of the local government are themselves not tied to an overarching strategy and have not been prioritized in the Afghan campaign.[19] As the Afghan war drags on and its costs mount, COIN has been treated like a fad. Still, dismissing COIN in Iraq and Afghanistan as having outlived its usefulness is shortsighted, at least as concerns the lessons that COIN can provide for the legal shortcomings of waging postmodern war. Even if the current unpopularity of the COIN operation in Afghanistan leads to abandoning the term, we should build the legal framework and continue to utilize the COIN tools when appropriate in future conflicts. Change the language, perhaps, but develop and maintain the legal framework. This volume serves as an opportunity to step back from the day-to-day and reflect on the challenges of shaping a legal framework for COIN and other forms of postmodern asymmetric warfare.[20]

We should also work toward a legal framework for COIN because COIN will likely continue to have value in postmodern conflicts. Rapid changes in insurgencies—including external support, virtual audiences, and global objectives—continue to shape the future of COIN.[21] There is no one-size-fits-all COIN manual, and flexibility and adaptability are likely essential to successes in fighting the postmodern wars, whatever they are called.[22] COIN is capable of nuance, works under unified command, and requires civilian intelligence support. COIN can isolate insurgents from the population, use minimum levels of force necessary to achieve established objectives, and assure legitimacy of the COIN effort among the local population. History shows that COIN has value when conventional military tactics are not sufficient to achieve security objectives.

Future wars will be irregular, whether COIN or counterterrorism or something else. They will be a blend, a complex mix that will continue to change over time. COIN can provide principles and standards, sufficiently supported by law that will be useful when our soldiers are asked to stabilize a conflict zone or intervene in mass atrocities in the future. The main thesis of the volume thus should be broadly construed. Whether labeled as COIN or some other form of postmodern warfare, existing legal frameworks are not sufficiently nuanced or nimble enough to accommodate postmodern conflict forms. COIN is the emblematic case study for the authors of this book because COIN illustrates the insufficiencies of the conventional modern IHL legal regime and provides a broad context to engage the legal parameters of today's operational challenges, including peacekeeping and postconflict reconstruction, for example. We also believe that strengthening the foundational legal framework for conducting irregular postmodern warfare is an essential step in establishing normative principles that will go a long way in aiding in the development of strategic justification for postmodern wars.

III. DEFINITIONS

According to the U.S. Department of Defense, COIN is a subset of "stability operations" defined as "military and civilian activities conducted across the spectrum from peace to conflict to establish or maintain order in States and regions." The goal of stability operations is "to provide the local populace with security, restore essential services, and meet humanitarian needs."[23] The Defense Department defines counterinsurgency as "comprehensive civilian and military efforts taken to defeat an insurgency and to address any core grievances."[24] An insurgency is "the organized use of subversion and violence by a group or movement that seeks to overthrow or force change of a governing authority."[25] Insurgencies are distinct from traditional war in important ways. First, insurgencies are by their nature asymmetric. Insurgents lack the resources in military, political, and economic means and governance infrastructure that the COIN operations enjoy. Yet insurgents' lack of infrastructure and governance responsibility means that they may fight anywhere and everywhere, at any time. The battle is not about particular territory. Moreover, insurgents are not seeking to defeat the government militarily. Often the struggle is over who governs, not who fights best.[26]

Second, one of the most basic objectives in COIN is to establish an effective rule of law in host-nation institutions. Although based in law, the rebuilding of local institutions occurs not by lawyers, but by military personnel and contractors, in part through forceful suppression of insurgents. Building host-nation institutions according to law is, of course, part of developing the legitimacy of the local government, a critical ingredient for success in COIN. Thus, complying with law in postmodern/COIN military operations is integral to the success of the operation both to defeat an insurgency and establish the foundation for long-term stability.[27]

The next section of this introduction briefly places our thesis in historical context. Insurgencies have occurred throughout human history, of course, and so has COIN. Lessons from both inform postmodern COIN, though we believe that the inadequate legal framework for COIN in today's conflicts presents novel challenges for the United States and the international community. Today's insurgents may have global impact, in part because of ease of access to media and communications and the 24/7 news cycle. In addition, cyberspace itself has become a battle space, where asymmetric warfare proliferates. At the same time, the United States is only one of many states that have outsourced significant security functions to private companies, and those contractors are not accountable directly to the laws that bind state soldiers.

IV. COIN IN HISTORICAL AND DOCTRINAL CONTEXT

The term "counterinsurgency" (COIN) was invented during the response to the wars of national liberation from the end of World War II through the 1970s. Theorists as diverse as Mao Zedong, Che Guevara, David Galula, and Bernard Fall influenced this classical COIN theory.[28] Classical COIN presumes an insurgency challenging a functioning, albeit usually weakened, state. The COIN force supports the state and thus defeats the insurgency. COIN operations in Thailand, Colombia, and Sri Lanka generally followed the classical model. Even before the postmodern era, however, insurgencies developed after state failure sought not to challenge or replace a state, but to further destabilize a region or gain some control over ungoverned spaces. Examples include Chechnya, Somalia, and East Timor.[29] In Afghanistan, of course, the insurgency existed before the modern government was in place.

Postmodern insurgencies may be reactive. Rather than starting a revolutionary war, insurgents may simply resist the revolutionary change brought by outside forces. Consider Waziristan—where in addition to al Qaeda and Taliban fighters, the United States and its allies fight against local tribesmen who fight to preserve their traditional control over the area.[30] Countering this kind of postmodern insurgency requires separating local fighters from the outside insurgents and dealing with the transnational effects of globalized conflict, including their electronically generated worldwide audience.

Postmodern insurgencies are characterized in part by the impact of modern communications.[31] The battlefield may be dramatized and brought to life almost anywhere, in short order, for a worldwide audience. What would have been modest tactical operations in the past can now have strategic and almost immediate impact. Public perception of compliance with law, for example, will continue to have significant impacts on COIN and its postmodern variants, but postmodern communications make the legal environment subject to lawfare manipulation (discussed below) and to unpredictable change.

Much of postmodern COIN takes place in urban areas, where the insurgent hides among the civilian population. Military engagements are short and fleeting, and innocent bystanders are omnipresent. Insurgents use propaganda to scare the population and

fuel resistance to COIN forces. If COIN forces attempt to segregate the population from insurgents, they are criticized for disrupting neighborhoods and for their unlawful actions. The success of a postmodern COIN campaign is thus at least as dependent on popular support, inside and beyond the battlefield to a larger area of influence.

The point of departure for postmodern COIN is the *U.S. Army/Marine Corps Counterinsurgency Field Manual* (*Counterinsurgency Manual*) and its companion, the *U.S. Army Stability Operations Field Manual*,[32] published in 2006 in the midst of the wars in Iraq and Afghanistan. Both of these insurgencies previously would have been labeled as noninternational armed conflicts (NIACs) in IHL terms. As IHL experts know, however, there is little legal doctrine that governs NIAC, at least not in the form of agreed upon internationally enforceable norms. Although the U.S. Army/Marine Corps had internal doctrine for various forms of nontraditional warfare developed in modern times—operations other than war (OOTW), peacekeeping, peace enforcement, low-intensity conflict—the COIN Manual broke new ground in recognizing the uniqueness of each insurgency, noting the patterns that likely occur in postmodern insurgencies, and expressly downplaying the use of military force. The Manual states that COIN takes place "in a complex civilian environment... [where the] war, is in large part a war of ideas, the battle largely for one of perception."[33]

The postmodern environment is murky, no doubt. The COIN Manual refers to the "mosaic war" of COIN. In other words, doctrine shifts, as does the legal framework. The Manual suggests, for example, that force protection, one of the mainstays of modern doctrine and consonant with IHL, may make the force less secure; that "soft power" may be more effective than traditional weapons; that overwhelming force may be less effective and more risks to the force may need to be accepted.[34]

Postmodern COIN is a strikingly different approach than the modern kill-or-capture approach to warfare.[35] The objective is to win the population. Some insurgents will have to be killed or captured, but the focus of COIN is on building support for the local government. Nor is COIN focused only on military operations. It includes various reconstruction tasks, in support of winning the population. As the COIN Manual states, COIN operations focus on the security of the population; ensuring essential services; establishing governance structures; developing the economy and infrastructure; and communicating with the people.[36] Of course, ensuring civil security may require combat operations against insurgents, but those maneuvers, typically fought in close quarters inside an urban area, must avoid harming the population that the COIN force is trying to win over.

V. SHAPING A LEGAL FRAMEWORK

Until recently, law was only marginally involved in serious discussions of COIN theory and practice. Classical counterintelligence theory focused on tactics and the contexts in which COIN is pursued. The background law was IHL, embellished in some circumstances by customary international law and domestic laws of states engaged in COIN operations. Particularly over the last decade, however, legal concerns with COIN and irregular warfare with nonstate entities generally have manifested themselves in two distinct ways. First, adversaries have used traditional IHL constraints as a tactical weapon against COIN forces—"lawfare"—where alleged shortcomings in states' obligations to comply with legal limits on the use of military force are showcased by adversaries to weaken the COIN operations. Others have asserted that COIN forces violate HRL in their operations, and similarly use the supposed violations as lawfare. Second, considerable attention has been paid in debating the legal contours for postmodern warfare to the choice of a legal framework for military operations—is it criminal law or international human rights law, treating insurgents or terrorists as criminals, subject to arrest, prosecution, and imprisonment? Or is it IHL, where the insurgents or terrorists are combatants, amenable to targeting, subject only to the *jus en bello* principles of distinction, proportionality, and military necessity?

The authors of this volume believe that lawfare is an inevitable tactic in postmodern warfare and that the best response to it is to create and embrace a workable legal framework for COIN and other forms of postmodern war. Lawfare cannot be removed from the global lens on military operations against insurgents, but its effectiveness can be blunted significantly by adapting IHL and other laws to better fit the circumstances of postmodern war. Regarding the supposed choice between criminal law enforcement and IHL to govern the conduct of postmodern warfare, we reject the either/or dichotomy altogether. Instead, the authors offer a range of alternative configurations for how law might be tempered, adjusted, reshaped, and even created anew to provide a coherent and practically useful framework for COIN and other postmodern warfare. At the same time, COIN embraces criminal law processes as integral to defeating insurgents or other irregular fighters. Arrest, detention, and prosecution are always preferred over targeting and the kill/capture option. Indeed, building the host state criminal law institutions is a complement to the use of criminal law in defeating the insurgency and an adjunct to building the legitimacy of the government. When the use of lethal force is required in postmodern warfare, compliance with law harmonizes with the government's legitimacy and distinguishes the COIN forces from the insurgents.

The chapters in this book will show that in important respects, a COIN or other postmodern military operation confronts operational pitfalls that will benefit from rethinking traditional IHL doctrine. Consider targeting. Under IHL, the combatant's privilege leaves civilians harmed as a result of targeting with no protection and no legal recourse. In COIN, it is important to consider whether a military operation will create more insurgents by virtue of its being conducted than it will eliminate if carried out. COIN operators must simultaneously defeat an insurgency while protecting the civilian population and avoid furthering the insurgent cause. The IHL principle of distinction requires considering whether a target may be struck in view of the presence of civilian bystanders, but it does not take into account the broader context of the targeting decision. The sort of cost/benefit calculation that goes on in COIN operations is not part of IHL. One approach may be to leave IHL as it is, with its binary focus, to govern all uses of military force during armed conflict. The COIN trade-offs could be left to operational guidance, or policy. In our view, COIN and other forms of postmodern warfare will be more successful and more legitimate in the eyes of the affected populations if the requirements are memorialized in law.

A second example of an IHL standard in need of framing for postmodern warfare explored in this book is the principle of proportionality. Protocol I of the Geneva Conventions forbids indiscriminate attacks, defined as "an attack which may be expected to cause incidental loss of civilian life, injury to civilians, damage to civilian objects, or a combination thereof, which would be excessive in relation to the concrete and direct military advantage anticipated."[37] Focusing on "military advantage" may be an insufficient marker in postmodern warfare. In insurgencies, those in noncombat roles may be as integral to the success of the insurgents' cause as the fighters. Those who provide shelter, cook food, deliver messages, and the like create for the COIN force a dilemma, where the need to separate insurgents from the population requires dealing with civilians using military force. Is a civilian Internet service provider (ISP) or radio station that spreads the insurgents' propaganda directly participating in hostilities under IHL? Probably not. Are its personnel subject to targeting in COIN? To arrest and detention? What discretion should the COIN force have in making such determinations? To what extent should the chance that targeting or arresting the ISP or radio station owner will inflame the insurgency factor into making the operational decisions in COIN?

The realities of postmodern warfare have already been noted by the United States. The COIN Manual explains that "in COIN operations, [military] advantage is best calculated not in terms of how many insurgents are killed or detained,... [instead] the number of civilian lives lost and property destroyed needs to be measured against how much harm the targeted insurgent could do if allowed to escape."[38] Moreover, COIN operations "that cause unnecessary harm or death to noncombatants may create more resistance and increase the insurgency's appeal—especially if the populace perceives a lack of discrimination in their use."[39] The COIN Manual focuses on the identity of the individual insurgent, the future potential harm that insurgent may cause, and attempts to measure those subjective factors against potential civilian losses, including civilian support for the insurgency. Such fine calibration is not found in traditional IHL. Postmodern warfare may demand that self-interest on the part of COIN operators may be influenced by humanitarian concerns. In the context of COIN, as Sarah Sewell has written, "killing the civilian is no longer just collateral damage.... Civilian casualties tangibly undermine the counterinsurgent's goals."[40]

The U.S. COIN Manual states that "an operation that kills five insurgents is counterproductive if collateral damage leads to the recruitment of fifty more insurgents."[41] United States forces may thus be obligated to assume greater risk in order to minimize loss of civilian lives. The COIN proportionality principle may thus be stricter than in

IHL, and postmodern warfare calls for a vigorous effort to flesh out just what the evolving standards for proportionality or its replacement look like.

The trend in postmodern warfare in the last decade has been for these adaptations to traditional IHL to take the form of operational law, or policy add-ons to IHL. Indeed, operational law has outpaced treaty-based legal developments and case law. It is not surprising that commanders in the field and their legal advisers would be the first to come to grips with the need for a reshaped legal paradigm in combating postmodern warfare. Nor has it been unhealthy for the reshaping to take the form of operational law standards. Going forward, however, nation states should assume full accountability for reshaping the framework for conducting postmodern warfare and adopt a new set of principles and guidelines, through states' political processes and at their highest levels. Fortunately, as the authors of this volume demonstrate, the reshaping does not involve the rejection of IHL or criminal law, or their counterterrorism doctrines. The new standards harmonize with the old, embellish them for new settings, and such reshaping may emanate from domestic and/or international institutions and processes.

VI. LOOKING AHEAD

Postmodern warfare will be with us for the foreseeable future, in one form or another. The fact that COIN in Afghanistan has lost its luster, and that the United States and other Western states are presently disinclined to invest in protracted conflicts abroad does nothing to blunt the inevitability of continuing insurgencies, ethnic and religious conflicts, wars over energy and other natural resources, and the need for military support in humanitarian responses. The contexts will vary, and the legal framework for postmodern wars will also be contingent on context. Yet the agreement in principle to a framework for COIN and other forms of postmodern war will be an important step for those seeking to protect innocents during wartime.[42]

A. Roadmap for the Book

The four parts of our book that follow offer a range of legal and policy perspectives on the problems of COIN in particular and irregular warfare in general as twenty-first-century asymmetric warfare continues to evolve. The contributors offer analyses and prescriptions that are complementary in some instances and widely divergent in others. Their work is unified by the excellent scholarship and lucid insights.

Part I highlights a recent trend in irregular warfare—the tendency for the regulatory paradigms in armed conflict to converge, particularly IHL and HRL. In Chapter 1, Daphné Richemond-Barak assesses the role of the U.N. Human Rights Council in addressing violations of HRL, its original mandate, and IHL, an adopted task. Richemond-Barak worries that convergence of HRL and IHL in this instance could lead to misapplications of IHL and dilution of the respect shown to the Human Rights Council for its mandated HRL work. In Chapter 2, Evan Criddle looks closely at the potential for convergence of IHL and HRL in COIN, focusing on the traditional IHL principle of proportionality. Criddle develops and applies a relational theory of *lex specialis* to distinguish when IHL or HRL doctrines should apply to set the parameters for what is proportionate in COIN.

Two former U.S. Army JAGs frame the thesis in Part II—that COIN operations afford an opportunity to reunify IHL. Eric Jensen argues in Chapter 3 that the traditional IHL dichotomy between international and noninternational armed conflict causes confusion and, at times, a failure to apply core IHL protections to armed conflict. Jensen would apply the full range of IHL principles to all forceful activities by state sovereign forces. In Chapter 4, Geoffrey Corn accepts the noninternational armed conflict categorization as a starting place for his examination of who may be targeted in unconventional conflicts with nonstate actors. Corn criticizes an ICRC study on civilians directly participating in hostilities and the study's application of a continuous combat function test to identify civilians that may be targeted. Corn argues that the better point of departure is to treat all noninternational armed conflicts as involving hostilities between opposing armed groups whose members are military objectives.

Part III focuses on the tensions that exist between protecting civilians and risking harm to the state force in COIN and other forms of irregular warfare. In Chapter 5, Peter Margulies takes a fresh look at the supposed duty to risk state forces when, for example, a state chooses to use air power to minimize risk to its own troops. Reviewing IHL and military ethics, Margulies argues against those who assert that a state has a categorical duty to risk its forces even when IHL would not so require according to the principle of proportionality. Alternatively, Margulies maintains that traditional IHL principles may be most fair to civilians and soldiers and that duties to the state that exceed IHL requirements may blunt the decisiveness commanders need. In Chapter 6, Chris Jenks focuses on the ongoing COIN operation in Afghanistan and explores the allocation of risk between the military force and the Afghan civilian population. Jenks prompts further discussion of whether the U.S. COIN doctrine emphasis on risk tolerance by the force works as theorized in practice. Gregory McNeal also examines civilian casualty mitigation in U.S. COIN operations in Chapter 7. He evaluates U.S. practice and theoretical and practical lessons learned while placing the concerns about mitigating civilian losses in their larger geopolitical context.

Finally, Part IV addresses an assortment of new legal and policy problems that arise on unconventional battlefields. In Chapter 8, Boaz Ganor revisits the IHL problems that accompany conflicts with nonstate actors. He points out that the protections for civilians under IHL are exploited by insurgents and terrorists, where military commanders may thus be compromised in conducting their operations. In response, Ganor proposes a new international legal framework, "multidimensional warfare," where he defines four categories of actors involved in warfare, each of which has a different status in the conflict. In Chapter 9, Robert Chesney offers a compelling case study of the 2011 targeted killing by the United States of Anwar al-Awlaki in Yemen. Chesney puts aside the IHL, HRL, and U.S. constitutional questions associated with the Awlaki operation and focuses instead on the U.N. Charter and its rules governing the use of armed force in international affairs. Chesney concludes that consent by the Yemeni government or Article 51 of the U.N. Charter provided adequate authority for the United States to target Awlaki. In Chapter 10, Corri Zoli explores the Gaza flotilla incident in 2010 to further illustrate the ways in which asymmetric warfare challenges traditional IHL. Zoli explores the politicization of the incident, and she shows how the legal and policy shortcomings of postmodern warfare are especially attenuated when conflicts take place in a maritime environment.

War remains a complex political phenomenon, and wars will include conventional and postmodern challenges in an ever-changing environment. Postmodern wars will likely continue to be "wars amongst the people," as described in the COIN Manual. Rather than leaving the legal rules of the game for military officials to set as they choose in each future conflict, policy makers, scholars, and practitioners should work to shape a legal framework for COIN and other forms of postmodern warfare, based on careful reflection, lessons learned, and strategic considerations. The law should be out ahead of the evolution of postmodern warfare, because the legal framework can itself generate the norms for the conflict. Law can shape strategy, and can provide guidance to the armed forces that will be sorely needed in these challenging conflict environments. We believe that the contributions in this volume will illuminate the role of law in the wars of the future.

1

Converging Paradigms in Contemporary Armed Conflict: Framing Problems of COIN

1

THE HUMAN RIGHTS COUNCIL AND THE CONVERGENCE OF HUMANITARIAN LAW AND HUMAN RIGHTS LAW

Dr. Daphné Richemond-Barak*

Introduction

In the final stages of the Sri Lankan Civil War in 2009, Human Rights Watch publicly criticized the Human Rights Council ("the Council") for failing "to address serious allegations of violations of human rights and humanitarian law by government forces, focusing only on the abuses committed by the LTTE [Liberation Tigers of Tamul Eelam]."[1] Underlying this criticism was the assumption that the Council can and ought to address violations of international humanitarian law (IHL)—in addition to violations of human rights.

Viewed within the context of the Council's creation in 2006, this criticism comes as somewhat of a surprise. The Council was founded as a replacement body to the much-discredited U.N. Commission on Human Rights (the Commission), which had become, in the Secretary General's words, selective and politicized.[2] As the first U.N. body to have undergone a full-fledged reform, the newly created Human Rights Council was granted a new mandate, a new membership, and renewed legitimacy to address issues of human rights. It was well known at the time of the Council's creation that its predecessor, the Commission on Human Rights, had addressed issues of international humanitarian law. And yet, no mention of either IHL or armed conflict appears in U.N. General Assembly Resolution 60/251 establishing the new Council.

The absence of any reference to IHL in the resolution establishing the Council—coupled with the widely held assumption that the Council should address such matters (as in the case of Sri Lanka)—raises important questions about the Council's mandate, the line separating human rights and IHL, and the role of human rights bodies in the context of armed conflict.

Both the treatment of IHL by human rights bodies and the flaws of the Council have been debated at some length. This chapter's contribution lies in its analysis of the Council's involvement in IHL as a manifestation of the convergence of human rights law and IHL. Moreover, this chapter is the first academic piece which questions the mandate and expertise of the Council to deal with IHL while building on state practice.

This chapter begins by examining the circumstances of the Council's creation (section I) and its recent forays into international humanitarian law (section II). I then compare the treatment of IHL by the Council and by other human rights bodies. Unlike the Council, other human rights bodies facing similar challenges have not encroached so directly on the territory of IHL: they have generally been reluctant to address humanitarian law head-on, and have felt compelled to justify any interpretation or application of IHL (section III). While I am mindful of the growing convergence of IHL and human rights law, I argue in section IV that the Council has neither the mandate nor the expertise necessary to act as enforcer of IHL. Finally, I envisage the consequences of the blurring of the IHL/human rights divide by the Council (section V).

I. FROM THE COMMISSION ON HUMAN RIGHTS TO THE HUMAN RIGHTS COUNCIL

The Commission on Human Rights was established on February 16, 1946, by Resolution E/20 at the first meeting of the U.N.'s Economic and Social Council (ECOSOC). It was created pursuant to Article 68 of the U.N. Charter, which provides that "[t]he Economic and Social Council shall set up the commissions in economic and social fields and for the promotion of human rights, and such other commissions as may be required for the performance of its functions." Under the terms of Resolution E/20, the Commission on Human Rights was to submit proposals, recommendations, and reports to ECOSOC regarding *inter alia* the status of women, freedom of information, the protection of minorities, and the prevention of discrimination on grounds of race, sex, language, or religion.[3] The Commission's mandate was thus decidedly focused on states' compliance with human rights norms.

In its lifespan, the Commission was continuously discredited. It was accused of being bureaucratic, excessively political, and ineffectual, especially due to the participation of states with problematic human rights records who exploited the organization as a shield against scrutiny and condemnation. The 2005 *In Larger Freedom* report of the U.N. Secretary General condemned the Commission's lack of credibility and professionalism:

> In particular, States have sought membership of the Commission not to strengthen human rights but to protect themselves against criticism or to criticize others. As

> a result, a credibility deficit has developed, which casts a shadow on the reputation of the United Nations system as a whole.[4]

The Commission was also criticized for its partial and subjective treatment of human rights issues and for resorting to double standards—focusing on certain states while ignoring other, graver, violations of human rights in other states:

> The Commission's ability to perform its tasks has been overtaken by new needs and undermined by the politicization of its sessions and the selectivity of its work.[5]

Coming from the U.N. Secretary-General, such severe accusations carried unprecedented weight. They reflected a broad consensus among developed countries—and many of the United Nations' strongest financial backers—that the Commission had become a moral and political liability. The dismantlement of the Commission on Human Rights had become necessary for the sake of the U.N. system as a whole, and, of course, for the sake of human rights.

Throughout this process, the composition and competence of a replacement body were discussed. The Secretary General's *High Level Panel on Threats, Challenges and Change* (known as the High Level Panel Report) noted that the new Human Rights Council would deal with human rights and economic issues—but made no mention of IHL.[6] And Addendum 1 to the Secretary General's Report *In Larger Freedom*, which dealt specifically with the reform of the Commission on Human Rights and devoted an entire section to the mandate of the new Council, did not even raise the possibility that IHL should be part of its mandate.[7]

Similarly, the summary of the open-ended informal consultations held in Geneva ahead of the reform of the Commission on Human Rights—the main document recording discussions that took place prior to the creation of the Council—provides extensive information as to the role of the new human rights body.[8] While it describes views on the Council's function, agenda, status, and composition—it does not refer at any time to international humanitarian law. The possibility of including IHL as part of the new body's mandate was raised only once by the Organization of Islamic States, which stressed that the new body should "deal with serious human rights violations and situations involving non-compliance with international humanitarian law. Its mandate should be to oversee the implementation of international human rights and international humanitarian law."[9] This suggestion, however, failed to trigger any discussion on the matter.

The absence of any reference to international humanitarian law in the discussions leading up to the creation of the Council is all the more surprising considering the practice of its predecessor.[10] The past practice of the Commission on Human Rights in matters of IHL was not mentioned—let alone considered a point of reference—when delineating the mandate of the new body.

Naturally, therefore, when the General Assembly voted in favor of the creation of the Council in March 2006,[11] the draft resolution did not mention IHL or any related concept.[12] While the expression "human rights" appears over thirty times in the resolution, IHL (or war, or armed conflict) is not mentioned even once. According to Resolution 60/251:

> The Council shall be responsible for promoting universal respect for the protection of all *human rights* and fundamental freedoms...
> The Council should address situations of violations of *human rights*...
> The Council shall make recommendations to the General Assembly for the further development of international law in the field of *human rights*...
> The Council shall undertake a universal periodic review...
> of the fulfillment by each State of its *human rights obligations and commitments*...
> The Council shall contribute ... towards the prevention of *human rights violations* and respond promptly to human rights emergencies...[13]

As can be seen, the resolution is notably silent on matters of IHL, despite explicit and repeated references to human rights law. I argue, therefore, that IHL is outside the Council's mandate, and that the Council should tread carefully when considering such matters.

II. THE ACTIVITY OF THE HUMAN RIGHTS COUNCIL

In stark contrast with its constitutive document which is focused exclusively on matters of human rights, the practice of the Council has focused heavily on matters of international humanitarian law. In this section, I address three specific instances in which the Council became involved in situations of armed conflict since its creation in 2006: the war pitting Israel against Hezbollah in the summer of 2006, the ongoing conflict unfolding in the Darfur region of Sudan, and Somalia. Illustrative in nature, these examples do not intend to provide an exhaustive picture of the IHL-related activity of the Council. Other examples abound—such as the Council's treatment of the situation in the Congo,[14] in Sri Lanka (as mentioned in the introduction to this chapter),[15] or in Myanmar.[16] The Council has also addressed alleged violations of IHL by Israel outside the context of the Second Lebanon War.[17]

The Second Lebanon War provided the new Council with its first opportunity to opine on matters of international humanitarian law. In July 2006, the Council decided to hold its first special session to discuss the situation then unfolding in Lebanon, following the kidnapping and killing of Israeli soldiers.[18] During this special session, the Council adopted a resolution addressing the Israeli response to the Hezbollah attack.[19] The subject matter and very terms of the resolution were imbued with IHL. The Council reaffirmed "the applicability of the Geneva Conventions Relative to the Protection of Civil Persons in Time of War, of 12 August 1949, to the Occupied Palestinian Territory, including East Jerusalem, and to other occupied Arab territories" and expressed "deep concern at the breaches by Israel, the occupying Power, *of international humanitarian law* and human rights law in the Occupied Palestinian Territory."[20] Importantly, the Council also established a Commission of Inquiry responsible for investigating alleged systematic targeting and killing of civilians by Israel in Lebanon, examining the type of weapons used by Israel and the conformity of such use with international law, and assessing the extent and deadly impact of Israeli attacks on human life, property, critical infrastructure, and the environment.[21] All of these questions related in large part (if not exclusively) to international humanitarian law.

The mandate of the Commission of Inquiry on Lebanon was discussed in greater detail at a second special session held by the Council in August 2006. The Commission of Inquiry declared that it would interpret such mandate "broadly" and "in the light of the principles and rules of international law, international humanitarian law and international human rights law, and having in mind the need for the respect for human life and dignity in the face of the complex challenges posed in the context of armed conflict."[22] In its report, the Commission of Inquiry characterized the conflict between Israel and Hezbollah as "an international armed conflict to which conventional and customary international humanitarian law and international human rights law are applicable."[23] Such determination was made based on the commission's finding that "[t]he fact that the Lebanese Armed Forces did not take an active part in them neither denies the character of the conflict as a legally cognizable international armed conflict, nor does it negate that Israel, Lebanon and Hezbollah were parties to it."[24]

Overall, the Council conveyed a very clear message when it convened its first and second special sessions to discuss the conflict in Lebanon: the Council, like its predecessor, would continue to address situations of armed conflicts. While in line with the late Commission of Human Rights practice, the Council's focus on IHL in its very first resolutions nonetheless stands in stark contrast to the lack of discussion of humanitarian law ahead of the creation of the Council and the absence of any reference to IHL in its foundation document. Keeping in mind the short lapse of time between the adoption of that document and the Council's first special session (only a few months), the contrast is even stronger.

Were one inclined to interpret the Council's declarations on the Second Lebanon War as inadvertent references to IHL, the Council's subsequent treatment of Darfur suggests otherwise. In late 2006, the Council noted "with concern the seriousness of the human rights and humanitarian situation in Darfur" and called "on all parties to put an immediate end to the ongoing violations of human rights *and international humanitarian law*, with a special focus on vulnerable groups, including women and children, while not hindering the return of all internally displaced persons to their homes."[25] Regularly since then, the Council has reiterated its call to end violations of human rights *and* international humanitarian law in Darfur—usually placing an emphasis on the latter. It has, for example, expressed concerned regarding

> violations of human rights and international humanitarian law in Darfur, including armed attacks on the civilian population and humanitarian workers,

> widespread destruction of villages, and continued and widespread violence, in particular gender-based violence against women and girls, as well as the lack of accountability of perpetrators of such crimes.[26]

As a matter of fact, in most of the Council's decisions dealing with the situation in the Sudan, the Council has addressed human rights and IHL concerns on an equal footing.[27]

The Council's treatment of the situation in Somalia confirms this pattern—with resolutions typically condemning violations of human rights law *and* IHL, and calling on all parties involved in the conflict to respect human rights *and* humanitarian law.[28] Since 2009, the Council has referred to the situation in Somalia as an armed conflict,[29] repeatedly referring to human rights law and humanitarian law as the applicable legal frameworks. There is hardly a report of the Council's Independent Expert on Somalia that does not make reference to IHL.[30] Take, for example, the Independent Expert's recommendation to establish "an Independent International Commission of Inquiry into human rights and humanitarian law violations, or similar mechanism to investigate all abuses committed by all actors involved in the Somali conflict."[31] Perhaps more important, the Council has recently addressed IHL on its own (i.e., without any accompanying reference to human rights law), when it called on parties

> to adhere to the principles of international humanitarian law relating to the protection of civilians. The waging of hostilities in urban areas—as provoked by Al-Shabaab—inevitably brings with it huge risks to the civilian population, particularly if the principles of international humanitarian law of proportionality, targeting of only military objects and the requirement to take precautionary measures to avoid civilian casualties are disregarded. Both TFG and AMISOM have suffered many fatal casualties as a result of attacks by Al-Shabaab.[32]

The Council's joint treatment of IHL and human rights law, accompanied by occasional but increasingly frequent references to IHL alone, demonstrate the growing involvement of this body in matters of humanitarian law. The activity of the Council in the context of the Second Lebanon War, Sudan, and Somalia exemplifies a general trend in the body's treatment of IHL. Such treatment is by no means unintended or secondary to its treatment of human rights issues: the Council addresses international humanitarian norms in almost all of its resolutions, often (but not always) in tandem with human rights law.

That IHL has become a major part of the Council's workload is not contested—not even by the Council itself. On the contrary, the Council is well aware of the scope and implications of its forays into international humanitarian law, and at least on one occasion has actively advocated for an expansive interpretation of its human rights mandate. When addressing the United States' argument that the Council's Special

Rapporteur on Extrajudicial, Summary, or Arbitrary Executions lacks competence to address IHL matters, the Rapporteur noted that, if accepted, such argument "would have far-reaching consequences for the Council and for *its ability to contribute in any way to many of the situations that are currently most prominent on its agenda*."[33] The Council thus acknowledges its growing involvement in matters of IHL.

As a result of this involvement, the Council produces a significant body of law in IHL—an area of law not formally part of its mandate.[34] I examine below the consequences of humanitarian lawmaking at the hands of a human rights body. Before I do so, however, I would like to address—only to dismiss it—the argument that the Council must inevitably interpret and apply IHL given the close relationship between IHL and human rights law.

III. ENFORCEMENT OF INTERNATIONAL HUMANITARIAN LAW BY OTHER HUMAN RIGHTS BODIES

Defenders of the Council's treatment of IHL would argue that the unclear boundaries between human rights and international humanitarian law make the Council's forays into IHL inevitable. In addressing the validity of this argument, it is instructive to examine how other human rights bodies have contended with similar challenges. How have such bodies conceived of their role in interpreting, developing, and enforcing IHL? Have they, too, expanded their mandate to encompass IHL alongside human rights?

At times, other human rights bodies have indeed found themselves compelled to interpret or even apply norms of IHL. Most human rights bodies address matters of IHL—generally without an explicit mandate to do so.[35] President Theodor Meron of the International Criminal Tribunal for the Former Yugoslavia explains that human rights bodies turn to IHL for primarily two reasons: either because human rights law could address the relevant issues only indirectly or because IHL seems more pertinent to the situation considered. The analysis of the practice of human rights bodies (other than the Council) reveals that human rights bodies usually address IHL reluctantly. Whenever they do so, they take careful steps to justify the treatment of such issues and assert their competence by reference to the terms of their mandate. Such justifications and references are singularly absent from the Council's decisions and resolutions.

Among all human rights bodies, the European Court of Human Rights (ECHR) has probably demonstrated the greatest reluctance to engage in an analysis and application of IHL. Even when claims arose out of a situation of armed conflict (for example, in the context of the Turkish occupation of Cyprus in *Loizidou v. Turkey*[36] or the conflict between Transdniestrian separatists and the Moldovan Security Forces in *Ilascu and Others v. Moldova and Russia*[37]),[38] the ECHR has avoided making any reference to humanitarian law and has provided relief to victims on the basis of human rights law only.[39]

By carefully confining itself to the application and interpretation of the European Convention of Human Rights and protocols, the ECHR demonstrates a real awareness of the limits of its competence. This exclusively human rights approach has given rise to some criticism. In particular in situations where victims of armed conflicts turned to the ECHR for relief or when the pleadings themselves referred to IHL (as in *Banković*[40] in the context of the war in the Balkans or *Isayeva* in Chechnya[41]), scholars have deplored the ECHR's reluctance to apply IHL.[42]

The Inter-American Commission on Human Rights has shown more willingness to consider IHL. One such example is the well-known decision of the Inter-American Commission to examine questions of IHL in *Abella v. Argentina* (better known as the *La Tablada* case).[43] The case considered the circumstances surrounding the attack on members of the Argentinean armed forces by over forty armed persons, leading to a lengthy battle and many deaths among both groups. Attackers who survived filed a complaint with the Inter-American Commission on Human Rights for violations of the American Convention on Human Rights and IHL by members of the Argentinean armed forces. The Commission therefore had to determine whether it had competence to examine the IHL-related allegations. The Inter-American Commission began by characterizing the violent incidents between the attackers and the Argentine armed forces as a noninternational armed conflict, to which Common Article 3 of the Geneva Conventions applies.[44] But before proceeding to analyze whether any violation of this provision had been committed by Argentina, the Inter-American Commission decided to "clarify the reasons why it has deemed it necessary at times to apply directly rules of international humanitarian law or to inform its interpretations of relevant provisions of the American Convention by reference to these rules."[45] Only following a lengthy analysis of its mandate and the questions arising in the specific case did the Inter-American Commission conclude that it had the competence to apply international humanitarian law.[46]

Skeptics questioned whether the commission was justified in extending its mandate to IHL as it did in *La Tablada*. One scholar argues that most of the arguments presented by the Commission do not provide compelling authority for the unqualified application of IHL and that it might "have sufficed for the Commission to apply provisions of the American Convention interpreted in light of international humanitarian law."[47]

The question of whether the Inter-American Commission was justified in extending its mandate falls beyond the scope of this chapter. What matters for our purposes is that the commission felt compelled to assert its competence to deal with matters of IHL prior to embarking on a direct application of this body of law.

This has consistently been the approach of the Inter-American Commission on Human Rights. Two years following *La Tablada*, in a discussion of the violent situation in Colombia (namely, acts of violence carried out by Colombian state agents and organs and their proxies as well as by armed dissident groups such as the FARC), the commission noted that it "will *apply* international human rights norms, particularly the American Convention on Human Rights (...), and will also *refer* to international humanitarian law, where relevant."[48] Though it did apply humanitarian norms directly, the commission acknowledged that "the American Convention and other universal and regional human rights instruments were not designed specifically to regulate in detail internal conflict situations and, thus, they do not contain specific rules governing the means and methods of warfare."[49] The commission then proceeded to explain its use of IHL as *lex specialis* and as a source of "authoritative guidance" in nine detailed paragraphs.[50] The Inter-American Commission on Human Rights again reiterated the need to "refer to and consider pertinent provisions of international humanitarian law as the applicable *lex specialis*" when applying human rights law in the context of counterterrorism in subsequent reports and decisions.[51] The possible use of IHL as a tool of interpretation of human rights instruments was perhaps best spelled out by the commission in response to U.S. objections that the Commission lacked competence to address legal issues relating to the detention of suspected terrorists in Guantanamo Bay:

> [T]he Commission remains of the view that it has the competence and the responsibility to monitor the human rights situation of the detainees at issue and in so doing to look to and apply definitional standards and relevant rules of the international humanitarian law *in interpreting and applying the provisions of the Inter-American human rights instruments* in times of armed conflict.[52]

Following a similar pattern, the Inter-American Court of Human Rights refuses to apply humanitarian law directly. The landmark case in this respect is certainly *Las Palmeras*[53]—even though a number of earlier cases had already announced the Court's position.[54] The case arose out events that occurred in Las Palmeras, Colombia, where Colombian armed forces allegedly fired from a helicopter, injuring a child, and Colombian police detained a number of individuals, some of whom were extra-judicially executed by the Columbian police.[55] Colombia filed a number of preliminary objections contesting the jurisdiction of the Court, including two objections arguing that neither the Court nor the Commission had the jurisdiction to apply humanitarian law.[56] The Court admitted these two objections and held that it lacked competence to analyze whether actions committed by the Colombian police and armed forces complied with the Geneva Conventions.[57]

Again in *Bámaca Velásquez v. Guatemala*,[58] the Court held that IHL may be used as a tool of interpretation of human rights law—but not applied directly as such. The case related to the detention and alleged mistreatment of Bámaca Velásquez and other combatants of the Guatemalan National Revolutionary Unit at the hands of the Guatemalan armed forces. The Court first characterized the situation in Guatemala as a noninternational armed conflict[59] and held that "the relevant provisions of the Geneva Conventions may be taken into consideration as elements for the interpretation of the American Convention."[60] This position, maintained in following cases, suggests a conservative and careful interpretation of its competence by the Inter-American Court of Human Rights.

The case of other U.N. human rights bodies is also instructive. At the Human Rights Committee, which was established by the International Covenant on Civil and Political

Rights of 1966, no real interpretation or application of IHL has been noticed.[61] The Human Rights Committee has referred to IHL mostly in observations and recommendations made on states' reports—which are not binding—and not with respect to decisions regarding individual applications.[62] As noted above in the context of Israel's objections to the Human Rights Committee's treatment of IHL,[63] the Committee has referred to IHL when commenting on state parties' periodical reports. Not limited to Israel, this practice has also included in the past states such as Croatia, Bosnia and Herzegovina, and Colombia.[64]

Even the Human Rights Division of the U.N. Observer Mission in El Salvador, whose mandate includes humanitarian law,[65] has given priority to violations of human rights law and has elaborated some operational guidelines designed to apply when investigating violations of IHL.[66]

Unlike its counterparts, the Council does not feel the need to establish its competence when addressing substantive issues of IHL. The Council approaches questions of IHL in the same manner as it approaches human rights questions—that is, as if they are part of its mandate. At best, the Council "acknowledges that human rights law and international humanitarian law are complementary and mutually reinforcing"—a statement that precedes, for example, a resolution on the protection of cultural rights and property in situations of armed conflict.[67] In another resolution entitled *Protection of Human Rights of Civilians in Armed Conflict*, the Council noted that "all human rights require protection equally and that the protection provided by human rights law applies as a *lex specialis*," and that "conduct that violates international humanitarian law...may also constitute a gross violation of human rights."[68] This is as far as the Council has gone in justifying its continual treatment of IHL.

While surprising in its own right, the failure of the Council to explain its overreaching is even more puzzling when compared to the practice of other human rights bodies. No comparable encroachment upon IHL has been found on the part of other human rights bodies, which have either refrained from applying IHL altogether or taken great care to justify any such application. In contrast, the Council not only does not limit itself to the application of IHL as *lex specialis*, but regularly applies IHL in its own right.

The comparison between the Council and other human rights bodies reveals the weakness of the claim that the former has no choice but to apply IHL under the *lex specialis* doctrine. The experience of other human rights bodies shows that other courses of action are indeed available to the Council when contending with IHL matters.

IV. CHALLENGING THE COUNCIL'S MANDATE AND EXPERTISE TO ADDRESS IHL

The Council's treatment of IHL raises the question of whether the Council has the mandate or the expertise to engage in the interpretation, development, and application of international humanitarian law.

A. Challenging the Council's Mandate

A number of states have expressed their reluctance to have the Council (or its predecessor, the Commission on Human Rights) involved in situations of armed conflict. Back in the days of the Commission on Human Rights, Turkey and the United States argued that matters of international humanitarian law fell outside the body's mandate.

In the 1990s, Turkey challenged the Commission's treatment of the conflict between the Kurdistan Workers Party (PKK) and the Turkish security forces. Turkey had faced criticism from a number of U.N. human rights bodies and experts (including the Commission's Special Rapporteur on Extra-Judicial, Summary, or Arbitrary Executions) relating to the alleged killing of Turkish and Iraqi civilians of Kurdish origins following Turkey's entry into Northern Iraq in March 1995. The Permanent Mission of Turkey to the United Nations responded that the operation conducted by the Turkish armed forces in Northern Iraq did not fall within the competence of the human rights bodies and that, accordingly, special rapporteurs and working groups of the U.N. human rights system were not entitled to oversee state performance with IHL in armed conflicts. To Turkey's argument, the U.N. bodies and experts replied that "they considered themselves competent to monitor the implementation of international human rights law within the framework of their respective mandates, irrespective of whether those rights were codified in international human rights law or humanitarian law."[69]

A decade later, the United States challenged the treatment of IHL by the Commission on Human Rights. The United States first raised the argument in 2003 in connection with the detention of suspected terrorists in Guantanamo Bay.[70] That same year, the United States once again challenged the competence of the Commission (in particular that of the Special Rapporteur on Extrajudicial, Summary, or Arbitrary Executions) following the targeted killing of a terrorist in Yemen.[71] The United States claimed that the question of the legality of the strike fell outside the Commission's mandate, as the strike was carried out as part of an armed conflict.

The United States raised similar arguments after the creation of the Council in 2006 in connection with the execution of Al Qaeda operative al Yemeni. The United States argued that "inquiries related to allegations stemming from military operations conducted during the course of an armed conflict with Al Qaida do not fall within the mandate of the Special Rapporteur."[72] In particular, the United States dismissed the Special Rapporteur on Extrajudicial, Summary, or Arbitrary Executions' argument that "all major relevant resolutions in recent years have referred explicitly to that body of law" as imparting "upon the Special Rapporteur a mandate to consider issues arising under the law of armed conflict."[73] According to the United States, the Rapporteur's mandate did not provide it with the authority to opine on questions of IHL—and the fact the Rapporteur (both under the Commission and the Council) "may have reported on cases outside of his mandate (...) does not give the Rapporteur competence to address such issues."[74] Past practice, according to the United States, could not make up for the absence of a mandate in IHL.[75]

Interestingly, these objections extend beyond the Council and its predecessor. State practice reveals a certain level of objection to the treatment of IHL by human rights bodies, generally speaking. Take, for example, the arguments raised by the United States before the Human Rights Committee.[76] Or the U.S. claim that the Inter-American Court of Human Rights "lacks the jurisdiction or specialized expertise to apply" international humanitarian law,[77] and that the Inter-American Commission on Human Rights could not grant precautionary measures in the context of the detentions of suspected terrorists at Guantanamo Bay, as such detentions took place during an armed conflict.[78]

In the same vein, the United Kingdom has argued that the Committee against Torture lacks jurisdiction to address humanitarian law,[79] and Israel has repeatedly challenged the competence of human rights bodies to deal with situations governed by the law of armed conflict. Israel's argument has focused on the absence of mandate of U.N. human rights bodies to discuss the situation in the West Bank and Gaza—a situation which, Israel argues, is governed by IHL. In the context of the submission of Israel's second periodic report to the Committee on Economic, Social, and Cultural Rights, for example, it was noted that

> Israel has consistently maintained that the Covenant does not apply to areas that are not subject to its sovereign territory and jurisdiction... Accordingly, in Israel's view, the Committee's mandate cannot relate to events in the West Bank and the Gaza Strip, inasmuch as they are part and parcel of the context of armed conflict as distinct from a relationship of human rights.[80]

Israel has expressed this position on a number of occasions before the Human Rights Committee,[81] the Committee on the Elimination of Racial Discrimination,[82] and the Committee on the Rights of the Child.[83]

The objections of various states to the treatment of IHL by human rights bodies reveal a level of support for the view that human rights bodies lack competence to apply international humanitarian law—an element largely overlooked in contemporary analysis of the relationship between human rights law and IHL. In other words, it is not at all clear that states support the application of IHL by human rights bodies—and even in cases in which states do support such application, the mandates of the human rights bodies themselves might limit the application of IHL.

In contrast to the practice of certain states, arguments in favor of the Council's treatment of IHL have been made. Though the Council's mandate does not expressly cover IHL, it has been suggested that the competence to deal with humanitarian law could have been "inherited" by the Council from its predecessor.[84] In the decade or so before the Commission on Human Rights was dismantled and replaced by the Council, the former dealt with IHL on a number of occasions. This treatment dates from the 1990s and includes situations of armed conflicts in Kuwait,[85] Yugoslavia,[86] Rwanda,[87] and Israel.[88] One could perhaps attribute this shift in the Commission's practice to a change in the nature of international conflicts—involving more nonstate entities, often taking place within the territory of a single state, not always being characterized by the relevant state as an armed conflict to which IHL applies. Regardless of the reasons that led the Commission to address IHL more frequently, the Commission's practice exists and cannot be disputed.[89] The question, rather, is whether such IHL practice granted the Commission (and later, the Council) the authority to act both as guardian of human rights *and* enforcer of IHL.

The argument seems plausible. International law places much weight on the practice of states and international organizations. The behavior of essential actors in the international arena can affect the development of existing law and even contribute to the crystallization of new law. In this sense, the practice of the Commission on Human Rights could theoretically have shaped the contours of the body's competence—to the point where, following fifteen years of practice in this sense, the Commission might eventually have been legally and legitimately entitled to opine on matters of IHL.

Even assuming that the practice of the Commission did have the effect of expanding its own mandate—a question that this chapter does not seek to answer—could the newly created Council have benefited from such an expansion? As noted earlier, the replacement of the Commission on Human Rights with the Council in 2006 constituted the first (and, to date, the unique) case in which a U.N. body underwent full-fledged reform. Following the dismantlement of the Commission on Human Rights, a completely new body was established, with a completely new mandate. As part of the reform of the Commission on Human Rights, a serious examination of the roles and mandates of both the Commission and its successor-to-be was conducted. And yet, the discussions leading up to the creation of the Council did not include any mention of whether the mandate of the new body should cover human rights and humanitarian law. The absence of such a discussion, together with the clear language of the resolution establishing the Council, militate against a "transfer of competence" from the Commission to the Council—even assuming the Commission acquired such competence by its practice.

Some argue that the omission of IHL from the debates leading up to the creation of the Council shows how obvious the question was to all. Alston and others, for example, concede that "the issue was never broached" but claim that "it was simply assumed that the Council would, in this respect as in most others, maintain the practice followed by the Commission."[90] I find this argument unconvincing. The debates show the willingness of states to deeply reform the Commission and create an entirely new and distinct body, with different features, mechanisms, mandate, and aspirations. For Alston and others to attribute the fact that IHL was never "broached" to a mere "assumption" contradicts the circumstances in which the Council was established. Drawing such wide-ranging conclusions as to the Council's competence from the *absence* of debates on IHL seems far-fetched by any legal standard.

If anything, the absence of such debates indicate a consensus to the effect that the Council's mandate would in fact be limited to human rights.

This conclusion is supported by the practice of a few human rights bodies whose mandate covers both human rights law and humanitarian law. Take, for example, the U.N. Observer Mission in El Salvador of 1991 or the U.N. Verification Mission in Guatemala (1994).[91] Whenever human rights bodies have been expected to intervene in situations of armed conflict, their mandates have referred to humanitarian law—undermining the argument that IHL was *assumed* to form part of the new Council's mandate and was thus not included expressly in its mandate.

Another argument raised in favor of the treatment of IHL by the Council emphasizes the absence of mechanisms specifically designed to enforce IHL. As noted by a group of experts consulted by the International Committee of the Red Cross (ICRC) on IHL enforcement mechanisms, "[t]here is unanimous agreement that existing IHL implementation mechanisms suffer from a lack of use and from a lack of effectiveness."[92] As a result, human rights bodies have increasingly been regarded as an alternative:

> The consequence of the weaknesses in the traditional enforcement mechanisms of international humanitarian law is that the victims of many conflicts have been left with no recourse other than regional or international human rights bodies. Therefore, such bodies have found themselves in the situation where they are called upon to address violations committed in armed conflicts, and so have had to decide upon the appropriate law to apply in such situations.[93]

That victims view recourse to human rights bodies as the only means to obtain remedy for IHL violations does not imply that these bodies ought to act as enforcers of IHL. The argument that victims have a right to an effective remedy for a violation[94] cannot provide a justification for overstepping a body's mandate or jurisdiction. As a matter of fact, all experts taking part in the ICRC meetings on enforcing IHL "cautioned strongly against any blurring of the distinction between the two bodies of law and voiced concerns that increased consideration by regional human rights bodies in particular might lead to a fragmentation or lack of universality in the application of international humanitarian law."

Among the issues raised against the increased consideration of IHL by human rights bodies was the "lack of express competence to examine issues of international humanitarian law, except where some fundamental human rights and protections of international humanitarian law overlap."[95]

I question neither the lack of IHL enforcement mechanisms nor its implications. The Council itself has noted that denying it competence on matters of IHL would have "the effect of placing all actions taken in the 'global war on terror' in a public accountability void, *in which no public and transparent international monitoring body would exercise oversight*."[96] The potential drawbacks this institutional gap holds for ensuring compliance with IHL, however, cannot suffice to overcome the major legal hurdles which stand in the way of using the Council to deal with matters of IHL.

In yet another attempt at rationalizing the Council's singular treatment of IHL, it has been suggested that human rights treaty bodies might feel more reluctant to pronounce on violations of IHL than human rights bodies established through U.N. Charter mechanisms.[97] Nontreaty bodies, Noam Lubell notes, "do not have the same treaty restrictions and are therefore more easily able to refer directly to violations of IHL."[98] The lack of deference to a treaty may—or may not—explain the greater reluctance of treaty bodies to deal with IHL. Still, it may not explain why the Council should feel less bound by its mandate than human rights treaty bodies. It, too, was granted a mandate—and it does not matter whether such mandate was encapsulated in a treaty or, as in the case of the Council, in a U.N. General Assembly resolution.

Neither can the mandate issue be solved by the appointment of rapporteurs and experts well-versed in IHL. The *Code of Conduct for Special Procedures Mandate-Holders of the Human Rights Council*, adopted by the Council in 2007, provides in its Article 7 that

> [i]t is incumbent on the mandate-holders to exercise their functions in strict observance of their mandate and in particular to ensure that their recommendations do not exceed their mandate *or the mandate of the Council itself*.[99]

The Council's lack of competence, in other words, cannot be "cured" by conferring an IHL mandate to rapporteurs and experts.

Overall, the arguments raised to assert the competence of the Council to deal with IHL are unconvincing. Neither the past practice of the Commission on Human Rights nor the absence of a suitable mechanism capable of enforcing IHL, the lack of deference to a treaty, the right of victims to an effective remedy, or the appointment of IHL specialists by the Council can justify the Council's pattern of incursion into humanitarian law. This is particularly true given the Council's failure to provide much explanation of how it derives such authority (as shown in section III, above), and the Council's apparent lack of expertise when pronouncing on matters of IHL (as I elaborate below).

B. Challenging the Council's Expertise

Beyond the jurisdictional aspects of the treatment of IHL by human rights bodies, a number of scholars have questioned the expertise of these bodies to interpret and apply international humanitarian law. Though scholars rarely mention the Council in this context,[100] the questions that arose are similar: What are the consequences of IHL being enforced by a nonspecialized body? Does a human rights body have the required expertise and resources to investigate state behavior in situations of conflict?

The Council's membership consists of forty-seven member states, each of which elects a representative at the Council.[101] At the moment, only the President of the Council and one out of four members of the Bureau has experience in IHL.[102] This is the first time the Council is headed by someone with IHL experience since the Council was created in 2006.[103] In general, it can be said that most of the Officers of the Council are diplomats—traditionally, though not always, with experience in human rights law.[104] Even more rarely do the Officers of the Council have expertise in IHL.

Statistics provided in connection with other human rights bodies confirm this trend. Only about 16 percent of the judges have expertise in IHL at the European Court of Human Rights, along with 50 percent of the members at the Human Rights Committee and 20 percent of the members of the Inter-American Court and Commission.[105] The need for human rights bodies to employ more members with IHL expertise and train the members in humanitarian law has already been noted.[106] In the case of the Council, whose activity heavily focuses on IHL, this need has become an imperative.

The absence of IHL experts on the Council carries important implications. The lack of training and expertise in IHL among the Council's officers affect their perception of issues related to armed conflict. Daniel Reisner argues that, in international law, "what you see depends on where you sit."[107] Using Reisner's terminology—what do state representatives "see" when sitting on the Council? They "see" the situation through the eyes of diplomats and/or human rights experts sitting on the United Nations' main human rights body—not as IHL experts or military experts. Their interpretation and application of IHL necessarily reflects this viewpoint.[108]

Perhaps this explains why the treatment of IHL by the Council has proved inconsistent at best and counterproductive at worst. The first issue concerns the characterization of conflicts by the Council. That those inconsistent findings may arise when human rights bodies determine the existence of armed conflicts should not come as a surprise.[109]

The Council, for example, has characterized the situation in Somalia as an armed conflict,[110] but it has yet to recognize that the ongoing hostilities in Syria have risen to the level of an armed conflict (as of March 2012). I find this discrepancy difficult to justify.

The second issue has to do with the interpretation and development of IHL norms by a human rights body—which has the potential of weakening the fundamentals of humanitarian law. This danger has been highlighted by Yuval Shany.[111] According to Shany, the tendency of human rights bodies "to alter the basic attributes of the 'armed conflict' paradigm through the application of human rights law may result in problematic normative-overreaching" and "lead to a law-application exercise divorced from the actual conditions and the needs of warfare."[112] For this reason, Shany concedes that "[o]ne may regard with a certain degree of apprehension attempts by human rights bodies to assess the lawfulness of counterterrorist military operations *predominantly* pursuant to human rights standards."[113] The main reason for such apprehension is that the standards developed by human rights bodies, imbued with human rights law, may lead to counterproductive results in terms of compliance. States may regard these standards as out of sync with the reality of the battlefield and be less inclined to comply with them. Shany also expresses regret that Goldstone applied a "high standard of required precautions" in its analysis of the duty of effective warning owed to civilians located in areas where military operations take place.[114] As a result of this and other (mis)applications of IHL in the report, Shany notes, "IHL with its greater tolerance for operational mistakes committed during the fog of war is cast aside, and human rights law, which arguably imposes a more exacting standard of care, is selected as the principal legal framework for the imposition of liability."[115]

Laurie Blank echoes Shany's concerns.[116] She, too, deplores the concurrent application of IHL and human rights law in the Goldstone Report. In particular, Blank takes issue at the report's analysis of the attack on the al-Daya family home, which the report found did *not* violate IHL but violated Article 6 of the International Covenant on Civil and Political Rights (ICCPR).[117] This type of finding, she argues, undermines the delicate balance "between minimizing suffering in war and enabling effective military operations" on which IHL is premised.[118]

These criticisms highlight the danger of having a significant bulk of IHL interpreted, developed, and applied by a human rights body like the Council. Until reforms to address the lack of expertise in IHL are implemented, the integrity of IHL will continue to be harmed. These reforms would include training and properly selecting representatives,[119] appointing specialized rapporteurs, heads of commissions of inquiry, and other experts specialized in IHL, and establishing dialogue among various international and regional bodies on issues of common interest.[120] But none of these steps, no matter how feasible or beneficial, would cure the Council's lack of a mandate to address questions of IHL. Fixing the Council's mandate would require approval by the General Assembly—a step unlikely to take place. As such, the Council's lack of a mandate in IHL raises potentially insurmountable hurdles.

V. THE CONVERGENCE OF HUMAN RIGHTS LAW AND HUMANITARIAN LAW

The Council's treatment of IHL illustrates the ever-growing convergence between human rights law and humanitarian law. This convergence has also been referred to as *fusing*, *meshing*, *complementarity*, or *confluence*.[121] These expressions all describe a reality where peace and war have become blurred—in large part as states have had to contend with terror, nonconventional warfare, and the realities of the modern battlefield. Like the traditional paradigm between human rights law and IHL, other traditional distinctions, such as the distinction between civilians and combatants or the *suma divisio* between the *jus ad bellum* and the *jus in bello*, have also eroded as a result of these changes.[122]

Human rights law and humanitarian law have grown closer—upholding similar values but setting forth different standards. At the same time, these bodies of law increasingly compete with each other to govern situations which do not necessarily fall clearly within one or the other's sphere of competence.[123] That the Council has encountered difficulties in delineating the scope of application of human rights law in relation to humanitarian law thus appears legitimate.[124] And yet, as I explored in section III, the experience of other human rights bodies suggests that convergence is not necessarily inevitable—at least not to the extent witnessed in the practice of the Council.

Convergence between human rights law and humanitarian law, including of the type witnessed in the practice of the Council, can be attributed to a number of factors. The two fields share a common goal[125]—the protection of individuals, especially those most vulnerable. Human rights law was traditionally defined as the body of law protecting the individual in time of peace; while humanitarian law protects individuals from hostilities in time of war.[126] At their core was the recognition of "one and the same ideal."[127] The two bodies, however, are based on different assumptions: while human rights law aims at curbing the power of the government, IHL has different sensibilities with respect to the balance of power. IHL grew out of an arrangement between sovereigns, not as a body of law protecting the individual against the state.

Though each field's sphere of application was theoretically clearly defined, the question of their interaction quickly arose.[128] The International Court of Justice's advisory opinion in *The Legality of the Threat or Use of Nuclear Weapons* made clear that the relationship between the two bodies of law was far more complex than traditionally envisaged. It laid out, for the first time in such clear terms, the *lex specialis* doctrine:

> The Court observes that the protection of the International Covenant of Civil and Political Rights does not cease in times of war, except by operation of Article 4 of the Covenant whereby certain provisions may be derogated from in a time of national emergency. Respect for the right to life is not, however, such a provision. In principle, the right not arbitrarily to be deprived of one's life applies also in hostilities. The test of what is an arbitrary deprivation of life, however, then falls to be determined by the applicable *lex specialis*, namely, the law applicable in armed conflict which is designed to regulate the conduct of hostilities.[129]

The complexity of the relationship between human rights law and IHL created some confusion at the doctrinal level. This confusion has also reached beyond the legal realm, with journalists and politicians referring to human rights and humanitarian law interchangeably. Take, for example, references to the principle of proportionality—a principle "common" to both fields, but with a different substantive content.[130] While scholars face challenges trying to determine which of the human rights law standard or IHL standard of proportionality apply and when,[131] the wider public simply uses the term broadly to cover all situations. Human rights law and humanitarian law both enjoy mass awareness and popular appeal, and the distinction between the two often gets lost.

In this climate, the forays of the Council into human rights law have gone largely unnoticed.[132] In fact, as noted in the introduction to this chapter in the context of Sri Lanka, the Council is now *expected* to address violations of humanitarian law.

The absence of any effective mechanism capable of enforcing IHL has strongly contributed to this situation. The institutional vacuum pertaining to IHL is striking, especially when compared to the plethora of mechanisms existing to monitor and enforce human rights—whether charter-based or treaty-based.[133] The only existing IHL-specific enforcement mechanism is the International Humanitarian Fact-Finding Commission ("the IHFFC")—a permanent body established by Article 90 of Additional Protocol I. Since coming into existence in 1991, seventy-two states have recognized the competence of the IHFFC[134] and endowed it with the competence to inquire into allegations of grave violations of the Geneva Conventions and accompanying Protocols, exercise good offices, and make recommendations when appropriate.[135] But the ability of the IHFFC to carry out its functions exists only when both the state requesting the involvement of the IHFFC and the state against which the allegations are made have accepted the Commission's competence.[136] Thus limited by its voluntary jurisdiction, the IHFFC has not yet been called upon to exercise its functions.[137] The IHFFC has therefore failed to provide states with a mechanism capable of ensuring compliance with IHL.

The common roots and popular appeal of IHL and human rights law, combined with the failure of IHL mechanisms, has led the Council (as well as other human rights bodies) to act as enforcer of IHL. Human rights bodies are increasingly seen as an alternative, particularly by victims seeking redress for violations committed during armed conflicts.[138] But as I explained in section IV, the absence of any other available mechanism should not and cannot justify an extension of the mandate of the Council to all matters of IHL. No matter how laudable the goal of enforcing IHL, the automatic application of IHL by the Council finds legal support neither in theory nor in practice.

The experience of the Council raises the question of whether certain limits should be placed on the convergence of IHL and human rights law. Certain scholars view convergence as a positive development.[139] Others have expressed the need for caution, noting that the "common aim" shared by IHL and human rights law "does not wipe away the fact that they are finally different legal *corpora* and that, as a consequence, in their articulation it is impossible to confuse or blend their rules."[140]

Convergence, I would argue, is a question of degree. While "the momentum behind the complementary application of the law of armed conflict and human rights law is too powerful a trend to reverse,"[141] a clearer separation between IHL and human rights law is necessary *at the enforcement level*. If not at the theoretical level, the paradigm and the delicate equilibrium between human rights law and IHL should be maintained at the enforcement level.

The Council's lack of mandate to address matters of IHL and the resulting damage to IHL norms strongly advocate for such a solution. One could envisage the creation of a committee within the Council expressly entrusted with the task of advising the body on humanitarian law questions. This could actually solve both the expertise and the mandate issues, if such a solution were adopted by members of the United Nations.[142]

As this solution appears unlikely, the Council should exercise greater deference to its mandate when becoming involved in questions of IHL. And when such involvement is inevitable, the Council should tread more cautiously when interpreting and applying IHL. The ultimate function of the Council should remain ensuring respect for human rights, with only limited (and well-justified) forays into IHL when necessary.

The Council should implement these recommendations for a number of reasons. It must do so in order to avoid losing its credibility like its predecessor. As the United Nations' main human rights body, the Council must uphold the strictest human rights standards and ensure that it does not become, de facto, enforcer of IHL—a mandate it does not possess. A not-insignificant number of states and scholars have questioned the authority of human rights bodies to interpret and apply IHL (as distinct from their ability to use IHL, from time to time, as a tool of interpretation of human rights law). If the Council continues to deal with IHL matters without showing deference to the issues raised by these states and scholars, the normative weight of its IHL-related findings and resolutions will eventually be called into question. In addition, I have shown how the human rights–infused approach adopted by the Council has challenged the fundamentals of IHL. The Council must change its *modus operandi* in order to avoid causing any further damage to the integrity of humanitarian law and preserve its legitimacy.

Finally and importantly, the Council must make a point to preserve the delicate and important relationship between IHL and human rights law—at the risk of losing the support of states. As feared by Yuval Shany, states could easily come to disregard IHL altogether if they feel it sets unrealistic standards. Maintaining the fragile but important paradigm between IHL and human rights law is necessary to ensure that states continue to trust international institutions designed to promote and ensure respect for human rights. What weighs in the balance, eventually, is the human rights charter- and treaty-based system:

> [C]ompliance with IHL rules cannot be directly supervised by a human rights tribunal unless formally stated in its constitutive document... [b]ut good will needs to find its place within the limits of pre-existing law and, in the absence of international tribunals specifically addressed to care for the respect of IHL, enlarging the competence of human rights courts requires careful thinking if the mechanism aspires to maintain the support of states.[143]

How the Council contends with this challenge will affect its prestige and legitimacy as the main U.N. human rights body. In turn, this will reflect upon the normative weight of its decisions and resolutions. For the sake of human rights and IHL, let us hope that the Council begins to take full measure of its limitations.

Conclusion

The assumption that the Council can *and ought to* address violations of IHL—as Human Rights Watch assumed with respect to the war in Sri Lanka—must be reconsidered. As we have seen, IHL is not within the Council's mandate, despite ample opportunity to include IHL in the Council's founding resolution. Beyond the question of its mandate, having the Council deal with IHL raises an array of problems: Is it well equipped, trained, and able to investigate and assess states' compliance with norms of IHL? Should we maintain the long-held paradigm between human rights and IHL? That is, should human rights and IHL become inseparable rather than merely complementary?

In and of itself, the convergence of human rights law and humanitarian law at the doctrinal level has not been questioned in this chapter. Rather, this chapter has called for a reassessment of the relationship between the two bodies of law at the enforcement level—highlighting the singular treatment of IHL by the Council in comparison to the treatment of IHL by other human rights bodies. Neither the past practice of the Commission on Human Rights nor the absence of a suitable mechanism capable of enforcing IHL, the lack of deference to a treaty, the right of victims to an effective remedy, or the appointment of IHL specialists by the Council, can justify the Council's pattern of incursion into humanitarian law. Next time the Council addresses a situation of armed conflict, we ought to ask ourselves whether there is a point at which we should draw the line. This must be done to prevent the Council from losing its credibility and legitimacy, to maintain state support for IHL and human rights mechanisms, and to preserve the integrity of humanitarian law.

2

PROPORTIONALITY IN COUNTERINSURGENCY: RECONCILING HUMAN RIGHTS AND HUMANITARIAN LAW

Evan J. Criddle*

AT THE CLOSE of the twenty-first century's first decade, international law has yet to develop a coherent framework for regulating the use of force in counterinsurgency. The problem is not that international law lacks norms applicable to counterinsurgency operations. Rather, the problem is that international law contains multiple, conflicting standards for evaluating a state's use of force in counterinsurgency, and courts and publicists have yet to reach a consensus about how these standards relate to one another.[1]

Two discrete subfields of international law—international human rights law (IHRL) and international humanitarian law (IHL)—currently compete for supremacy in the counterinsurgency context. Each of these bodies of law aspires to safeguard human dignity during national security crises, but each seeks to honor human dignity in its own way. IHL authorizes states to target enemy fighters freely in support of military objectives, provided that collateral damage to civilians is not manifestly "excessive."[2] HRL, on the other hand, safeguards the universal demands of human dignity for all—even, and perhaps especially, during armed conflict—by prohibiting all casualties that are not strictly necessary to preserve human life.[3] Because international law does not contain clear choice-of-law rules to mediate conflicts between HRL and IHL, these two formulations of the "proportionality" principle have continued along parallel tracks without converging into a unified and theoretically satisfying standard. This fragmented legal framework has real-world costs not only because it compromises states' ability to demonstrate their adherence to international standards, but also because the difference between HRL's strict proportionality standard and IHL's more flexible standard is measured in military and civilian casualties.

How, then, should international law reconcile IHL and HRL in counterinsurgency? Some courts, military lawyers, and legal scholars have argued that international law should distinguish the respective domains of IHL and HRL based on the *nature of the threat* to national security (i.e., ordinary disturbances versus armed conflict) and the corresponding *type of operation* conducted (i.e., law enforcement versus military). According to this logic, HRL's strict proportionality standard would apply whenever a state is able to address internal unrest through traditional law enforcement tools. When internal disturbances spark genuine armed conflict, on the other hand, IHL's proportionality standard would displace HRL. I refer to this approach in the discussion that follows as the "operational theory" of *lex specialis*. Although the operational theory has its detractors, it has emerged as arguably the leading conceptual framework for reconciling IHL and HRL.

In this chapter, I argue that the operational theory is normatively unattractive as a choice-of-law rule because it ignores a critical factor in assessing a state's legal authority to use force: the *character of the relationship* between particular states and the persons they target with coercive force. A general theme running throughout contemporary international law is the principle that states bear special duties of care toward their own nationals and foreign nationals over whom they have asserted control. Under HRL, states are obligated to guarantee a variety of basic civil, political, social, and economic rights to their citizens and resident foreign nationals. IHL likewise requires states to honor the dignity of prisoners of war and other detained foreign nationals by providing basic medical attention, food, shelter, and other necessities of life. These affirmative duties of care between states and persons over whom they have asserted control are different in kind from states' obligations toward extraterritorial foreign nationals over whom they have not asserted control. The latter obligations are better described as quasi-tort duties of noninterference rather than affirmative fiduciary duties of care. Individual states do not ordinarily bear duties under international law to *protect* and *fulfill* human rights for foreign nationals outside their control, although they do bear duties to *respect* the human rights of other peoples by refraining from military aggression and other harmful acts that would treat foreign nationals as mere means to their own ends. Because the operational theory ignores the relational character of state obligations under international law, it leads to counterintuitive and morally troubling results, including the proposition that states waging counterinsurgency may kill their own people under some circumstances where lethal force is not strictly necessary to preserve equal freedom for all. Given these results, international lawyers would do well to question whether the operational theory in fact offers the best interpretive framework for conceptualizing a state's legal authority to use force in counterinsurgency.

This chapter outlines an alternative theory for reconciling HRL and IHL in counterinsurgency. The starting point is Immanuel Kant's insight that a state's legal authority to use force is best understood in relational, deontological terms. When states conduct counterinsurgency operations against their own people, they bear a fiduciary obligation to ensure that their use of force is strictly necessary to preserve fundamental legal order for all. A state may not treat its people as mere objects that it may oppress or destroy for its own benefit or for the benefit of particular groups or individuals. States engaged in internal armed conflict and belligerent occupation must comply with HRL's strict proportionality requirements during counterinsurgency, irrespective of the nature of the threat to national security or the operational character of the state's response. Conversely, because states do not ordinarily assert control over foreign nationals operating beyond their borders, their duties toward foreign nationals in traditional international armed conflicts and transnational armed conflicts with nonstate actors are generally duties of noninterference rather than affirmative responsibilities to protect. As such, states engaged in self-defense are entitled to privilege the lives of their own soldiers over those of a foreign aggressor (as reflected in IHL but not HRL) to the extent consistent with *jus ad bellum* and IHL's general humanitarian principles of distinction, necessity, and proportionality.[4] This relational theory of *lex specialis* offers a principled framework for distinguishing the respective domains of HRL and IHL, while lending support for the human rights-focused proportionality jurisprudence of the European Court of Human Rights (ECHR) and the Inter-American Court of Human Rights (Inter-American Court).

I. TWO PARADIGMS FOR PROPORTIONALITY

Over the past decade, legal scholars have observed that counterinsurgency operations straddle an active "fault line" where IHL and HRL overlap with potentially destabilizing effects for international law as a normative system.[5] In this part, I trace the fissures along this fault line and explore efforts by international tribunals and publicists to resolve tensions between IHL proportionality and HRL proportionality in counterinsurgency.

A. International Humanitarian Law

Traditionally, IHL has been concerned primarily with international armed conflict between states rather than noninternational armed conflict (NIAC) between state and nonstate actors, the primary context for counterinsurgency operations. The Geneva Conventions prescribe detailed rules to protect civilians and combatants captured within international armed conflicts. For example, states engaged in international armed conflict are required to distinguish combatants from civilians.[6] When attacks on military targets pose a risk of collateral civilian casualties, states must take "all reasonable precautions to avoid losses of civilian lives"[7] and must ensure that unintended civilian casualties are not "excessive in relation to the concrete and direct military advantage anticipated."[8] These principles of distinction and proportionality do not generally shield combatants from the use of force within an international armed conflict; rather, states remain free to target enemy combatants without subjecting themselves to legal censure, provided that their chosen means and methods of attack are lawful and the targets are not incapacitated by injury or attempting to surrender.[9] IHL thus safeguards civilians in international armed conflict from excessive state force without according comparable protection to combatants.

Although the Geneva Conventions speak less clearly regarding the use of force in NIAC, they offer some support for extending IHL proportionality to counterinsurgency. Additional Protocol I to the Geneva Conventions, which focuses on the protection of victims of international armed conflicts, extends IHL proportionality to internal struggles against colonialism and apartheid.[10] Additional Protocol II, which addresses the protection of victims of NIAC, has a similarly limited field of application, addressing only conflicts involving dissident armed groups "under responsible command" that "exercise... control over... territory" and are capable of conducting "sustained" military operations and complying with IHL.[11] As a result, Additional Protocol II does not apply to most contemporary insurgencies, which are networked rather than hierarchical, opportunistic rather than sustained, and itinerant rather than territorially entrenched.[12] Nonetheless, commentary accompanying Additional Protocol II notes in passing that the proportionality norm is a general principle with broader application.[13] Thus, states must observe "general principles relating to the protection of the civilian population which apply irrespective of whether the conflict is an international or an internal one," including "the principle of distinction and the principle of proportionality which only intervenes when it is not possible to ensure the total immunity of the population."[14]

Some authorities have asserted that IHL proportionality applies to all armed conflicts, international and noninternational alike, as a matter of customary international law (CIL).[15] For instance, in *Prosecutor v. Tadić*, the International Criminal Tribunal for the Former Yugoslavia reasoned that principles of discrimination from the law of international armed conflict represented "customary rules" that apply equally "in civil strife."[16] Similarly, in its landmark 2005 study on customary IHL, the International Committee of the Red Cross (ICRC) asserted that "gaps in the regulation of the conduct of hostilities in Additional Protocol II have largely been filled through State practice, which has led to the creation of rules parallel to those in Additional Protocol I, but applicable as customary law to non-international armed conflicts."[17] Although some legal scholars and military lawyers have embraced the ICRC's vision of IHL proportionality analysis as CIL for noninternational conflicts, this conclusion is by no means universally accepted and remains empirically underdeveloped.[18]

B. International Human Rights Law

HRL presents a different paradigm for evaluating the proportionality of a state's use of force within internal conflicts. Under HRL, the "right to life" is widely described as a *jus cogens* norm that does not admit derogation even in public emergencies such as a violent riot or insurrection. The human right to life does not prohibit all use of lethal force by states; rather, it imposes a requirement of justification: states may not use lethal force unless they can show that this extraordinary measure is "absolutely necessary" to protect life or legal order.[19] Thus, states may take life only if they first take appropriate precautions to avoid or minimize casualties and only if nonlethal measures such as arrest or incapacitation would be likely to impose disproportionate costs.[20] This strict necessity requirement for the use of lethal force places HRL in stark conflict with IHL, which requires only that civilian casualties not be "*excessive* in relation to the concrete and direct military advantage anticipated."[21]

HRL's proportionality principle differs from IHL in another respect: it takes into account threats not only to ordinary civilians but also to armed belligerents. While military casualties are irrelevant to IHL's proportionality inquiry, HRL requires states to consider all potential casualties—lawful combatants, noncombatant fighters, and ordinary citizens alike—when planning and executing counterinsurgency operations.[22] A state violates the right to life if it uses lethal force where nonlethal measures would do, even if those killed were participating directly in armed combat against the state. As the ECHR has explained, a law enforcement or military operation must be "planned and controlled by the authorities so as to minimise, to the greatest extent possible, recourse to lethal force."[23] When states do employ lethal force, they bear a continuing obligation to "take appropriate care to ensure that any risk to life is minimised."[24] HRL thus imposes a significantly more protective standard than IHL for determining whether a state's use of force is adequately proportional.

To be sure, these formal differences between HRL proportionality and IHL proportionality do not always counsel different results on the ground. In the heat of an active firefight, for example, both IHL and HRL would allow counterinsurgents to take the lives of enemy insurgents who are participating actively in the confrontation in order to preserve their own lives and the lives of innocent civilians. Conversely, neither IHL nor HRL would authorize the type of indiscriminate attacks against innocent civilians that prompted the U.N. Security Council's recent condemnation of the Libyan counterinsurgency. Differences between the IHL and HRL proportionality standards become more troubling, on the other hand, in contexts where they could prompt counterinsurgents to develop substantially different rules of engagement, depending upon the applicable legal regime. The paradigmatic example involves an unarmed fighter who temporarily takes leave from an organized insurgent group in order to plow his fields away from the battlefield. While most authorities suggest that IHL would permit counterinsurgents to use lethal force freely against such an insurgent,[25] HRL would not authorize lethal force if the insurgent could be apprehended without substantial risk of injury to counterinsurgents.[26] Here the difference between HRL and IHL would constitute a material conflict of laws necessitating recourse to choice-of-law principles.

Unfortunately, such conflicts between IHL proportionality and HRL proportionality arise frequently in counterinsurgency operations. Whenever a civil uprising leads to armed conflict, states must decide whether to adhere to the strict proportionality standards of HRL or employ the more flexible standards of IHL. Uncertainty regarding the international legal standards for counterinsurgency operations has compromised the efforts of compliance-minded states to develop principled rules of engagement. How this conflict between IHL and HRL ultimately gets resolved will have profound implications for the relative safety of countless numbers of counterinsurgent soldiers, insurgent fighters, and innocent civilians in future conflicts.

C. Reconciling IHL and HRL: Current Approaches

The International Court of Justice (ICJ) took a first stab at reconciling IHL and HRL in its 1996 *Advisory Opinion on Nuclear Weapons*.[27] The Court acknowledged the potential for conflict between IHL and HRL but suggested that there could be no true conflict between HRL and IHL in practice because "whether a particular loss of life" violated the human right to life could "only be decided by reference to the law applicable in armed conflict and not deduced from the terms of the Covenant itself."[28] In short, the ICJ concluded that IHL proportionality would supply the relevant proportionality standard whenever states cross the threshold of armed conflict.

Following the *Nuclear Weapons* decision, the Inter-American Commission on Human Rights adopted a similar approach in *Abella v. Argentina*.[29] Under review was the Argentine government's response to an armed attack by militants against military barracks in the province of Buenos Aires.[30] The Commission reasoned that if the incident was a mere "disturbance," HRL would govern the proportionality of the state's response; if the incident constituted "armed conflict," on the other hand, the Commission could look to IHL for the relevant standard. The basis for this bifurcated approach, according to the Commission, was that the American Convention on Human Rights simply did not address the special context of armed conflict.[31] Concluding that the *Abella* incident could not "be properly characterized as a situation of internal disturbances,"[32] the Commission held that IHL would supply the relevant proportionality standard for determining whether a state had honored international law.[33]

Since *Nuclear Weapons* and *Abella*, international and regional courts have challenged the idea that HRL proportionality does not prescribe rules applicable to the use of force in armed conflict. In the 2004 *Palestinian Wall* case, for instance, the ICJ observed that IHL and HRL might apply concurrently in some contexts, forcing courts to "take into consideration both [of] these branches of international law, namely human rights law and, as *lex specialis*, international humanitarian law."[34] While the ICJ did not clarify what it would mean for a court to "take" both IHL and HRL "into consideration," some experts construed the decision as recognizing that IHL and HRL sometimes operate concurrently, with each body of law acting as an interpretive guide for the other.[35] Both the ECHR and the Inter-American Court have embraced this approach, applying HRL proportionality to cases arising within their jurisdiction pursuant to their path-dependent, treaty-based mandates. In a series of cases involving internal conflicts in Chechnya, Colombia, Guatemala, and Turkey, these courts have applied ordinary HRL proportionality to counterinsurgency operations, notwithstanding the fact that the operations occurred within theaters of active armed conflict.[36] In so doing, they have implicitly rejected the view espoused in *Nuclear Weapons* and *Abella* that the HRL's strict proportionality principle is inapplicable to armed conflict.

Illustrative of this approach is the ECHR's 2005 case, *Isayeva, Yusupova and Bazayeva v. Russia* (*Isayeva*).[37] During Russia's October 2001 military campaign against insurgents in Chechnya, the Russian military established a temporary "humanitarian corridor" to facilitate the departure of civilians from Grozny, Chechnya's besieged capital. As civilian refugees were departing, the Russian military bombed a segment of the humanitarian corridor where militants were believed to be attempting escape. Sixteen civilians were killed, including two Red Cross workers.[38] In reviewing this episode for compliance with the European Convention on Human Rights, the ECHR declined to apply IHL as the relevant *lex specialis* and instead looked to HRL for the applicable proportionality standard. The Court stressed that Russia's "use of force" against Chechen insurgents "must be no more than 'absolutely necessary'" and that the Russian military "must take appropriate care to ensure that any risk to life is minimised."[39] Although the Court recognized that "the situation that existed in Chechnya at the relevant time called for exceptional measures," it nonetheless condemned the particular bombing under review because Russia was unable to show that this measure was strictly necessary under the circumstances.[40]

In the wake of *Isayeva* and related decisions of the ECHR and Inter-American Court, the fault line between IHL and HRL in counterinsurgency has become increasingly unstable. Human rights scholars generally have embraced the idea "that international human rights standards are, in some manner, applicable alongside humanitarian law during military" operations, providing more exacting standards for the use of force in NIAC.[41] In contrast, some experts in the law of armed conflict have dismissed HRL as a "law enforcement paradigm" that is ill-tailored to armed conflict, where IHL is *lex specialis*.[42] At present, it remains unclear whether one of these approaches will emerge as the definitive legal standard for assessing proportionality in counterinsurgency.

II. A RELATIONAL THEORY OF THE USE OF FORCE

Given the persistent tensions between IHL and HRL, how *should* international law define the proportionality principle in counterinsurgency?

To answer this question in a principled fashion, the international community needs a legal theory capable of clarifying the basis for a state's authority to use force. In the discussion that follows, I provide a rough outline for such a theory and consider the theory's general implications for international regulation of counterinsurgency operations. The theory I develop takes as its starting point the republican insight that central features of both HRL and IHL reflect a fiduciary relationship between states and persons who are subject to their control. Although this brief chapter is not the place for a detailed defense of the relational character of state sovereignty (something I and others have undertaken elsewhere[43]), the discussion that follows suggests a variety of ways in which a relational theory of state legal authority to use force might illuminate the normative basis for key features of both IHL and HRL and mediate conflicts between them.

A. The Fiduciary Character of Sovereignty

The idea that states serve as agents or trustees on behalf of their people has deep roots in many legal and political traditions, but until recently it has not received sustained attention as a normative theory of international law. The fiduciary conception of sovereignty merits further exploration in international legal theory, however, because it offers an attractive foundation for international law's increasing emphasis on human dignity as the central concern of state sovereignty.

As a general matter, fiduciary obligations arise in contexts where one person (the fiduciary) assumes discretionary power of an administrative nature over the legal or practical interests of another (the beneficiary),[44] thereby rendering the beneficiary vulnerable to the fiduciary's potential abuse of power. In relationships that bear these characteristics, private law intervenes to ensure that the fiduciary exercises her discretionary power reasonably. Under the paradigmatic duty of loyalty, fiduciaries are forbidden from engaging in self-interested transactions without their beneficiary's informed consent. Where a fiduciary has multiple beneficiaries, the fiduciary is obligated to exercise her powers reasonably and evenhandedly for all. The fiduciary must also take proper care to ensure that she does not squander her beneficiary's interests through arbitrary administration or neglect.

In previous writings, Professor Evan Fox-Decent and I have argued that Kant's theory of fiduciary relations offers a compelling philosophical justification for attributing fiduciary obligations to public institutions and officials.[45] According to Kant, all persons have an innate right to as much freedom as can be reconciled with others' freedom. The law's role is to honor individual dignity by enshrining legal rights within a regime of equal freedom such that no person has the capacity to dominate or instrumentalize another.[46] Fiduciary obligations safeguard legal persons from domination by ensuring that the fiduciary cannot treat her beneficiary as a mere means to her own ends. The paradigmatic example for Kant is the parent-child relationship. Children have an "innate (not acquired) right to the care of their parents until they are able to look after themselves... without any special act being required to establish this right."[47] As long as children depend upon their parents for care, their innate right to equal freedom dictates that parents cannot arbitrarily "destroy their child"; rather, they bear an obligation to tend to their child's interests, treating their child as a person endowed with dignity to whom they "cannot now be indifferent."[48]

Extending this relational account of legal obligation to public law, states are subject to fiduciary obligations under a Kantian theory of right to the extent that they assume control over the legal or practical interests of their people. All public institutions and officials—legislative, executive, and judicial—exercise discretionary powers of an administrative nature over the public's legal and practical interests. Just as parents bear a duty to tend to the legal and practical interests of children they bring into the world, so too the states must protect their people against domination (the threat of arbitrary interference) and instrumentalization (exploitation as a means to another person's ends). A state's beneficiaries for these purposes are persons who stand in a continuing relationship with the state by virtue of the state's continuing assertion of control (e.g., citizenship, residency, detention). For ease of exposition, I will refer to this category of persons as a state's "people."

Conceptualizing the relationship between a state and its people in fiduciary terms clarifies the basis for a state's authority to wield sovereign coercive powers under international law. According to Kantian legal theory, the primary purpose of public law and public institutions is to safeguard human dignity by liberating individuals from a condition where each person would be subject to the threat of arbitrary coercion from others (*dominium*).[49] States may use force to resolve disputes between private parties within their jurisdiction as necessary to preserve legal order, which Kant defines as a regime of secure and equal freedom for all. On the other hand, states may not use their coercive powers to instrumentalize individuals as mere means to the ends of public officials, rather than as ends in themselves. The relational theory of state legal authority also harnesses the republican insight that a state's mere capacity to use force arbitrarily (*imperium*) imperils liberty and undermines human dignity. A primary purpose of public law, in Kantian legal theory, is to reconcile states' authority to use coercive force with human dignity by guaranteeing that states employ force only on a nondiscriminatory basis for the benefit of their people and only where such action is strictly necessary to guarantee a regime of secure and equal freedom.

A state's fiduciary relationship with its people has at least two dimensions. A state's primary function is to furnish a regime of secure and equal freedom for its people. Secondarily, a state serves as an agent for its people in international relations. The mere fact that a state has the *power* to affect the interests of foreign nationals beyond its borders does not entail a corresponding *right* to do so, nor does it follow that a state bears fiduciary duties toward all persons everywhere. To the extent that states have the power to extend their influence extraterritorially, international law satisfies the principles of nondomination and noninstrumentalization by imposing tort duties of noninterference and criminal prohibitions such as the crime of aggression to prohibit states from using or threatening to use force unilaterally beyond their borders. Only when a state asserts ongoing control over the legal or practical interests of individuals beyond its borders does the fiduciary principle apply, triggering legal duties such as the *jus cogens* prohibitions against slavery and torture. The fiduciary principle thus links a state's legal authority to use coercive force to its correlative obligation to respect human dignity, honoring the principles of nondomination and noninstrumentalization.

The fiduciary principle's relational conception of state sovereignty resonates with the emerging "responsibility to protect" norm in international legal and political discourse.[50] On both accounts, states bear an obligation to secure their people from "avoidable catastrophe" such as "mass murder" or "starvation."[51] The fiduciary principle suggests, however, that a state's responsibility to protect applies more broadly than simply the prevention of mass atrocities. States must take care to safeguard their people from all abuses of public power, including foreign aggression, public corruption, and arbitrary deprivations of life and liberty. To respect the dignity of their people, states must also employ deliberative decision-making processes and offer reasoned justifications for their use of public powers. The fiduciary principle thus provides a general normative account of international law that is congenial to the responsibility to protect, anchoring this emerging norm in the relational fiduciary duties that govern states' relationships with their people.

B. The Fiduciary Character of Human Rights

When viewed from this perspective, human rights come into focus as relational legal duties emanating from the fiduciary character of state sovereignty. In previous writings, Professor Fox-Decent and I have argued that human rights are essential correlates of a state's fiduciary duty to secure conditions of nondomination and noninstrumentalization for its people.[52] Human rights are constitutive of sovereignty in the sense that a state may not violate the human rights of its people without undercutting its own claim to act as a sovereign entity under international law.

The fiduciary principle frames human rights in dynamic, relational, and institutional terms. Human rights respond to the threats that arise within particular relationships between public institutions and the people they serve. The relational character of human rights informs states' obligations to protect civil and political rights; while a state must endeavor to secure its own people from violence, it does not ordinarily bear a comparable responsibility unilaterally to protect persons beyond its borders who are not otherwise subject to its control.

One implication of the fiduciary principle is that states may derogate from their treaty-based human rights obligations when such action is strictly necessary to preserve legal order for their people.[53] During genuine public emergencies such as an invasion, riot, or insurgency, a state may find it necessary to restrict some personal liberties such as the freedom to travel in order to safeguard the public's basic right to secure and equal freedom. Such emergency measures limit personal freedoms, but they are consistent with the fiduciary theory of human rights to the extent that they are strictly necessary to preserve a regime of secure and equal freedom for all.

Some human rights norms are never suitable for derogation under the fiduciary theory. International law characterizes these norms as *jus cogens*—peremptory prohibitions that states may not transgress under any circumstances. Violations of peremptory norms are never consistent with a state's fiduciary obligations because such actions exploit the state's subjects as mere instruments of state policy, not as persons entitled to be treated as the state's coequal beneficiaries. Under the fiduciary principle, therefore, public institutions and public officials that engage in slavery, summary execution, or torture cannot claim to have acted under the aegis of sovereign authority.[54]

A significant contribution of the fiduciary theory is that it lends support for international and regional courts' context-sensitive formulation of the "right to life." Leading human rights conventions such as the International Covenant on Civil and Political Rights, the American Convention on Human Rights, and the European Convention on Human Rights identify the right to life as a peremptory norm from which states may not derogate even during public emergencies.[55] This characterization of the right to life as nonderogable is apt on the fiduciary theory, but only because courts have not defined the right to life as an unqualified entitlement to freedom from lethal force, but rather as a relational right that is correlative to the state's duty to act reasonably and nonarbitrarily to protect its people. The right to life does not prohibit a state from killing an armed insurgent who is participating directly in hostilities, at least where such action is strictly necessary to preserve legal order for others. When a state uses lethal force to prevent an individual from killing others subject to its care, it preserves a regime of equal freedom by ensuring that no private party has the capacity to deprive another of life unilaterally. The fiduciary principle thus provides a theoretical framework congenial to HRL's prevailing definitions of the right to life.

C. The Fiduciary Character of International Humanitarian Law

Central features of IHL, like HRL, can be understood as reflecting a relational conception of state authority. Perhaps the clearest example is Common Article 3 of the Geneva Conventions, which identifies the minimal standards of care states must observe in NIAC with regard to "[p]ersons taking no active part in hostilities, including members of armed forces who have laid down their arms and those placed *hors de combat* by sickness, wounds, detention, or other cause."[56] Once a state gains effective administrative control over incapacitated belligerents in NIAC, it must treat them "humanely, without any adverse distinction founded on race, colour, religion or faith, sex, birth or wealth, or any other similar criteria."[57] Humane treatment includes abstaining from "violence to life and person" such as "cruel treatment and torture," "outrages upon personal dignity" such as "humiliating and degrading treatment," and "the passing of sentences and the carrying out of executions without previous judgment pronounced by a regularly constituted court."[58] States also bear an affirmative obligation to collect and care for the wounded and sick with the aid of humanitarian organizations such as the International

Committee of the Red Cross.[59] Each of these principles addresses states' duties to treat persons subject to their unilateral administrative authority as human beings endowed with dignity.

Many features of the law of international armed conflict likewise fit congenially within a fiduciary theory of sovereignty. IHL currently safeguards human dignity through prohibitions against aggression[60] and indiscriminate attacks,[61] protections for civilians in conflict zones,[62] and duties of humane treatment for prisoners of war (POWs)[63] and wounded, sick, and shipwrecked combatants.[64] States that detain foreign combatants assume fiduciary obligations to provide for their basic welfare, including the obligation to refrain from engaging in torture, summary execution, and humiliating and degrading treatment.[65] States need not, of course, grant detainees in armed conflict the full spectrum of civil, political, social, and economic rights that their own citizens enjoy. Even within international armed conflicts, however, states must honor the fundamental dignity of all foreign nationals who are subject to their control. These humanitarian features of IHL comport with the relational character of state legal authority under international law.

D. The Relational Theory as an Agenda for Progressive Development

Although key features of both HRL and IHL dovetail with the relational theory's normative account of state sovereignty, neither HRL nor IHL fully satisfy the fiduciary principle in their current form. In previous writings, Professor Fox-Decent and I have advocated modest reforms to HRL in order to harmonize current legal doctrine with the fiduciary character of state sovereignty.[66] This chapter does not afford the space for a comprehensive evaluation of comparable asymmetries between the fiduciary theory's normative account and IHL. In some respects, IHL's protections to individuals in international armed conflict may exceed states' fiduciary obligations.[67] In other respects, IHL likely does not provide sufficient safeguards for human dignity to satisfy the fundamental principles of nondomination and noninstrumentalization. For present purposes, however, it will suffice to observe that the relational account of a state's legal authority to use force furnishes a normative agenda for the progressive development of both HRL and IHL.

III. A RELATIONAL THEORY OF *LEX SPECIALIS*

The relational theory suggests a new approach for resolving tensions between HRL and IHL, including current debates over the appropriate proportionality standard for counterinsurgency operations. Under the relational theory, a state's legal authority to use force depends upon the character of its relationship with the individuals it targets. Once a state asserts control over a class of persons, it assumes a concomitant responsibility to protect them and provide for their basic welfare. The greater the degree of control a state asserts over its people, the greater its responsibility to protect. Thus, when uncertainty arises regarding the scope of a state's legal authority to use force under international law, state decision makers, international judges, and other international actors should resolve interpretive ambiguities by following the standard from either HRL or IHL that most closely tracks the relational duties that flow from the character of a state's relationship with particular groups or individuals.

A. The Operational Theory of *Lex Specialis*

Some scholars and judges have characterized IHL as *lex specialis*, arguing that two factors—the severity of a threat to national security and the type of operation necessary to address the threat—are what distinguish the *lex specialis* of IHL from HRL as *lex generalis*. According to this approach, HRL regulates the use of force under ordinary circumstances but defers to IHL whenever hostilities cross the threshold of "armed conflict." A state in an internal armed conflict would therefore be free to target the hypothetical insurgent-farmer away from the battlefield with lethal force—even if the insurgent-farmer could be captured without a substantial risk of injury to counterinsurgents. The operational context of armed conflict, combined with the insurgent-farmer's status as a civilian participating directly in hostilities, would authorize the state's use of lethal force under IHL. This focus on the nature of the threat and the type of operations conducted by the state to address that threat (whether law enforcement, military, or something in between) has become the preferred choice-of-law model for military lawyers in the United States and abroad.

Professor Geoffrey Corn, a former senior law of war expert in the U.S. Judge Advocate General's Corps, has offered a sophisticated, pragmatic defense of this operational theory of *lex specialis*.[68] Professor Corn argues that applying HRL's proportionality standard to NIAC would be "operationally debilitating" for soldiers on the battlefield who must be "ready, willing, and able to kill on demand."[69] Soldiers cannot be trained to apply HRL proportionality, he suggests, without introducing uncertainty that would jeopardize soldiers' safety and undermine their ability to achieve military objectives.[70] States must have the latitude to use "overwhelming combat power at designated times and places against an enemy" in order to demonstrate their "full spectrum dominance" effectively to enemy leaders.[71] Furthermore, IHL's proportionality standard already takes human dignity into account, and it does so in a way that "will not be expected to compromise mission effectiveness or place the forces in significant danger."[72]

These arguments for applying IHL proportionality would be persuasive if we were to assume (as does IHL generally) that a state engaged in armed conflict may always attach greater weight to the lives of its own soldiers than to those of enemy fighters and innocent civilians. Under a fiduciary theory, however, this assumption does not hold in contexts of internal armed conflict and belligerent occupation (the traditional domains of counterinsurgency) because a state bears duties of loyalty and care for all of its people, military and civilian alike, as equal beneficiaries of state action. Hence, a state may not use force against its people in a manner that would unnecessarily sacrifice the lives of some for the benefit of others. Any collateral damage to civilians likewise must be strictly necessary to preserve legal order for all. Nor may a state treat domestic insurgents in internal armed conflict as mere objects that can be destroyed in the absence of an imminent threat to legal order. Under the relational theory of state legal authority, IHL's proportionality standard is poorly tailored to the fiduciary duties that states bear toward their people.

B. The Relational Theory of *Lex Specialis*

The relational theory offers an attractive alternative approach for conceptualizing the "subject matter" of IHL/HRL choice-of-law analysis. Rather than defining the appropriate proportionality standard according to the type of *operation* conducted by a state (i.e., law enforcement or military), the relational theory counsels resolving interpretive ambiguities by applying the legal norm that best tracks the *relationship* between a state and those it would target with force. Whenever a state uses force against groups or people over whom it has asserted control, HRL's stricter proportionality standard constitutes the applicable *lex specialis*. Conversely, when states use force against those beyond their domain of asserted control, IHL proportionality applies. This relational approach to the *lex specialis* principle arguably casts the recent jurisprudence of the ECHR and the Inter-American Court in its best light.

Under the relational theory of *lex specialis*, HRL proportionality would govern a state's use of force in internal armed conflict, the paradigmatic context for counterinsurgency operations. States conducting internal armed conflict must take precautions to avoid unnecessary casualties to domestic insurgents. One specific implication of this standard, as we have seen, is that counterinsurgents in internal armed conflict would be required to apprehend rather than summarily kill the hypothetical unarmed insurgent-farmer who is found plowing his fields while temporarily away from the battlefield. States operating within their own territorial jurisdiction must take care, as well, to prevent all civilian casualties that can be avoided through the adoption of reasonable precautionary measures (HRL proportionality); it is not enough for counterinsurgents to ensure merely that civilian casualties are not "excessive" in relation to military objectives (IHL proportionality). The relational theory of *lex specialis* thus supports adherence to HRL proportionality in internal armed conflict.

HRL would also supply the applicable proportionality standard for states conducting counterinsurgency operations within contexts of belligerent occupation. Because an occupying state asserts control over the legal and practical interests of persons within occupied territory, the fiduciary principle requires an occupier to respect the basic human rights of occupied peoples during belligerent occupation regardless of whether the resident population has consented to the occupation.[73] Occupying states must afford "the civilian population" the "maximal safeguards feasible under the circumstances."[74] Although a state need not give persons within an occupied territory all of the benefits bestowed upon its own nationals, it must refrain from colonialist exploitation and other forms of domination. The relational theory of *lex specialis* supports these features of the law of occupation and suggests that states must honor the heightened protection of HRL's strict proportionality standard when they use force within an occupied territory.

Of course, in the transnational battlefields of twenty-first-century counterinsurgency, the planning and execution of insurgent attacks frequently extends beyond the jurisdiction of a single country's territory. In Afghanistan, for example, counterinsurgency operations are conducted not only by Afghan security forces but also by a coalition of foreign militaries with the consent of the current Afghan government. Counterinsurgency operations in Afghanistan are transnational in the additional sense that they are not confined to the territory of Afghanistan alone; for years, Taliban insurgents have gathered in, and mounted attacks from, the mountainous tribal regions of neighboring Pakistan. To the extent that coalition forces in Afghanistan conduct counterinsurgency operations at the behest of the government of Afghanistan or with the consent of the government of Pakistan, they arguably accede to the host states' fiduciary obligations to respect the human rights of Afghan nationals, whether these nationals are in Afghanistan or across the border in Pakistan. Thus, coalition officials coordinating attacks in Pakistan would bear an obligation under HRL to take all available precautions to minimize civilian casualties.

On the other hand, the fiduciary principle does not require states to apply HRL's stricter proportionality standard to all armed conflicts. A state defending its people from foreign aggression or terrorist attacks launched from abroad does not ordinarily assume a continuing control-based relationship with—and accordingly bears no fiduciary obligations toward—foreign belligerents. Instead, a state's primary obligation as fiduciary in such contexts is to protect the secure and equal freedom of its own people. As such, a state defending its people from armed attack originating abroad may target foreign fighters in order to repel ongoing attacks or prevent future attacks, provided that its use of force is consistent with *jus ad bellum* and applicable IHL constraints, including the principles of necessity, distinction, and proportionality. Under the relational theory of *lex specialis*, for example, IHL proportionality would apply to Israel's military response to Hezbollah rocket attacks during the July 2006 Israel-Lebanon conflict, irrespective of whether Hezbollah militants were best characterized at the time as state or nonstate actors. Plausibly, IHL's proportionality standard would also govern the United States' military strike against al Qaeda leader Osama bin Laden in Abbottabad, Pakistan—an action conducted by U.S. Navy Seals without Pakistan's foreknowledge or advance consent. In sum, to the extent that conflicts of law arise between IHL and HRL, the relational theory would give primacy to IHL proportionality in traditional international armed conflicts and at least some transnational armed conflicts involving nonstate actors, enabling the state as fiduciary to employ reasonable measures to protect its own people from attack.

Although the relational theory supports applying HRL proportionality to internal armed conflict and belligerent occupation, it does not necessarily follow that *all* HRL norms should apply as *lex specialis* in these contexts. When IHL and HRL come into tension, states should apply the norm that most closely tracks their relational duties toward persons subject to their coercive powers. Some norms of IHL—including, perhaps, customary standards for humane detainee treatment—might offer the best fit for states' relational obligations in internal armed conflict and belligerent occupation. With respect to the current debate over proportionality in counterinsurgency operations, however, the relational theory supplies a strong normative argument for HRL's stricter standard.

C. Case Study: The *Targeted Killings* Case

A brief case study from the Israeli-Palestinian conflict illustrates the relational theory's application to contemporary counterinsurgency operations. In 2005, the Supreme Court of Israel took several steps toward the relational theory in its controversial decision, *Public Committee Against Torture in Israel v. Israel* (*Targeted Killings*).[75] At issue in *Targeted Killings* was the Israeli military's practice of targeting suspected terrorists in the West Bank and the Gaza Strip with lethal force to prevent terrorist attacks in Israel. Between the start of the second *intifada* in February 2000 and the end of 2005, the Israeli military killed nearly three hundred suspected terrorists and approximately one hundred fifty innocent civilians and wounded hundreds of others in military strikes designed to prevent future attacks in Israel.[76] At the request of two human rights groups, the Israeli Supreme Court, sitting as a High Court of Justice, agreed to review the military's targeted killing policy and issue an advisory opinion clarifying the legal standards applicable to lethal strikes in the West Bank and Gaza.

The Court stressed at the outset of its opinion that Israel's targeted killing operations were conducted within the context of a longstanding belligerent occupation. The "normative system" applicable to this relational context "is complex," the Court observed, because it implicates not only IHL, but also HRL and "fundamental principles of Israeli public law, which every Israeli soldier 'carries in his pack' and which go along with him wherever he may turn."[77] Although the Court characterized IHL as the relevant *lex specialis* for this "international armed conflict,"[78] its proportionality inquiry did not end with IHL's traditional standard. Instead, it played with several different formulations of the proportionality principle. At one point, the Court suggested that targeting civilian terrorists would be permissible only if "innocent civilians nearby are not harmed" (a standard even stricter than HRL proportionality).[79] Almost in the same breath, however, the Court reasoned that collateral damage to innocent civilians would be lawful if the military observed an appropriate "balance" between "military advantage" and "civilian damage" (roughly, IHL proportionality).[80] Ultimately, the Court appeared to settle on a legal standard for targeted killing that closely tracked HRL proportionality: Israel could target terrorists in the West Bank and Gaza only where "there is no other less harmful means" to neutralize the threat.[81] Thus, the Court required the Israeli military to conform their targeted killing operations to the strict proportionality standard commonly associated with HRL.

To be sure, the *Targeted Killings* opinion is hardly a model of clarity, and critics are free to debate whether the Court was aware of the distance between its strict formulation of the proportionality principle and IHL's looser standard. It is entirely possible that the Court unintentionally conflated IHL and HRL, mischaracterizing the former in terms typically associated with the latter. A more charitable reading of the decision is possible, however. The Court clearly recognized that "the question of the legality of [targeted killing] according to customary international law is complex" and multilayered, implicating overlapping bodies of international and domestic law.[82] The Court also recognized that human dignity constitutes an overarching principle that "feeds the interpretation of international law, just as it feeds the interpretation of internal Israeli public law."[83] Guided by these concerns, the Court might have fashioned its proportionality standard based on a determination that IHL's proportionality standard required interpretive adaptation for the special relational context of belligerent occupation. Whether the Court adopted the HRL proportionality standard *qua* HRL or simply viewed the standard as a context-specific formulation of IHL, it appears to have concluded that a strict necessity standard best fit Israel's limited occupation of the West Bank and Gaza. This charitable reading of *Targeted Killings* reveals the Supreme Court feeling its way toward a context-sensitive approach to the use of force, one that takes both IHL and HRL into consideration while resolving conflicts based on relational context and a fundamental commitment to human dignity.

Had the Court in *Targeted Killings* employed the relational theory of *lex specialis*, its legal analysis would have been more coherent and persuasive. The starting point for analysis, once again, is the character of the relationship between Israel and suspected terrorists targeted in Judea, Samaria, and Gaza. In *Targeted Killings* and earlier decisions, the Court has held that these territories should be viewed as a single territorial unit under Israeli occupation, notwithstanding Israel's formal pull-out from Gaza in 2005.[84] Although Israel does not wield all of the public powers traditionally associated with sovereignty, its continued assertion of control, including the right to intervene militarily for security purposes, constitutes a form discretionary power that triggers the fiduciary principle.[85] As the ICJ observed in *Palestinian Wall*, Israel's assertion of control and its human rights obligations within the West Bank and Gaza are interrelated; Israel must honor the human rights set forth in the International Covenant on Civil and Political Rights "in light of the powers available to it" in the occupied territories.[86] Israel cannot, therefore, use lethal force in the West Bank and Gaza unless (1) targeted killing strikes are strictly necessary to guarantee legal order, and (2) appropriate precautions are taken in advance to avoid or minimize injury to innocent civilians. To the extent that HRL and IHL conflict in this context, HRL's stricter proportionality standard constitutes a minimum baseline of humane treatment that applies for the duration of Israel's occupation, irrespective of whether terrorist attacks from the West Bank and Gaza would otherwise constitute "armed conflict" for purposes of IHL.

Conclusion

This chapter has outlined a relational theory of state legal authority to use force in an effort to resolve tensions between IHL and HRL and lay the groundwork for a more coherent approach to counterinsurgency regulation. The relational theory supports the recent jurisprudence of the ECHR and the Inter-American Court by affirming that a state may use lethal force in internal armed conflict and belligerent occupation only if it "take[s] appropriate care to ensure that any risk to life is minimized."[87] Even when civil unrest crosses the armed-conflict threshold, a state must refrain from using force against its people except where strictly necessary to secure legal order. Conversely, the relational theory supports applying IHL proportionality primarily within its traditional domain of international armed conflict; states acting as agents for their people in the international realm may legitimately use force to defend their people from foreign aggression, subject to IHL principles of military necessity, distinction, and proportionality. On this relational theory of *lex specialis*, it is the character of a state's *relationship* with its target, not the severity of the threat to national security or the nature of the state's response to that threat, which dictates which proportionality standard applies.

Whether states will accept a broader role for HRL in counterinsurgency operations will depend, in part, on the perceived legitimacy of HRL's proportionality principle. But it will also depend on whether states perceive that adherence to HRL's proportionality standard will enhance or undermine their chances for success on the battlefield. With respect to this latter concern, recent developments in counterinsurgency theory offer grounds for cautious optimism. Increasingly, military strategists have recognized that success in counterinsurgency operations depends crucially upon the support of the embattled populace, and that public support, in turn, turns upon counterinsurgents' ability to minimize fatalities.[88] The U.S. military's *Counterinsurgency Manual* reflects this view when it observes that an "operation that kills five insurgents is counterproductive if collateral damage leads to the recruitment of fifty more insurgents."[89] Rules of engagement should therefore "address lesser means of force when such use is likely to create the desired effects and joint forces can do so without endangering themselves, others, or mission accomplishment."[90]

To implement this population-centric vision of counterinsurgency, the United States has experimented in Afghanistan with new rules of engagement that are more protective of civilian life than IHL proportionality.[91] While U.S. forces may defend themselves from insurgent attack, tactical directives instruct that lethal force should be employed only "where it is determined no other options are available to effectively counter the threat"—a standard that mirrors HRL.[92] In some respects, these new directives for counterinsurgency are even *more* protective of life than HRL. For instance, counterinsurgents are instructed to refrain from using lethal force against insurgent fighters when innocent civilians are present or if it is unclear whether innocent civilians are present in order to avoid the slightest risk of civilian casualties.[93] These unprecedented efforts to avoid fatalities reflect a growing appreciation among military strategists that success in counterinsurgency often depends on states counterintuitively using *less* force against insurgents and accepting *greater* risks to their own soldiers.[94]

The relational theory of *lex specialis* resonates with the new population-centric emphasis in counterinsurgency theory. HRL proportionality arguably tracks public perceptions regarding the legitimate use of force more closely than IHL proportionality. In an age of instant media and precision munitions, states are likely to face increasing public pressure to comply with HRL's strict-necessity standard. When states can minimize civilian casualties by abstaining from the use of lethal force, this option is not merely a moral imperative; it is also a strategic imperative, because it conserves the reservoir of public trust that is essential to sustain counterinsurgency. States that manage their political capital wisely by adhering to HRL proportionality may be more likely to achieve their military objectives.[95] While it would be premature to draw firm conclusions regarding the long-term prospects for HRL proportionality in counterinsurgency, there is mounting evidence that adherence to the HRL standard is not as politically unrealistic or operationally debilitating as some critics have suggested in the past. The prospects for state acceptance of the relational theory of counterinsurgency, including broader application of HRL proportionality in internal armed conflict and belligerent occupation, thus appear stronger now than ever before.

2

Reunifying the Laws of Armed Conflict: Noninternational Conflict and COIN Operations

War is a challenge to law, and the law must adjust. It must recognize that the old wineskins of international law, domestic criminal procedure, or other prior frameworks are ill-suited to the bitter wine of this new warfare. We can no longer afford diffidence. This war has placed us not just at, but already past the leading edge of a new and frightening paradigm, one that demands new rules be written. Falling back on the comfort of prior practices supplies only illusory comfort.[1]

3

REUNIFYING THE LAW OF ARMED CONFLICT IN COIN OPERATIONS THROUGH A SOVEREIGN AGENCY THEORY

Eric Talbot Jensen*

AS THE ABOVE quote from the 2010 D.C. Circuit case of *Al-Bihani v. Obama* reflects, the U.S. government's response to the war on terror and the resulting counterinsurgencies have led many governments, practitioners, and academics to debate the continuing validity of the current legal paradigm. Many, such as Judge Brown, have concluded that the law of armed conflict (LOAC) is not well suited for this type of warfare and that the LOAC should be amended to accommodate modern conflicts, such as counterinsurgency.

The roots of this concern lie in the modern bifurcation of the LOAC into rules that apply in international armed conflicts (IAC) and a subset of those rules which apply in noninternational armed conflicts (NIAC). This bifurcation began as early as the beginning of the twentieth century and achieved formal codification in the 1949 Geneva Conventions.[2] The bifurcation is created in Articles 2 and 3 that are common to all four Conventions and then extended in the two 1977 Protocols to the 1949 Geneva Conventions.[3]

This bifurcation was intended to add protections to the victims of armed conflict and encourage greater compliance with the LOAC across a wider range of conflicts. History, however, shows that this bifurcation has had little effect, if any, on nonstate compliance with the LOAC[4] and has mainly acted to limit states which seek to be compliant. Further, as illustrated by the case of the U.S. response to the war on terror and its subsequent counterinsurgencies, the application of law has been almost exclusively focused on conflict classification. If a state calls an armed conflict an IAC, it is bound by one set of duties and authorities, and if it calls it a NIAC, it is bound by another. Further, if it avoids calling a conflict an armed conflict at all, it can claim its use of armed forces is not governed by the LOAC, thus potentially creating the "no law" zone the United States sought with regard to terrorists and insurgents in the post–9/11 conflicts.[5]

In addition to the U.S. experience, recent events in Colombia,[6] Russia,[7] and Mexico[8] demonstrate the failure of the bifurcation. Because the law focuses on how the conflict is classified—whether as an IAC, a NIAC, or not as armed conflict at all—states are able to determine the law that applies as a matter of policy, rather than as a matter of fact.

Modern practice by states leads to the conclusion that the international community's focus on conflict classification to determine which law applies is misplaced and does not facilitate application of fundamental LOAC protections. Rather than using the type or existence of armed conflict as the gauge for LOAC applicability, states should apply the full LOAC every time they utilize their armed forces as state agents to apply sovereign force, including in counterinsurgencies. This turns the focus from what a state chooses to call a conflict, to the forces a state chooses to use to deal with a conflict. Application of the LOAC to all forceful activities by state sovereign forces is drawn from the historical development of the LOAC and will provide a more solid foundation upon which to place the LOAC, diminishing the potential for political manipulation of the law and for incentivizing noncompliance.

Applying this author's sovereign agency theory of LOAC applicability will immediately accrue benefits not currently in place in modern armed conflict. First, applying the full LOAC every time a state uses its armed forces will mean that the starting point for humanitarian protections is always the most robust of possible alternatives. Second, it will provide clarity for armed forces, making them more efficient and effective. Third, it will not require significant changes to what many states already do. History has shown that applying the full LOAC to all forceful activities of a state's armed forces is a manageable approach, though it has only been done as a matter of policy to this point. For the sovereign agency theory of LOAC applicability to truly overcome the problems inherent in the conflict classification paradigm, it must be accepted as a matter of law.

For the purposes of this argument, applying the "full LOAC" means a state should apply those customary provisions that normally apply during IAC as well as any conventional obligations imposed by a state's specific treaty obligations. This includes the application of human rights law as appropriate.[9]

Section I briefly describes the current paradigm of LOAC applicability based on conflict characterization, including a brief historical review of the bifurcation of the LOAC into provisions regulating NIAC and IAC. It further demonstrates that the bifurcation has not been effective in curbing either the violence against victims of armed conflict or in promoting LOAC compliance by participants in armed conflict. Rather, it has become a political tool to manipulate the applicable law and leads to a lack of clarity on the battlefield. Section II expands on the argument made above that states should apply the LOAC to their use of armed forces to apply sovereign force and outlines some benefits of the sovereign agency theory, including historical examples. Finally, section III analyzes some recent developments that have positioned states to make just such a transition in the law and offers a way forward to complete the transition.

I. THE FAILURE OF CURRENT BIFURCATED PARADIGM

"The terms 'international' and 'non-international' conflict import a bipartite universe that authorizes only two reference points on the spectrum of factual possibilities. The terms are based on a policy decision that some conflicts... will be insulated from the plenary application of the law of armed conflict—even though such conflicts may be more violent, extensive and consumptive of life and value than other 'international' ones. The terms are, in effect, a sweeping exclusion device that permits the bulk of armed conflict to evade full international regulation. This exclusion is not one that comports easily with the manifest policy of the contemporary law of armed conflict, which seeks to introduce as many humanitarian restraints as possible into conflict, without judgments about its provenance, its locus, or about the justice of either side's cause."[10]

By the early nineteenth century, states recognized basically two forms of armed conflict: armed conflict between two or more states, and civil wars.[11] Interstate conflict, or what has become known as IAC, invoked all the principles of the laws of war as they were then understood. In contrast, during civil wars, states did not feel constrained to apply such international rules, and the treatment of opposing fighters was considered a matter of domestic concern. This difference "was based on the premise that internal armed violence raise[d] questions of sovereign governance and not international regulation."[12]

However, the international community, led by the International Committee of the Red Cross (ICRC), soon recognized the plight of victims of civil wars or noninternational armed conflicts. As early as the 1912 IXth International Conference of the Red Cross meeting in Washington, D.C., the ICRC argued for extending some LOAC rights to victims of civil wars. This initiative was not well received by the majority of the participants who felt that "the Red Cross Societies have no duty whatever to fulfil [*sic*] toward rebel or revolutionary troops, which the laws of [a] country can only consider as criminals."[13]

The ICRC continued to advance the idea of codifying protections for victims of noninternational armed conflicts through the subsequent decades,[14] until the aftermath of World War II. At the 1949 Diplomatic Conference, which would ultimately produce the revised modern Geneva Conventions, the ICRC once again raised its request to apply the full LOAC to NIACs. While the majority of states remained opposed to such a sweeping measure, they agreed to have the issue sent to a Mixed Commission, tasked with examining articles that were common to all four proposed Conventions. Within these "common" articles were those that determined the applicability of the LOAC. The Mixed Commission created Article 2, which accepted the traditional approach of applying the Conventions to armed conflicts between sovereign states.[15] Common Article 3 then provided a set of minimum protections to "conflict[s] not of an international character occurring in the territory of one of the High Contracting Parties."[16] Though a step in the right direction according to the ICRC, it was a minimal one as these provisions in Article 3 represented only the "most rudimentary principles of humanitarian protection."[17]

Perhaps as important as the actual protections provided by Common Article 3, its creation signified the beginning of the application of the LOAC to noninternational armed conflicts and the bifurcation of the LOAC between IACs and NIACs. In the two decades following the adoption of the 1949 Geneva Conventions, the majority of conflicts throughout the world were NIACs. In its assessment of these NIACs, the ICRC determined that "[h]owever useful Article 3 had proved, it had been shown to have numerous loopholes, and this made it no longer possible to ensure sufficient guarantees to the victims in question."[18] The ICRC then pushed for greater clarification of the rules that would apply to armed conflicts and convened a Conference of Government Experts in 1971, urging for the dissolution of the bifurcation of LOAC applicability and applying the full LOAC to NIACs as well.[19]

The ICRC forwarded its proposals to states, but found that the majority of states preferred to maintain the distinction between IACs and NIACs. As a result, the ICRC abandoned the "single protocol" approach[20] and proposed two separate protocols to the 1977 Conference of State Parties—one dealing with IAC and one with NIAC. These two proposals provided the basis for two Additional Protocols to the 1949 Geneva Conventions, one dealing with IACs, which has come to be known as Additional Protocol I (AP I),[21] and the other dealing with NIACs, which has come to be known as Additional Protocol II (AP II).[22]

The legal effect of the promulgation of the two Protocols was the cementing of conflict classification as the standard for LOAC application. The Protocols divided the application of the law into two categories and assigned rights and responsibilities within them, effectively requiring a threshold question regarding conflict characterization in every discussion of applicable law. As Emily Crawford has observed, "characterization of the conflict is crucial to determining what level of protection is provided for combatants and civilians."[23]

The Additional Protocols were widely ratified, though there were some notable exceptions such as the United States.[24] Though widely ratified, the Protocols have not been widely invoked, particularly AP II. In fact, since ratification the vast majority of claims that AP II is applicable to an armed conflict have come from international bodies or third-party states and not from the state within whose borders the conflict is occurring.[25]

Instead of being an effective means of extending LOAC protections to noninternational armed conflicts, such as counterinsurgencies, states normally try extremely hard to avoid the applicability of these Protocols to their conflicts.[26] Recent examples include Israel in its conflict with the Palestinians[27] and Mexico in its current fight against narcotics trafficking.[28] As will be discussed below, in these cases, just as with the United States in the war on terror mentioned at the beginning of this chapter, the state is trying to select the applicable portion of the bifurcated LOAC in order to choose law applicability.

This means that the bifurcated system has become manipulable[29] by states for political purposes, allowing them to choose how they intervene in the affairs of another state as a means of ensuring that particular provisions of law will or will not apply to the conflict. As Stewart put it:

> States and non-state actors have proved equally willing to favour or fabricate accounts of foreign participation in internal conflicts for their own wider political

> gain. As a result, the characterization of armed conflicts involving international and internal elements, and the applicable law that flows from that characterization, are frequently "the subject of fierce controversy of a political nature."[30]

While this type of manipulation of the law for political purposes is certainly not a new phenomenon, with regard to the LOAC, it demonstrates that the bifurcation of applicable law has not worked. This is especially true in the case of counterinsurgencies where the domestic government under attack has incentives to neither legitimize the insurgents nor provide them with any other protections than those mandated under domestic law.

Instead of accomplishing the desired goals of protecting victims and encouraging state compliance, the bifurcation of the LOAC has had the opposite effect. Rather than increasing protections in armed conflict, it "frustrates the humanitarian purpose of the law of war in most of the instances in which war now occurs."[31] Clearly, in light of all of these concerns, it is time to reexamine the bifurcated paradigm of LOAC application.

II. THE SOVEREIGN AGENCY THEORY OF LOAC APPLICABILITY

> *"What is inhumane, and consequently proscribed, in international wars, cannot but be inhumane and inadmissible in civil strife."*[32]

Given the ineffectiveness of LOAC bifurcation at providing protections for individuals on the modern battlefield, states should reconsider the LOAC applicability paradigm. One solution would be a return to the roots of the application of sovereign force and combatancy—the principle of agency. When a state deploys its military to an armed conflict, it imbues those forces with agency and exempts them from the individual consequences of traditional criminal activities, such as murder and destruction. Under a sovereign agency theory of LOAC applicability, as long as a member of the military is acting as the state's agent and taking advantage of this immunity, the full provisions of the LOAC should apply, including the protections for victims of armed conflict.

A. Sovereignty and the Development of the LOAC

Rules regulating warfare have existed since the beginning of history,[33] but have not been applied universally or consistently.[34] For example, in the seventeenth century Grotius wrote:

> I saw prevailing throughout the Christian world a license in making war of which even barbarous nations should be ashamed; men resorting to arms for trivial or for no reasons at all, and when arms were once taken up no reverence left for divine or human law, exactly as if a single edict had released a madness driving men to all kinds of crime.[35]

In response to this scene, Grotius authored one of the seminal works in international law in an attempt to right this uncontrolled culture of violence.

When European powers signed the Treaty of Westphalia, they essentially solidified states as sovereigns and the primary actors in the international community.[36] This also empowered states with the authority and responsibility to monopolize violence through standing armies and navies.[37] The process of establishing that monopoly on violence created a system of agency between sovereign and soldier. As Rousseau indicated, the soldier was not viewed as an individual but as an agent of his sovereign until such time as he could no longer fight or had laid down his arms. Then, he reverted to his status as an individual and was treated as such.[38]

In response to this agency relationship, the laws and customs of warfare grew to regulate the use of force by the sovereign's armies and navies.[39] Soldiers and sailors acting as the sovereign's agents were granted certain privileges, such as combatant immunity, and correspondingly required to comply with certain duties, such as carrying out the sovereign's orders and staying within the bounds of those orders. Failure to remain within those bounds could result in individual criminal responsibility.[40] Compliance with those orders generally granted immunity for otherwise illegal acts.

As this agency relationship became more generally accepted, states began to reciprocally recognize the agents of other states and grant reciprocal individual immunity for sovereign-directed acts by other sovereign's militaries. For example, Article 57 of the Lieber Code which was written for the American Civil War, states "[s]o soon as a man is armed by a sovereign government and takes the soldier's oath of fidelity, he is a belligerent; his killing, wounding, or other warlike acts are not individual crimes or offenses."[41] Contrast that with a provision from the same code, but dealing with fighters not associated with the sovereign:

> Men, or squads of men, who commit hostilities, whether by fighting, or inroads for destruction or plunder, or by raids of any kind, without commission, without being part and portion of the organized hostile army, and without sharing continuously in the war, but who do so with intermitting returns to their homes and avocations, or with the occasional assumption of the semblance of peaceful pursuits, divesting themselves of the character or appearance of soldiers—such men, or squads of men, are not public enemies, and, therefore, if captured, are not entitled to the privileges of prisoners of war, but shall be treated summarily as highway robbers or pirates.[42]

Because the tradition, practice, and reciprocity that evolved from the granting of agency to a sovereign's military revolved around interstate conflicts, states had not allowed the rules to apply to other types of armed conflict. In fact, it was the bifurcation of the modern LOAC that unhinged the foundation of these rules in sovereign agency and refocused the LOAC applicability analysis on conflict classification. A return to an agency-based LOAC applicability will resolve some difficulties that have resulted from the LOAC bifurcation paradigm and promote the broader application of humane regulation of armed conflict.

B. Sovereign Agency in Counterinsurgency

Counterinsurgencies represent one of the clearest examples in the catalogue of contemporary armed conflict of the need for the sovereign agency theory of LOAC applicability—a need reiterated repeatedly in the U.S. post–9/11 experience. In response to the transnational terrorist attacks against the United States on September 11, 2001, the Bush administration argued that the ensuing armed conflict did not fit neatly within the current bifurcated paradigm of LOAC applicability. The conflict with al Qaeda was neither an IAC, because there were not two states at war with each other, nor a NIAC, because it was not a traditional civil war and because of the transnational nature of al Qaeda.[43] The U.S. decision was highly contested amongst allies, academics, and practitioners.[44]

If the same situation occurred under a sovereign agency paradigm, once the United States deployed its armed forces to use violence against al Qaeda and the Taliban, the full LOAC would apply. For those members of the military called upon to apply that law in the conflict in Afghanistan, such clarity would have been a welcome relief.

Similarly, in the case of Mexico and its fight against drug trafficking,[45] there has been no clear statement of applicable law. However, Mexico has deployed almost fifty thousand military and police forces against the "forces" of the cartels, which some estimate at around one hundred thousand.[46] The resulting deaths exceed the death totals in Afghanistan for the same period.[47] Despite this, Mexico has not conceded that this conflict is an "armed conflict" and has not agreed to apply the AP II provisions to the situation.[48] Under the sovereign agency theory, it would be clear which law applied to the military's use of force.

Israel demonstrates another difficult situation under the bifurcated LOAC paradigm that is easily solved by the sovereign agency theory. Israel has resisted the claim that the conflict with Palestinians is either an IAC or a NIAC. The government and Supreme Court have described the conflict in various ways,[49] making arguments which are rooted in conflict classification for LOAC applicability. At least some of this discussion is generated by a fear of Israel providing international legitimacy to its enemies, if the conflict is given a formal categorization.[50] Under the sovereign agency theory, the full LOAC would apply to Israel's deployment of its forces and would be triggered by the deployment of the military, not the government's decision on conflict classification.

These examples illustrate situations in which the current LOAC applicability paradigm does not provide the battlefield protections it is intended to provide. Transition to a sovereign agency theory where the military is governed by the LOAC any time it is used as the sovereign's agent to use force would provide clarity to an area of international law and benefit states in many practical ways.

C. Benefits of the Sovereign Agency Theory

Adopting the sovereign agency theory of LOAC applicability will provide a number of immediate benefits. First, it would guarantee that the most robust baseline of battlefield protections would be the default in every armed conflict. The law that applies in current situations of IAC is the most extensive and provides the most detailed and robust protections for both victims of and participants in armed conflict. Because it is this body of law that would apply under the sovereign agency theory, every time a sovereign used its forces to apply sovereign force, this extensive and robust body of legal regulation would apply. This applicable law would include all the customary rules on weapons, attacks, targeting, and even detention, as well as all of a state's conventional law obligations such as arms control, weapons prohibitions, and other pertinent treaty obligations.

The debates concerning the detention of persons in Guantanamo would have been unnecessary had the United States been applying the sovereign agency theory. Because they were being held by the military, they would have been treated in compliance with the Geneva Conventions, a position that only officially came about in 2006,[51] after the *Hamdan* decision by the U.S. Supreme Court.[52]

Second, application of the sovereign agency theory would provide much needed clarity to participants in armed conflict. Under the current paradigm, members of the military have to wait for their civilian governments to determine the categorization of the armed conflict before they know what law will apply to their operations.[53] This lack of clarity inevitably causes confusion and consternation for members of the armed forces. For example, in the reports filed by several U.S. Army legal offices, it was noted that there was a lack of legal direction from the civilian authorities on what law to apply during the early stages of Operation Iraqi Freedom.[54] This was also the case in Afghanistan where the United States initially took the view that the Geneva Conventions did not apply to their military operations but later clarified that the conflict was considered an armed conflict to which the LOAC applied.[55]

This dilemma is echoed by the U.K. Ministry of Defence's *Manual of the Law of Armed Conflict*. In discussing what law applies to armed conflicts generally, the manual states:

> There is thus a spectrum of violence ranging from internal disturbances through to full international armed conflict with different legal regimes applicable at the various levels of that spectrum. It is often necessary for an impartial organization, such as the International Committee of the Red Cross, to seek agreement between the factions as to the rules to be applied.[56]

If a third party is required to resolve applicable law, it seems obvious that there exists a lack of clarity that inevitably places the armed forces in an untenable situation of not knowing which legal standards to apply during hostilities. Furthermore, if this decision of applicable law is a matter of negotiation between parties, it will invariably be politicized and prone to manipulation based on policy considerations, rather than made as a legal determination, as it was in the case of the United States in Afghanistan. In the interim, military forces on the ground are left with few legal answers.[57]

As further evidence, in the Tadic Jurisdictional Appeal decision, the International Criminal Tribunal for the Former Yugoslavia (ICTY) characterized the conflict in the Former Yugoslavia "at different times and place as either internal or international armed conflicts, or as a mixed internal-international conflict."[58] Many counterinsurgencies will fall into a similar classification problem. Of course, this type of a post-hoc determination about conflict classification is completely unhelpful to military forces facing a deployment to the conflict zone. If trained jurists, such as those sitting on the ICTY, have to struggle with these questions years after the conflict and with a clear view of the facts and they still find that the conflict in question was of different types at different times, how can one expect even the most well-meaning governments to be able to discern a clearer answer in advance and to adequately prepare even the most professional armed forces to apply the correct LOAC provisions in the appropriate ways?

Acceptance of the sovereign agency theory would provide much needed clarity and simplicity. Militaries already train to the full LOAC standard and then adjust from those standards to meet other mission requirements. Knowing in advance that regardless of the mission, militaries would be applying the IAC standards would greatly increase the efficiency of that training and the effectiveness of its application in armed conflict.

These are just a few of the benefits that would accrue from the sovereign agency theory of LOAC applicability. But they represent a significant paradigm shift in LOAC protections, particularly in counterinsurgencies where states are dis-incentivized to apply robust protections to the insurgents.

D. Issues

A decision to apply the sovereign agency theory is not without issues. As discussed earlier, states have resisted the application of international law to circumstances other than conflict between sovereigns.[59] Whether because it legitimizes the opposing armed group, or prevents the state from manipulating the law to their benefit (as discussed above), states will view the sovereign agency theory as surrendering some of their sovereignty over domestic issues. For example, certain weapons limitations treaties, such as the Chemical Weapons Convention,[60] might preclude the use of certain weapon systems that otherwise only apply to IACs.

Additionally, if states adopted the sovereign agency theory and applied the full LOAC during an internal armed conflict, the military's detention of opposing fighters would be governed by the LOAC rather than by domestic criminal law. This would be particularly difficult if a state's military forces were working side by side with the state's law enforcement forces, such as in Mexico currently.[61] But both of these problems can be minimized by state action through rules of engagement[62] or other appropriate policy remedies.

Perhaps the most vexing cost is the lack of reciprocity with nonstate actors. A state's decision to apply the sovereign agency theory will have no effect on whether the nonstate actors it opposes will do the same. In fact, it is probable that nonstate actors will be more likely to see a state's unilateral declaration of compliance with the LOAC as a new point of advantage that can aid an asymmetrically disadvantaged opponent.[63] On the other hand, states which are already committed to be compliant will merely have to be compliant in a broader range of conflicts. Further, there is some evidence that nonstate actors can be incentivized to become compliant in certain circumstances.[64]

Though there are certainly issues with the adoption of the sovereign agency theory, these are neither insurmountable nor serious enough to prevent its implementation. In fact, as the next section will demonstrate, some states are already moving in that direction.

III. THE WAY AHEAD

> *"It is all war, whatever its cause or object, and should be conducted in a civilized way... There is no distinction from a military view between a civil war and a foreign war until after the final decisive battle."*[65]

Assuming a state recognizes the benefits listed above, is it really possible for that state to unilaterally embrace the sovereign agency theory and adopt the full LOAC when it is engaged in bitter insurgency such as the United States faced in Vietnam in the 1970s or in Afghanistan in the 2000s?

This section will discuss two influential factors: first, that it is a manageable approach by states to make the change and will not cause as much inconvenience as might be anticipated; and second, in addition to state practice already moving in this direction, both scholarly opinion and international jurists are embracing the underlying notion of sovereign agency.

A. A Manageable Approach for States

The role of militaries has changed dramatically since the 1940s, as have the types of armed conflicts most prevalent today.[66] States more fully recognize the need for clarity across the increasingly diverse types of missions that they assign to their armed forces today: including fighting nonstate organized armed groups and insurgencies,[67] conducting counterdrug operations against narcotics traffickers,[68] dismembering transnational criminal business networks,[69] forcefully separating belligerents, and implementing peace agreements.[70]

The United States has faced many of these challenges and has used its military in a number of nontraditional roles, including peace operations, disaster relief, humanitarian aid, and support for counterdrug operations.[71] These missions have been doctrinally termed Operations Other Than War,[72] highlighting their nontraditional nature and distinguishing them from more traditional IAC. In fact, the majority of U.S. operations since World War II have not been officially recognized as international armed conflicts.[73]

The prevalence of nontraditional conflicts, such as counterinsurgencies, has presented the United States with the question of how to apply the law in these cases. Two examples may be instructive.

Despite being involved in a ferocious counterinsurgency in Vietnam, the United States made the decision to treat captured Viet Cong as prisoners of war in compliance with the Geneva Conventions.[74] The Department of Defense (DoD) also promulgated a Directive which applied the law of war to operations occurring in Vietnam.[75] The successor to the DoD Directive also existed on 9/11[76] and was reissued on May 9, 2006. In the latest edition, the United States promulgates a policy that "[m]embers of the DoD Components comply with the law of war during all armed conflicts, however such conflicts are characterized, and in all other military operations."[77] It is significant to note that this policy was restated during a time when U.S. military forces were actively engaged in two simultaneous counterinsurgencies—Iraq and Afghanistan. No doubt, the benefits discussed above, including increased clarity, may have played a role in that decision. Regardless, the practical effect of the DoD policy is that the United States is already implementing sovereign agency theory. If the United States can do so in those circumstances, such a policy does not seem unmanageable for any nation.

Other nations are following suit. In the recent *Customary Humanitarian Law Study*, the ICRC analyzed state practice and articulated its analysis of what principles of the LOAC could be considered customary.[78] The study found that most of the customary provisions of IAC concerning targeting and the treatment of the victims of armed conflict are already being applied by states in NIAC.[79] The practice of these states, in combination with the practice and policy of the United States, indicates that a transition to agency theory would be very manageable and is already underway.

B. International Jurisprudence and Scholarship

International jurisprudence and scholarly argument are not as persuasive in this area as actual state practice, but when the two align, it is a strong predictor of the movement of the law.

Recent decisions in international tribunals have also expressed dissatisfaction with the bifurcation of the LOAC and have been slowly eroding the differences between IAC and NIAC. For example, in the Tadic Jurisdiction Appeal, the Appeals Chamber for the International Criminal Tribunal for the Former Yugoslavia (ICTY) held that customary rules governing internal conflicts include:

> [P]rotection of civilians from hostilities, in particular from indiscriminate attacks, protection of civilian objects, in particular cultural property, protection of all those who do not (or no longer) take active part in hostilities, as well as prohibition of means of warfare proscribed in international armed conflicts and ban of certain methods of conducting hostilities.[80]

Antonio Cassese, who was then president of the ICTY, concluded that "there has been a convergence of the two bodies on international law with the result that internal strife is now governed to a large extent by the rules and principles which had traditionally only applied to international conflicts."[81]

Many scholars also agree that the meaningful difference in LOAC applicability between IAC and NIAC is dwindling. Perhaps the most profound statement on the growing convergence between the IAC and NIAC is the International Institute of Humanitarian Law's *Manual on the Law of Non-International Armed Conflict*.[82] The manual "is a guide for behaviour in action during non-international armed conflict. While not a comprehensive restatement of law applicable in such conflicts, it nevertheless reflects the key principles contained in that law."[83] An analysis of these "key principles" shows a distinct similarity to the provisions governing IAC. The convergence of these two bodies of law demonstrates that the continued bifurcation is unnecessary and a different paradigm is warranted.

Along the same lines, in its *Customary Law Study*,[84] the ICRC finds that numerous provisions of Protocol II are customary international law and apply in all armed conflicts.[85] These provisions are also a part of the IAC regime, demonstrating the close links between the two.

C. Further Actions

With states' armies applying the agency theory as a matter of policy, and with that policy supported by the jurisprudence of international tribunals and the writings of eminent scholars, it is easy to envision the way ahead. While now only done as a matter of policy or out of some rationale such as convenience or the desire for clarity, states need to embrace the agency theory of LOAC applicability and apply the full LOAC as a matter of law. When states proclaim that they accept a legal obligation to apply the full LOAC to every employment of their armed forces to a mission where those armed forces are expected to use violence, the LOAC bifurcation paradigm will be set aside, and the sovereign agency theory will be ready to fill the gap.

This transition could be initiated by states already applying the theory as a matter of policy, such as the United States. The United States could call for a Convention and propose a revision of the Geneva Conventions to accomplish this purpose. Alternatively (and far more likely to succeed), each individual state could make a unilateral decision to embrace the sovereign agency theory and make public declarations to that effect.[86] This could be followed by incorporation of this decision into the state's domestic law.

Once a number of states made this choice, others could be invited to follow the same pattern. State pressure, particularly from allies or friends, is an effective means of encouragement to hesitant states. Ultimately, individual state practice is the most effective mechanism to accomplish the end of the LOAC bifurcation. If this course of action occurs, however, AP I and AP II would have to be significantly revised or abrogated in order to remove the codification of the LOAC bifurcation.

Conclusion

The current LOAC applicability paradigm requires a state to classify the conflict and then determine which law applies based on that determination. These decisions by states have often been open to manipulation in order to accomplish policy objectives. The recent actions by a number of states confirm the truth of this. It is time for the international community to rethink the current paradigm and select a more effective and principled basis for LOAC applicability.

The sovereign agency theory is that more effective basis. The application of the LOAC to all activities by state sovereign forces during armed conflict is a much better means of protecting the victims of armed conflict and will provide a much more solid foundation upon which to place the LOAC. The benefits of doing so include greater clarity for those actually applying the force, as well as making the law less manipulable.

The transition would not be too difficult and is supported by the practice of states as well as international jurisprudence and scholars. For states such as the United States, it would merely require the commitment to do, as a matter of law, what they are now doing as a matter of policy. In a time of armed conflict where NIACs, such as counterinsurgency in Afghanistan, dominate the legal landscape, it is vital to increase the protections granted by the LOAC to all individuals on the modern battlefield.

4

TWO SIDES OF THE COMBATANT COIN: UNTANGLING DIRECT PARTICIPATION IN HOSTILITIES FROM BELLIGERENT STATUS IN NONINTERNATIONAL ARMED CONFLICTS

Geoffrey Corn*

Introduction

Determining who qualifies as a lawful object of attack in the context of contemporary military operations against nonstate belligerents[1] is an increasingly demanding challenge. While it is axiomatic that only persons who qualify as either enemy belligerents or civilians taking a direct part in hostilities fall into this category, the nature of the nonstate operatives has blurred the line between civilians protected from deliberate attack and belligerent operatives subject to attack.[2] Indeed, many scholars of military strategy contend that this blurring is the result of deliberate tactics employed by nonstate operatives in an effort to offset the military superiority of their nation-state military opponents by adding tremendous complexity to the target decision-making process.[3]

The target identification and selection process is never easy; uncertainty is an element of any armed conflict. Threat identification has always been a core task of U.S. armed forces, a task intended to not only ensure the lawful use of combat power but to also maximize the tactical and operational effects of that use while mitigating the risk to civilians.[4] However, when engaged in combat operations against nonstate belligerents whose physical characteristics are almost always indistinguishable from the civilian population, the task of threat identification becomes exponentially more complex. Furthermore, the consequence of error is exacerbated by the risk of alienating the civilian population, a consequence fundamentally inconsistent with the core tenet of counterinsurgency (COIN) strategy—protection of that population.[5]

The difficulty in distinguishing the protected (civilians) from the unprotected (belligerents and civilians taking a direct part in hostilities) does not, however, warrant a fundamentally different targeting paradigm in counterinsurgency operations, form of a noninternational armed conflict (NIAC), than in conventional international armed conflicts (IACs).[6] The integrity of the target legality framework depends on the recognition of opposing belligerent groups in any armed conflict. This recognition facilitates implementation of the principle of distinction by allowing belligerent forces to segregate those they encounter into two distinct groups: those presumed hostile and therefore subject to immediate attack, and all others (civilians) presumed nonhostile and therefore protected from immediate attack.[7] As will be discussed in this chapter, neither of these presumptions is absolute; both may be rebutted based on the nature of the interaction with friendly forces. However, these presumptions add a modicum of clarity to an increasingly uncertain operational environment, clarity derived from the very nature of armed conflict and essential to protect both civilians and belligerents from the effects of overzealous or unjustifiably hesitant targeting authority.

This all may seem obvious, but in reality there is an increasing tendency to treat all nonstate actors as merely a conglomeration of civilians who take a direct part in hostilities. This trend gained momentum following publication by the International Committee of the Red Cross (ICRC) of the Interpretative Guidance on the Meaning of Direct Participation in Hostilities (DPH Study)—although, ironically, the DPH Study rejects this position.[8] Indeed, the derivative effect of considering all such nonstate actors as civilians is that no one is genuinely protected by the distinction obligation.

The trend appears to be the result of the combined effect of the lack of an explicit definition of a combatant that is applicable to NIAC, and the DPH Study's endorsement of the concept of "continuous combat function" (CCF) as a means of establishing direct participation in hostilities (which results in a loss of civilian protection from attack).[9] The effect of the CCF concept has made it more convenient to analyze the legality of attacking nonstate actors through the DPH methodology than to assess whether such actors fall into a category of presumptively targetable belligerents subject to attack no differently than their IAC counterparts.

This chapter will challenge that approach to targeting categorization in NIAC as flawed. In so doing, it will argue that it is critical to acknowledge that NIAC—any NIAC—involves hostilities between opposing armed belligerent groups whose members are presumptive military objectives. To support this argument, the chapter begins by discussing the law of armed conflict's (LOAC's) categorization of civilians and belligerents (combatants in IAC), and how a lack of an explicit treaty definition of combatant in the NIAC context is an obstacle to acknowledging analogous categorization in NIAC.

The chapter then explores organizational membership and how subordination to command and control is the fundamental difference between belligerents and civilians in any armed conflict. It will explain the difference between status- and conduct-based targeting and why a focus on conduct to assess belligerent status is merely a permutation of traditional status recognition analysis.

The chapter then contrasts that approach by examining why the use of conduct undermines the extension of the DPH rule to define enemy belligerent forces. These problems result in the DPH Study's problematic and arguably schizophrenic imposition of a minimum force requirement even when targeting those engaged in CCF. As the chapter details, the utility of CCF is not in assessing which civilians are taking a direct part in hostilities, but in determining when an individual appearing to be a civilian is in fact a belligerent operative of an armed organized group.

The chapter will then address why treating all nonstate opposition personnel as civilians taking a direct part in hostilities—even when applying the CCF concept—provides these operatives with an unjustifiable windfall and conflates law and rules of engagement. The chapter concludes with a proposal of how to reconcile the DPH Study with status-based targeting presumptions to maintain the distinction integrity. Acknowledging that NIAC involves armed hostilities between competing belligerent groups is a critical first step, and the DPH Study makes an important contribution to this acknowledgment. However, this must be accompanied by an additional acknowledgment: *all* belligerent operatives—those involved in IAC and NIAC—are subject to status-based targeting authority.

I. BACKGROUND

A. LOAC Categorization of Civilians and Combatants

Combatant. For a layman, the meaning of this term probably seems obvious: a soldier who fights for her country. However, in the LOAC lexicon, nothing could be further from the truth. While it might be tempting to invoke the ubiquitous "you know it when you see it" Supreme Court definition of pornography for the term "combatant," operational reality and legal definition appear severely attenuated where this term is concerned.[10] The mere fact that a definition for "combatant" remained purely customary until finally defined in the 1977 Protocol Additional I to the Geneva Conventions of 1949 (AP I) indicates the extent of legal uncertainty and complexity associated with this term.[11]

Ironically, it was not widespread discomfort with the uncertainty of the term "combatant" that motivated adoption of an express definition in AP I.[12] Instead, it was the need to define "civilian"—a definition central to the protection established for civilians at the core of AP I's targeting regime—that provided the motivation. Because AP I adopted a negative definition of "civilian" (all individuals who are *not* combatants), it was necessary to provide an explicit definition of "combatant."[13] Although it took until 1977 for the development of a positive treaty definition of "combatant," there existed prior to this date a general customary understanding of the term. This customary understanding was based on two intertwined concepts: privileged belligerent and prisoner of war.[14]

In 1899, The Hague Convention II and Annexed Regulations—the first comprehensive multilateral treaty regulating land warfare—defined individuals lawfully authorized to participate in hostilities.[15] According to Article 1 of the Regulations:

> The laws, rights, and duties of war apply not only to armies, but also to militia and volunteer corps fulfilling the following conditions:
>
> 1. To be commanded by a person responsible for his subordinates;
> 2. To have a fixed distinctive emblem recognizable at a distance;
> 3. To carry arms openly; and
> 4. To conduct their operations in accordance with the laws and customs of war.
> In countries where militia or volunteer corps constitute the army, or form part of it, they are included under the denomination "army."[16]

It is clear from this provision that regular armed forces, or "armies," were considered ipso facto lawfully authorized to engage in hostilities. An identical version of this definition was included in the 1907 revision of the Regulations.[17] This provision reflected the general understanding of the time that individuals authorized by their state to participate in armed hostilities—most obviously members of the regular armed forces but also militia and volunteer personnel properly connected to the command and control structure of the regular armed forces—were lawful belligerents.[18] Because the treaty recognized the lawful authority of these state forces to participate in hostilities, all individuals falling within this Hague definition were considered lawful belligerents: belligerents because the duties of war required them to engage in hostilities; lawful because the treaty recognized their legal right to do so.

The four-part test for determining when associated militias qualified as lawful belligerents subsequently influenced the development of a treaty definition of "prisoner of war" (POW).[19] Although neither the 1899 nor 1907 Hague Regulations included an express POW definition, both these treaties implied that any individual satisfying the definition of "belligerent who falls into enemy hands" would be a POW.[20] The linkage between the lawful belligerent definition and POW status became express in 1929, when the first treaty devoted to the protection of POWs came into force. Article 1 of the 1929 Geneva Convention Relative to the Treatment of Prisoners of War included within the protection of the treaty all captured individuals who met The Hague Regulation definition of "lawfully qualified belligerent."[21] When updated in 1949, the drafters incorporated the Hague "lawful belligerent" definition verbatim as the "POW" definition. According to Article 4 of the 1949 Geneva Convention Relative to the Treatment of Prisoners of War (GPW):

> A. Prisoners of war, in the sense of the present Convention, are persons belonging to one of the following categories, who have fallen into the power of the enemy:
>
> (1) Members of the armed forces of a Party to the conflict, as well as members of militias or volunteer corps forming part of such armed forces.

(2) Members of other militias and members of other volunteer corps, including [*sic*] those of organized resistance movements, belonging to a Party to the conflict and operating in or outside their own territory, even if this territory is occupied, provided that such militias or volunteer corps, including such organized resistance movements, fulfil the following conditions:
(a) that of being commanded by a person responsible for his subordinates;
(b) that of having a fixed distinctive sign recognizable at a distance;
(c) that of carrying arms openly;
(d) that of conducting their operations in accordance with the laws and customs of war.[22]

This undisputed interrelationship between the Hague definition of "lawful belligerent" and the GPW definition of "POW" led to reliance on Article 4 of the GPW as the controlling definition of "lawful belligerent," or "combatant." Nonetheless, it was clear that nothing in Article 4 of the GPW purported to define the term "combatant."[23] Instead, Article 4 simply relied on the antecedent Hague definition of "lawful belligerent" as the core for its "POW" definition. However, in the absence of an express definition of "combatant," it became common practice to equate the GPW definition of "POW" with that of "combatant"—a practice difficult to question considering the origins of the "POW" definition.[24] What remained uncertain, however, was whether the implied combatant definition of Article 4 of the GPW (and by implication, Article 1 of the Hague Regulations) was exclusive, or whether it only defined combatants legally authorized to engage in hostilities. In other words, what is the proper characterization of a "fighter" without legal privilege? Can this fighter be considered a de facto combatant without privilege? Or by exclusion, is this individual a civilian engaging in conduct inconsistent with his presumptive inoffensive status?[25]

The lack of an express "combatant" definition in the various treaties developed to regulate armed conflict made it impossible to answer this question with certitude. Additional clarity emerged in 1977. In that year, AP I provided the first express treaty definition of "combatant."[26] As indicated by the ICRC Commentary to the treaty, the GPW left the definition of this critical term to inference: "In the Third Convention, which deals only with the protection of prisoners of war, and not with the conduct of hostilities, this combatant status is not explicitly affirmed, but it is implicitly included in the recognition of prisoner-of-war status in the event of capture."[27]

In order to provide a more certain definition of "combatant," Article 43 (2) of AP I provides that "members of the armed forces of a Party to a conflict (other than medical personnel and chaplains covered by Article 33 of the Third Convention) are combatants, that is to say, they have the right to participate directly in hostilities."[28] Simple and direct, this definition reveals its roots in both the Hague Regulations and the GPW. Combatants include all members of the armed forces (with the exception of medical personnel and chaplains—individuals technically considered noncombatant members of the armed forces because of their limited function of caring for and ministering to the wounded and sick). As combatants these individuals are legally authorized to participate in hostilities. Who falls into the category of "armed forces"? Article 43 (1) answers this question by defining "armed forces":

> The armed forces of a Party to a conflict consist of all organized armed forces, groups and units which are under a command responsible to that Party for the conduct of its subordinates, even if that Party is represented by a government or an authority not recognized by an adverse Party. Such armed forces shall be subject to an internal disciplinary system which, inter alia, shall enforce compliance with the rules of international law applicable in armed conflict.[29]

Accordingly, armed forces include the organized armed units of a party to an armed conflict, and all such individuals are combatants vested with the legal authority to engage in armed hostilities. Perhaps of equal importance, this definition was considered exclusive, meaning that *only* members of the armed forces (as defined by Article 43 (1)) qualify as combatants within the meaning of international law. This conclusion is emphasized by the Commentary discussion of Article 43:

> All members of the armed forces are combatants, *and only members of the armed forces are combatants*. This should therefore dispense with the concept of "quasi-combatants," which has sometimes been used on the basis of activities related more or less directly with the war effort. Similarly, any concept of a part-time status, a semi-civilian, semi-military status, a soldier by night and peaceful citizen by day, also disappears. A civilian who is incorporated in an armed organization such as that mentioned in paragraph 1, becomes a member of the military and a combatant throughout the duration of the hostilities....[30]

As noted above, the explicit definition of "combatant" and "civilian" as the exclusive legal categorization of individuals in armed conflict was directly linked to AP I's codification of the principle of distinction, requiring combatants to constantly distinguish between those individuals qualifying as lawful objects of deliberate attack, and civilians protected from being made the deliberate object of attack, necessitated this express definition.[31] The definition of "combatant" therefore established two competing targeting presumptions: combatants are presumed to be lawful objects of attack, whereas all other individuals (civilians by exclusion) are protected by a rebuttable presumption of inoffensiveness with an accordant immunity from deliberate attack.

A combatant—an individual who according to AP I is granted the legal privilege to participate in hostilities—must therefore be a member of the armed forces or a member of a paramilitary organization associated with the armed forces operating under traditional command and subject to the military unit's disciplinary structure.[32] It is clear that civilians (even when authorized by a state to be present in the conflict area and therefore entitled to POW status upon capture) are not combatants pursuant to AP I because they are not fully integrated into a military command, control, and disciplinary structure.[33] While Article 43 therefore added clarity to the categorization of individuals associated with armed conflict, two aspects of the definition call into question the extent of its impact on the broader question of targeting status outside the context of interstate hostilities. First and most obviously, because the definition is embedded in AP I, it is limited to situations of international (interstate[34]) armed conflicts. Additional Protocol II (AP II), AP I's sister treaty supplementing the law applicable to NIAC, included no analogous definition.[35] Second, the definition of "combatant" included only individuals fighting on behalf of a state authority. Article 43 simply did not address individuals fighting on behalf of nonstate entities.[36]

The inapplicability of Article 43's "combatant" definition to NIAC has contributed to what the authors assert is a false conclusion that "combatant" is an alien concept outside the context of IAC. Instead of focusing on the lack of an analogous definition in AP II, the more appropriate focus in assessing targeting status in NIAC is the relationship between the principle of distinction and the division between civilians and de facto combatants.[37] As noted above, Article 43 was included in AP I as the essential predicate to implementing this principle. Accordingly, the underlying premise of Article 43 is that in order to facilitate the distinction process, it is essential to establish a clear dichotomy between combatant (privileged or unprivileged) and civilian. Without this dichotomy, the efficacy of this critical principle will inevitably be diluted. In essence, if everyone is a civilian, then no one is genuinely protected by the distinction obligation, for government forces will inevitably blur the line between "enemy" and "civilian."

The fact that Article 43's definition of combatant is tethered to the Hague definition of "qualified belligerent" likely explains why a similar article was not included in AP II. By linking the definition of "combatant" with legal qualification to participate in hostilities, the definition became incompatible with the law of NIAC, where by definition only the government forces may lawfully use force. From the inception of Common Article 3, the first LOAC provision developed specifically to regulate NIAC, states adamantly opposed even the suggestion that nonstate belligerents were vested with legal privilege to engage in hostilities against state forces.[38] Although the ICRC pressed for such a development, because it would prohibit states from punishing their own citizens for taking up arms against lawful authority, the proposal was dead on arrival.[39] State opposition to granting dissident and insurgent belligerent forces legal privilege to engage in hostilities has been a constant feature of the LOAC even during an era of amalgamation of IAC and NIAC. Indeed, the inapplicability of legal privilege for nonstate operatives in NIAC is today the most significant genuine difference between the law of IAC and NIAC.[40] Accordingly, applying Article 43's definition of "combatant" to NIAC was and remains incompatible with this fundamental difference in the nature of these two categories of armed conflict.

Nonetheless, what is most significant about AP I's definition of "combatant" is the segregation of individuals into two distinct categories for purposes of facilitating implementation of the principle of distinction. This segregation effectively sets the conditions for belligerents (members of opposing armed organized groups) to distinguish between lawful objects of attack and individuals protected from attack.[41] In the context of IAC, two considerations made it totally logical to link the definition of "combatant" to the regular armed forces and militias properly integrated with those forces. First, armed forces had customarily been considered combatants. Second, IAC by its very nature almost always involves hostilities between two or more nation-state militaries.[42] Neither of these considerations extends to NIAC, especially the classic internal armed conflict.[43] Internal conflicts had customarily involved hostilities between government armed forces and dissident armed forces or other organized belligerent groups (or even between competing organized nonstate belligerent groups). Designating all of these belligerents as combatants ran afoul of the assumption that state armed forces acted pursuant to lawful authority, whereas dissident forces acted without such authority.[44] Furthermore, NIAC by its nature did not involve hostilities between the regular armed forces of two states, but between regular state armed forces and nonstate belligerent groups.[45]

Nonetheless, the recognition of opposition belligerent groups is an essential aspect of all conflict regulation. Nothing about the nature of NIAC undermines the importance of this recognition in relation to implementing the principle of distinction. Indeed, while there is no analogous combatant definition in AP II, that treaty does impose an explicit distinction obligation to "parties" to a NIAC. According to Article 13:

1. The civilian population and individual civilians shall enjoy general protection against the dangers arising from military operations. To give effect to this protection, the following rules shall be observed in all circumstances.
2. The civilian population as such, as well as individual civilians, shall not be the object of attack. Acts or threats of violence the primary purpose of which is to spread terror among the civilian population are prohibited.[46]

This prohibition against making civilians the object of attack in the context of NIAC necessarily implies that there must be other individuals falling outside this protection who are subject to a presumption of targetability. How are such individuals characterized? If they fall within a presumptive lawful attack authority, they cannot properly be considered civilians, because civilians benefit from the inverse presumption protecting them from deliberate attack. Are they combatants? The absence of a combatant definition in AP II might suggest a negative answer to this question. Certainly, it is difficult to sustain an argument that they are combatants within the meaning of the AP I "combatant" definition—an individual legally privileged to participate in hostilities. However, the obvious division between lawful and unlawful objects of attack reflected in Article 13 indicates that organized nonstate belligerent operatives are in effect combatants without privilege, individuals subject to presumptive lawful attack authority by virtue of their status, yet lacking any international legal authority to participate in hostilities against the state.[47]

The Commentary to Article 13 corroborates the conclusion that Article 13 does indeed reflect the assumption that civilians are distinct from the organized armed groups engaged in NIAC.[48] These groups are normally not characterized as combatants (for the reasons explained above), but instead as "parties to the conflict." For example, in emphasizing the obligation that belligerent parties in NIAC take measures to ensure compliance with Article 13, the Commentary notes:

> Each *party* should, in good faith, design such measures and adapt them to the specific circumstances, bearing in mind the means available to it, and based on the general principles relating to the protection of the civilian population *which apply irrespective of whether the conflict is an international or an internal one*.[49]

The Commentary also acknowledges that distinguishing between true civilians and members of armed nonstate groups may often be difficult in the NIAC context, but that this does not alter the fundamental protection of the civilian. Interestingly, the term combatant slips into the Commentary discussion:

> It cannot be denied that in situations of non-international armed conflict in particular, the civilian population sometimes shelters certain *combatants*, and it may be difficult to ascertain the status of individuals making up the population. However, we must point out that if the mere presence of some individuals not protected under paragraph 3 of this article were to permit an attack against a whole group of civilians, the protection enjoyed by the civilian population would become totally illusory.[50]

These Commentary references all point to the same conclusion: although the treaty law of NIAC does not include an express "combatant" definition, NIAC involves hostilities between state armed forces and nonstate de facto combatants.

The alternative inference derived from the absence of a definition of "combatant" in treaties regulating NIAC is that the concept is exclusive to IAC. Accordingly, NIAC involves hostilities between state armed forces and civilians directly participating in hostilities. This interpretation is fatally flawed. As will be explained in more detail below, it distorts the fundamental lines of authority and obligation historically associated with armed conflict. By treating members of armed belligerent groups as civilians directly participating in hostilities—even when applying an expansive definition of direct participation endorsed by the DPH Study—the authority of government forces to engage and subdue these individuals is diluted. Furthermore, the refusal to segregate the population into de facto combatant and civilian compromises the principle of distinction.

These inevitable consequences of failing to acknowledge nonstate belligerents in the context of NIAC motivated the ICRC to conclude that the legal authority to attack members of nonstate belligerent groups in NIAC is identical to the authority to engage combatants in IAC.[51] While not using the term "combatant" to characterize members of such groups, it is clear that the ICRC understands that the authority to target nonstate belligerent operatives is not based exclusively on a DPH test. Instead, the DPH rule operates as a limitation on the protection afforded to civilians in NIAC no differently than it does in IAC. However, this rule is inapposite to members of armed belligerent groups, for such individuals are not properly considered true civilians, but instead "belligerents" subject to lawful attack by virtue of their connection and role within the group. Accordingly, unlike civilians who take a direct part in hostilities, the legality of attacking such individuals is derived from their membership status, and not their individual conduct assessed on a case-by-case basis. This conclusion is reflected in the DPH Study. According to Part II:

> While it is generally recognized that members of State armed forces in non-international armed conflict do not qualify as civilians, treaty law, State practice, and international jurisprudence have not unequivocally settled whether the same applies to members of organized armed groups (*i.e.* the armed forces of non-State parties to an armed conflict). Because organized armed groups generally cannot qualify as regular armed forces under national law, it might be tempting to conclude that membership in such groups is simply a continuous form of civilian direct participation in hostilities. Accordingly, members of organized armed groups would be regarded as civilians who, owing to their continuous direct participation in hostilities, lose protection against direct attack for the entire duration of their membership. *However, this approach would seriously undermine the conceptual integrity of the categories of persons underlying the principle of distinction, most notably because it would create parties to non-international armed conflicts whose entire armed forces remain part of the civilian population.* As the wording and logic of Article 3 GC I-IV and Additional Protocol II (AP II) reveals, civilians, armed forces, and organized armed groups of the parties to the conflict are mutually exclusive categories also in non-international armed conflict.[52]

If the ICRC is correct (consistent with the thesis of this chapter), what then distinguishes civilians from nonstate belligerent operatives[53] in the context of NIAC? As noted above, the answer cannot be found in any positive treaty provision applicable to NIAC. Instead, the answer is revealed by considering the fundamental difference between belligerents and civilians in any armed conflict: subordination to belligerent command and control.

B. Role of Organizational Membership and Subordination to Command and Control

Armed conflict involves the contest between belligerent opponents organized into military units acting to achieve tactical, operational, and strategic objectives. The very purpose of the organized nature of belligerent groups is to ensure that individuals act to achieve these collective military goals. The inherent obligation of members of such groups to obey the orders of military and civilian superiors is consistent with this purpose. Command structure, rank, and the duty of obedience all reflect a simple axiom of belligerent groups: members of these groups act as agents of the group leadership to achieve military goals, not as individuals.[54]

From the inception of treaty-based recognition of belligerents in the 1899 Hague Regulations, the centrality of this subordination to command authority has defined the concept of belligerent.[55] As noted above, this was originally reflected in the requirement that individuals operate on behalf of the state—the only entity authorized within the Westphalian system to authorize lawful belligerent conduct.[56] However, it is also clear that this was not in and of itself sufficient to qualify as a belligerent. In addition, belligerents were those individuals integrated into the command and control structure of a military unit for the purpose of engaging in armed hostilities. These latter elements of belligerent status are reflected in the requirement that the individual carry arms openly and conduct operations in accordance with the laws of war—both of which imply exercising a belligerent function.[57] Accordingly, subordination to the authority of belligerent group leadership for the purpose of engaging in hostilities at the direction of that leadership is the true sine qua non of belligerent status. As the DPH Study notes:

> Organized armed groups belonging to a non-State party to an armed conflict include both dissident armed forces and other organized armed groups. Dissident armed forces essentially constitute part of a State's armed forces that have turned against the government. Other organized armed groups recruit their members primarily from the civilian population but develop a sufficient degree of military organization to conduct hostilities on behalf of a party to the conflict, albeit not always with the same means, intensity and level of sophistication as State armed forces.[58]

This language indicates another acknowledgment that NIAC involves hostilities between armed belligerent groups—in the terminology of Common Article 3, "Parties" to an armed conflict.[59] If this is true (as the author believes it is), then it seems logical that the key factor in determining belligerent status—particularly for nonstate operatives—cannot be the state subordination element. Instead, the key factor is membership in the belligerent organization exercising a function historically associated with belligerent operatives, namely participating in hostilities. This is precisely the equation proposed by the DPH Study to distinguish between a genuine civilian and a member of a nonstate belligerent group:

> Consequently, under IHL, the decisive criterion for individual membership in an organized armed group is whether a person assumes a continuous function for the group involving his or her direct participation in hostilities (hereafter: "continuous combat function").[60]

With the exception of dissident armed forces (whose "status" as belligerents is revealed by their membership in an armed force formerly loyal to government authority), determining NIAC belligerent status is unquestionably complex. Absent the prospect of combatant immunity, there is little incentive for nonstate operatives to wear uniforms or other insignia that distinguishes them from the civilian population.[61] As a result, it is often difficult to determine the difference between civilians and nonstate belligerent operatives based on outward appearance. Instead, the difference will almost inevitably turn on the nature of the individuals' conduct. When that conduct indicates the individual is acting as an agent of the belligerent group leadership for the purpose of engaging in hostile functions, it in effect establishes belligerent status.

Herein is the ultimate complexity of acknowledging belligerent status in NIAC: while it is this status that subjects the operative to the scope of targeting authority identical to that applicable to a combatant in IAC, determining that status inevitably requires assessment of individual conduct.[62] Focusing on relevant indicators to determine belligerent status is nothing new. This has always been an aspect of the execution of military operations during armed conflict, what military operators would call "threat recognition."[63] What complicates this process in the NIAC context is the reality that the indicators of threat recognition (belligerent status) are rarely as concrete and overt as in the traditional IAC context. The evolving nature of NIAC, which today includes armed hostilities between states and transnational nonstate groups, further exacerbates this complexity.[64] It is critical, however, to distinguish between targeting authority triggered by status and targeting authority triggered by conduct, even when conduct is the key "threat recognition" factor resulting in status. Thus, while conduct is the key analytical indicator of status, this does not equate to conduct-based targeting authority—the type of authority utilized to respond to a genuine DPH situation.[65]

This conduct-based status determination is the true significance of the CCF concept endorsed in the DPH Study when applied in the NIAC context.[66] In essence, in NIAC, CCF is a methodology to assess belligerent status and thereby trigger status-based targeting authority.[67] As a method of assessing status, CCF is therefore merely a threat recognition equation that leads to a status determination triggering accordant targeting authority. It is not a method for determining when a civilian's conduct justifies a hostile response because of DPH (conduct-based targeting authority).[68] CCF is therefore merely the threat recognition criterion that is used to confirm belligerent status, a conclusion reflected in the DPH Study:

> Continuous combat function requires lasting integration into an organized armed group *acting as the armed forces of a non-State party to an armed conflict*. Thus, individuals whose continuous function involves the preparation, execution, or command of acts or operations amounting to direct participation in hostilities are assuming a continuous combat function. An individual recruited, trained, and equipped by such a group to continuously and directly participate in hostilities on

> its behalf can be considered to assume a continuous combat function even before he or she first carries out a hostile act. This case must be distinguished from persons comparable to reservists who, after a period of basic training or active membership, leave the armed group and re-integrate into civilian life. Such "reservists" are civilians until and for such time as they are called back to active duty.[69]

This test provides a logical and workable method to trigger status-based targeting authority in NIAC. While it is obviously less clear-cut than reliance on uniforms to distinguish combatants from civilians, it accounts for two realities. First, as the DPH Study emphasizes, members of organized belligerent groups in NIAC (individuals who are presumed hostile to friendly forces and therefore subject to a presumption of targetability) are a distinct category from civilians.[70] Second, the lack of any real incentive for members of these groups to wear distinctive uniforms or insignia necessitates reliance on their patterns of conduct as the principal threat recognition criterion used to assess when they are part of an "enemy" party to the conflict.[71]

The CCF test has, however, contributed to the flawed inference that nonstate operatives cannot be characterized as belligerents, but must instead be treated as civilians.[72] The CCF focus on individual conduct is interpreted by some as simply indicating a special class of civilians taking a direct part in hostilities. This leads to the erroneous conclusion that CCF-derived targeting authority is analogous to what military operators know as conduct-based rules of engagement.[73] While this error will be addressed in more detail below, it erroneously conflates the focus on conduct to assess membership (thereby triggering status-based targeting authority) with the focus on individual conduct to determine when a civilian loses presumptive protection from attack.

It is true that the ultimate outcome of both of these equations is a justified use of force. However, the impact of competing presumptions associated with status versus conduct-based targeting authority indicates there are important, albeit subtle, second- and third-order consequences resulting from this erroneous interpretation. The conclusion that use of force is permitted against both a civilian directly participating in hostilities and a member of an armed belligerent group therefore does not justify interpreting CCF in NIAC as a test to determine when a civilian is directly participating in hostilities as opposed to when the civilian becomes a member of a belligerent group. It is the thesis of this chapter that CCF in NIAC establishes belligerent group status, and not simply a temporary loss of civilian protection from attack. This is because integration into the belligerent forces of a nonstate group is functionally synonymous with integration into a combatant force in IAC. Indeed, the DPH Study seems to emphasize this consequence of CCF in NIAC:

> [I]t distinguishes members of the organized fighting forces of a non-State party from civilians who directly participate in hostilities on a merely spontaneous, sporadic, or unorganized basis, or who assume exclusively political, administrative or other non-combat functions.[74]

It is obviously significant that the term "distinguishes" is used in this passage, for it confirms that individuals who engage in a CCF in NIAC must be distinguished from civilians, even those civilians who take a direct part in hostilities.

II. WHY THE USE OF CONDUCT TO ESTABLISH STATUS UNDERMINES THE EXTENSION OF THE DPH RULE TO DEFINE ENEMY BELLIGERENT FORCES

It is clear that both organizational status and DPH—whether assessed on a case-by-case basis or pursuant to the CCF test—result in lawful targetability.[75] This raises an obvious question related to targeting individuals hostile to a friendly force: why does it matter whether forces apply status- or conduct-based targeting authority? This question has certainly become more difficult to answer in the wake of the CCF test, which mitigates the consequence between status- and conduct-based targeting authority. However, treating CCF as a form of civilian conduct—triggering targeting authority—dilutes the authority to address threats to friend forces emanating from organized belligerent groups. Such dilution should be unacceptable—to both belligerents and, perhaps surprisingly, civilians.

All battlefield-targeting authority falls into two broad categories: status- and conduct-based. Status-based targeting authority is, as described above, triggered by the determination that a proposed object of attack is a member of an opposition belligerent force.[76] In contrast, conduct-based targeting is based on the determination that an individual presumed inoffensive is engaged in conduct hostile to the friendly force.[77] Accordingly, status-based targeting has always typified the traditional conception of armed hostilities between armed forces, whereas conduct-based targeting reflects more of a constabulary function performed by the armed forces in the context of armed hostilities.

This dichotomy is reflected in the traditional DPH equation. Prior to the advent of the CCF test, DPH was an extremely restrictive concept. Only when a civilian engaged in conduct that would result in actual and immediate harm to an armed force did he lose legal protection from being made the deliberate object of attack.[78] This test was and remains premised on a critical presumption: civilians are inoffensive and deviation from that condition is an exceptional situation.[79] Accordingly, because the evidence supporting attack must be sufficient to prove a gross deviation from the presumed standard of behavior, the burden of risk is placed squarely on the responding armed force.[80]

The allocation of risk resulting from status-based targeting decisions is fundamentally different. The determination of hostile status triggers the opposite presumption from that afforded to civilians: that members of the opposition group represent an actual and ongoing threat of hostility to friendly forces. A status determination triggers the authority, derived from the principle of military necessity, to take "those measures not forbidden by international law which are indispensable for securing the complete submission of the *enemy* as soon as possible."[81] It is critical to recognize that this authority is not confined to the individual. Reference to "the enemy" indicates (in accordance with the customary nature of armed hostilities) that it is the opposition organization, and not just individual operatives of the organization, that is the legitimate objective of submission.[82] Accordingly, the law authorizes resort to deadly force immediately upon the belligerent status determination. Perhaps more important, because members of opposition belligerent groups qualify as lawful military objectives as a result of their status, the status determination triggers the legal authority to employ force against them as a measure of first resort, with no requirement to exhaust less than lethal means of subduing the threat. This reflects an axiom of targeting law: the burden of risk associated with armed hostilities is placed squarely on members of armed belligerent groups for as long as their status remains extant. They remain subject to attack unless and until they take affirmative action to rebut the presumption of threat, specifically by surrender or by being made combat ineffective as the result of wounds or some other disability.[83]

The CCF test certainly represents an important step forward in reconciling the concept of direct participation with the reality of contemporary armed conflicts. However, it confuses these competing but complementary presumptions because it straddles a line between status- and conduct-based targeting authority, blurring the resulting operative targeting presumptions.[84] While CCF indicates the individual is a member of an enemy belligerent group, the test is clearly conduct oriented, and the consequence of satisfying the test embedded within the DPH rule.[85] As a result, a CCF determination results in the loss of presumptive protection from attack, just as actually firing a weapon at a military force would result in the loss of protection for a civilian acting individually.[86] However, because it is associated with a rule resulting in the loss of *civilian* protection, it risks becoming the exclusive test for assessing the targetability of individuals not incorporated into an armed force subordinate to state authority and control—namely nonstate belligerent groups in NIAC. This is highly problematic because it effectively renders all nonstate actors civilians, who in turn, benefit from presumptive protection from attack.

It is notable that the ICRC recognized in the DPH Study that such an interpretation is unjustified, and that it diminishes the protection of the civilian population at large.[87] This recognition is absolutely appropriate. As the ICRC notes, the principle of distinction depends on the recognition of opposing belligerent groups, for such recognition facilitates targeting categorizations during armed conflict.[88] In effect, this is a concession that the legal authority for targeting such groups is based on a fundamentally different presumption than that applicable to civilians directly participating in hostilities—even those falling within the CCF test. Nonetheless, the DPH Study itself reflects an ongoing uncertainty as to the exact effect of the CCF concept, which contributes to the broader uncertainty as to the presence of nonstate belligerent groups. The most profound example of this is Section IX of the DPH Study.[89]

III. INTERPRETIVE GUIDANCE SCHIZOPHRENIA

A. CCF and the Minimum Force Requirement

Perhaps the most significant practical distinction between status- and conduct-based targeting authority is the effect of the principle of proportionality.[90] Because a determination that an individual is a member of an enemy belligerent group triggers status-based targeting, the LOAC has always permitted the attacking force to employ deadly force as a measure of first resort. In other words, status indicates the individual is a lawful military objective: a person, place, or thing subject to lawful attack.[91] This authority is in no way qualified by the rule of proportionality. Instead, the only limitation on the selection of methods and/or means of warfare to attack a military objective (beyond the initial targetability determination) is the prohibition against the calculated infliction of unnecessary suffering.[92]

It could be argued that the unnecessary suffering prohibition implicitly imposes a proportionality obligation. This is, however, incorrect. The foundation of the rule prohibiting unnecessary suffering is the bar against the employment of weapons (means) or tactics (methods) against a lawful object of attack *calculated* to cause superfluous injury or suffering.[93] What is superfluous? This has unquestionably been a vexing LOAC question. However, the nature of military doctrine, training, tactics, and operations all indicate that causing death, as a measure of first resort, is not considered legally superfluous.[94]

In fact, deliberately inflicting injuries that cannot be treated—for example permanent loss of sight or hearing—is far more likely to be considered to run afoul of this rule than killing the same enemy belligerent.[95] This might seem oxymoronic, but it is a reflection of the customary understanding that employing methods and means of warfare with the objective of causing death as a measure of first resort is consistent with the fundamental authority derived from the principle of military necessity to take all measures necessary to bring about the prompt submission of an enemy organization.

This conclusion is reinforced by one of the most important rules in AP I: Article 51(5).[96] Commonly called the proportionality rule, Article 51(5) prohibits attacks on lawful military objectives when the commander anticipates that the attack will produce excessive collateral damage or incidental injury.[97] It is clear that the beneficiaries of this rule are not the objects of attack (military objectives), but instead victims of collateral effects, namely civilians.[98] The benefit afforded to the objects of attack is purely gratuitous—an otherwise lawful attack against military objectives is prohibited not because of the disproportionate effect it will have on them but because of the desire to protect civilians and/or civilian property in their vicinity.[99]

No analogous proportionality protection is afforded to lawful objects of attack, including enemy belligerent forces.[100] Accordingly, it is legally impossible to employ excessive force against the deliberate and lawful object of violence in armed conflict; those objects may be attacked with whatever amount of force is considered necessary to bring about their immediate submission.[101] Perhaps more important, it is common practice to use overwhelming force against such enemy objectives in order to influence the subsequent behavior of enemy leadership and other enemy forces.[102] Of course, once the enemy belligerent is rendered hors de combat by wounds, sickness, or capture, he no longer qualifies as a lawful object of attack.[103] But the protection afforded to such an individual is in no way derived from a proportionality requirement; it is simply the result of the fact that the individual is no longer capable of engaging in belligerent conduct as an agent of the enemy belligerent leadership.[104]

If CCF is merely a threat-recognition methodology resulting in a belligerent status determination, then individuals engaging in CCF should be subject to attack without any proportionality-based constraint. The DPH Study, however, suggests a contrary conclusion. In Section IX, the DPH Study indicates that the use of force directed against individuals falling within the CCF definition is subject to such a constraint.[105] However, even the purported source of this constraint reveals the invalidity of this proposition.

As noted above, *jus in bello* proportionality does not protect enemy belligerent operatives who qualify as objects of deliberate attack.[106] As a result, Section IX is unsupported by any positive or customary LOAC obligation. This is essentially conceded in the DPH Study when it cites the general principle of humanity as the source of the constraint. According to Section IX:

> In the absence of express regulation, the kind and degree of force permissible in attacks against legitimate military targets should be determined, first of all, based on the fundamental principles of military necessity and humanity....
>
> While it is impossible to determine, *ex ante*, the precise amount of force to be used in each situation, considerations of humanity require that, within the parameters set by the specific provisions of IHL, no more death, injury, or destruction be caused than is actually necessary for the accomplishment of a legitimate military purpose in the prevailing circumstances....
>
> In sum, while operating forces can hardly be required to take additional risks for themselves or the civilian population in order to capture an armed adversary alive, it would defy basic notions of humanity to kill an adversary or to refrain from giving him or her an opportunity to surrender where there manifestly is no necessity for the use of lethal force. In such situations, the principles of military necessity and of humanity play an important role in determining the kind and degree of permissible force against legitimate military targets.[107]

This application of the principle of humanity may seem appealing, but it is fundamentally inconsistent with the scope of authority established by the principle of military objective. This principle, as explained above, allows for or restricts deliberate attack authority based on presumptions, presumptions that are inherently overbroad and underinclusive.[108] Nonetheless, these presumptions provide for a certain degree of clarity and consistency in an otherwise chaotic environment.

It is clear that Section IX was motivated by the reality that application of the CCF concept would expand the scope of lawful targeting based on a determination of DPH. This likely produced discomfort: deliberate-attack authority applicable to individuals who perform a CCF but who, at the moment when force is applied against them, may not be identifiable as belligerents based on conduct which creates an imminent risk to friendly forces. It also seems clear that the ICRC understood the minimal authority for applying a proportionality qualifier to belligerents. In justifying this purported limitation, the DPH Study begins by conceding that it will rarely apply to a lawful combatant in the IAC context:

> In classic large-scale confrontations between well-equipped and organized armed forces or groups, the principles of military necessity and of humanity are unlikely to restrict the use of force against legitimate military targets beyond what is already required by specific provisions of IHL.[109]

The discussion then shifts the focus to the civilian engaged in DPH, revealing its palpable discomfort with subjecting such a civilian to targeting authority analogous in scope to that applicable to the traditional combatant:

> The practical importance of their restraining function will increase with the ability of a party to the conflict to control the circumstances and area in which its military operations are conducted, and may become *decisive* where armed forces operate against selected individuals in situations comparable to peacetime policing. In practice, such considerations are likely to become particularly relevant where a party to the conflict exercises effective territorial control, most notably in occupied territories and non-international armed conflicts.
>
> For example, an unarmed civilian sitting in a restaurant using a radio or mobile phone to transmit tactical targeting intelligence to an attacking air force would probably have to be regarded as directly participating in hostilities. Should the restaurant in question be situated within an area firmly controlled by the opposing party, however, it may be possible to neutralize the military threat posed by that civilian through capture or other non-lethal means without additional risk to the operating forces or the surrounding civilian population.[110]

As will be explained below, restricting this asserted targeting authority qualification to this type of civilian engaging in DPH (ostensibly in the IAC context) is arguably defensible. However, in a classic manifestation of the proverbial "slippery slope," the DPH Study then extends the qualification to the NIAC nonstate belligerent:

> Similarly, under IHL, an insurgent military commander of an organized armed group would not regain civilian protection against direct attack simply because he

temporarily discarded his weapons, uniform and distinctive signs in order to visit relatives inside government-controlled territory. Nevertheless, depending on the circumstances, the armed or police forces of the government may be able to capture that commander without resorting to lethal force.[111]

This one paragraph reveals the inherent schizophrenia of Section IX. It is simply incompatible with the DPH Study's implicit (if not explicit) recognition that CCF in the NIAC context results not in limited targetability, but instead targeting based on belligerent status. In essence, Section IX is an attempt to "re-civilianize" the NIAC nonstate belligerent by cloaking that operative with a proportionality protection.[112] However, imposing a minimum necessary force obligation on government forces engaged in armed conflict against nonstate belligerents is inconsistent with the concession that CCF produces belligerent status, and with the overall concept of status-based targeting presumptions.

This proposed application of a unitary minimum necessary force qualifier to both civilians engaging in DPH in the IAC context (to include in an occupied area) and nonstate belligerents is problematic on a broader level. By applying a force limitation rule ostensibly justified vis-à-vis civilians (even those taking a direct part in hostilities) to nonstate belligerents, Section IX confuses the effect of CCF in the NIAC context. This confusion has, in the opinion of the authors, contributed to the more widespread misconception that nonstate belligerents are merely civilians engaged in DPH who remain civilians while they are so engaged (albeit subject to use of force). However, the mixing of armed conflict and law enforcement response authority reflected in the above extract from Section IX exposes the danger of this premise.[113] In short, continuing to characterize these belligerent operatives as civilian leads to an inevitable but impermissible outcome: the dilution of state targeting authority and a corresponding lessening of the protection of the civilian population writ large.

This outcome is not only unjustified and inconsistent with status-based targeting principles, it is arguably perverse, for it vests an unprivileged belligerent with greater protection than the privileged counterpart. As noted above, the DPH Study at least implicitly concedes CCF in NIAC results in belligerent status—a method of identifying members of organized armed groups (a party to an armed conflict).[114] However, applying the Section IX minimum force qualification to these belligerent operatives would grant them protection from being made the object of attack with the use of deadly force as a first resort.[115] No analogous protection applies to lawful combatants, who by virtue of that status are subject to such risk so long as they are capable of acting as agents of enemy leadership.[116]

Why the ICRC would propose such a minimum force qualification is understandable. By conceding that CCF results in belligerent status for NIAC nonstate operatives, the DPH Study conceded a broad scope of targeting authority for government forces. However, the CCF concept itself reflects the reality that threat identification in this context is far more complex than in traditional IAC (because clear objective criteria of belligerent status are rarely available).[117] Accordingly, the risk of error in the status decision is obviously increased vis-à-vis a nonstate operative. Imposing a minimum force qualification to targeting authority would obviously mitigate that risk by requiring the attacking commander to forego the use of deadly force wherever and whenever feasible.

This anomaly was apparently not lost on the ICRC. Instead of accounting for it by drawing a bright line between belligerent targeting authority and the authority to use force in response to a civilian taking a direct part in hostilities, Section IX used it as an opportunity to open another front: extending the minimum necessary force obligation to all enemy belligerents.[118] The DPH Study cites Jean Pictet in support of this extension. However, in the same footnote, the DPH Study concedes that:

> During the expert meetings, it was generally recognized that the approach proposed by Pictet is unlikely to be operable in classic battlefield situations involving large-scale confrontations (Report DPH 2006, pp. 25f., 78) and that armed forces operating in situations of armed conflict, even if equipped with sophisticated weaponry and means of observation, may not always have the means or opportunity to capture rather than kill (Report DPH 2006, p. 63).[119]

Indeed, opposition to this interpretation as fatally flawed is reflected in subsequent scholarly critiques by members of the DPH working group, which notes that it is inconsistent with the widespread and customary understanding of status-based targeting authority.[120]

Why would the ICRC muddy the proverbial waters with this overreaching? The most obvious reason, as noted above, is that it was considered necessary to offset the inherent expansion in risk to presumptive civilians produced by the CCF concept. However, by asserting the constraint applied to NIAC belligerents, the proponents of Section IX produced an anomaly that could only be eliminated by asserting the minimum necessary force requirement applies to *all* enemy belligerents, even traditional combatants. In this regard, it must be emphasized the DPH Study's concession that application of this constraint will often be impracticable in traditional force-on-force combat in no way suggests that the obligation is inapplicable in that context. Instead, Section IX injects another feasible precaution obligation into the use of combat power in any armed conflict. Unfortunately, unlike all other such obligations, which are intended to mitigate the collateral impact on civilians as the result of attacking a lawful objective, the beneficiary of this asserted obligation is the lawful and deliberate object of attack itself.

B. Conflating of Law and Policy

There may be—and indeed often are—policy and operational rationale for imposing a minimum force requirement on government forces combating nonstate belligerent groups. Indeed, the history of armed conflict is replete with examples of commanders imposing restraints on their forces in order to achieve some tactical, operational, or strategic objective.[121] But couching the requirement as a legal constraint at best conflates *lex lata* and *lex ferenda*.[122] At worse, the conflation, while cloaked in terms of humanity, will add uncertainty in the already complex NIAC environment and, ultimately, result in increased civilian casualties. The conflation also confuses the differing considerations through which the use of force is limited: legal, policy, and operational.

While the term "rules of engagement" (ROE) is commonly misunderstood to mean the LOAC, the ROE are a subset of the LOAC. The LOAC forms the outer, and legal, bounds of permissible conduct in hostilities. Those parameters are further reduced, not legally, but through a host of policy considerations, including political and diplomatic.[123] Military commanders also impose restraints and limitations on their forces.[124] The result of limiting conduct due to policy and operational considerations is the ROE. But to contend that because the ROE are a subset of the LOAC and that the ROE contain limitations on the use of force means that the limitations derive from the LOAC are sophism.

The practice of policy- and operationally-based limitations on what the LOAC would otherwise permit is not limited to operations against unconventional threats in a NIAC. But however broadly they may be applied in the spectrum of conflicts, these considerations do not justify transforming policy constraints into a general legal obligation. The ability to deliberately attack enemy belligerents with the full force of combat power available for mission accomplishment—an authority that implicitly allows the use of deadly force as a measure of first resort—is an essential aspect of armed hostilities between organized belligerent groups. Indeed, the ability to mass the effects of combat power at the decisive place and time often contributes to accelerating enemy capitulation, thereby sparing many enemy belligerents who might otherwise be subject to a loss of life even if a minimum necessary force obligation were applied.

Admittedly, the complexity borne of government forces operating in a NIAC—where the need to use deadly force as opposed to minimum force is unclear and will vary from person to person—cannot be eliminated. But this dilemma is not newly discovered. Indeed, as this chapter has acknowledged, there are circumstances when a minimal force requirement may be appropriate, as in the case of the civilian who directly participates in hostilities. Again, such targeting is based on conduct and not a presumption of offensiveness stemming from membership status in a belligerent armed group. While there may be a legal and policy basis for requiring government forces to employ the least necessary force in certain circumstances in a NIAC, to extend a minimum force requirement to include members of belligerent armed groups is both logically and operationally inconsistent, dilutes the concept and protections of a civilian, and further incentivizes noncompliance with the LOAC.

IV. Maintaining Distinction Integrity in the Application of the CCF Test

There is a rational way to reconcile Section IX with status-based targeting presumptions: maintain the distinction integrity. Acknowledging that NIAC involves armed hostilities between competing belligerent groups is a critical first step in doing so, and the DPH

Study makes an important contribution.[125] However, this must be accompanied by an additional acknowledgment: *all* belligerent operatives—those involved in IAC and NIAC—are subject to status-based targeting authority. This acknowledgment must then lead to the critical conclusion that Section IX cannot properly be applied to any belligerent, including those identified by their CCF in a NIAC.

Recognizing a bright line between genuine civilians—even those who deviate from their presumptive inoffensiveness by taking a direct part in hostilities—and belligerents will effectively preserve the critical distinction between conduct- and status-based targeting. Only belligerents will be subjected to status-based targeting. This is a necessary aspect of armed conflict, because that broad targeting authority is premised on the conclusion that the object of attack is not acting as an individual agent, but instead executing the will of belligerent leadership. It is obvious that this presumption is at times overbroad. It is a necessary (and at times unfortunate) aspect of armed conflict that enemy belligerents may be killed when capture may have been a viable alternative, or even when they present no genuine actual risk to the attacking force. This is a consequence of status-based targeting authority derived from the presumption of threat triggered by the status determination. That authority terminates only when the enemy belligerent is effectively separated from this subordinate agent-derived presumption through incapacity or surrender.[126]

This status-based targeting authority is triggered by the determination that an individual is a member of an opposition belligerent group. It is generally assumed that this determination will be relatively uncomplicated in the IAC context due to the expectation that enemy belligerents will wear distinctive uniforms. However, even during IAC this is not always true. One need only consider the Taliban armed forces whose appearance was generally indistinguishable from the civilian population, or unconventional forces in Iraq who discarded their uniforms in order to appear to be civilians, as examples of this reality.[127] However, NIAC will almost always involve hostilities between at least one party to the conflict unlikely to don distinctive uniforms (as an asymmetrical tactic to capitalize on opposition concerns over targeting civilians in error or because of the absence of any legal incentive to distinguish themselves from the population). As a result, conduct will increasingly become the key threat/status identification criterion. This much is acknowledged by the DPH Study's discussion of CCF in the NIAC context.[128]

Properly understood, CCF in NIAC is therefore neither synonymous with DPH, nor a means of determining when a civilian is engaged in DPH. Instead, CCF is a method for determining belligerent status, with all the consequences that flow therefrom. Whether CCF produces the same outcome in IAC is unclear from the DPH Study. CCF unquestionably results in loss of protection from attack. However, because IAC involves hostilities between lawful combatants (who must satisfy legally defined qualification requirements to obtain that status[129]), CCF does not result in lawful belligerent status.[130] In contrast, because the concept of unprivileged belligerent in the context of IAC is generally rejected (although not universally), individuals engaging in a CCF in IAC must by default remain in the category of civilian, albeit with a loss of protection from attack and lack of combatant immunity.[131]

Applying a proportionality constraint to protect these individuals is therefore fundamentally different from applying it to their NIAC counterparts. This is because in the IAC context, CCF is not a status determination equation, but instead a conduct-based criterion leading to the loss of protection from attack.[132] Nonetheless, the individual who loses that protection, while subject to attack, remains in the category of civilian. Because of this, it is not illogical to assert that any response to such a threat must be qualified by a minimum necessary force limitation. This is because as civilians these individuals are not subject to status-based targeting, but continue to benefit from the use of force limitations inherent in conduct-based targeting authority.[133] First, use of force is justified only when the individual's conduct is sufficient to rebut the presumption of inoffensiveness (the apparent purpose of the CCF test).[134] Second, the sole purpose of a responsive use of force is to restore the civilian to a condition of inoffensiveness, and not to permanently disable the individual's ability to execute the will of enemy belligerent leadership, or to influence the future conduct of the enemy belligerent leadership.[135] Accordingly, if measures short of deadly force can effectively achieve that limited purpose, resort to deadly force as a measure of first resort would exceed the conduct-based authority triggered by the DPH.[136]

A. Organized Belligerents: Agents of Leadership Will

Armed conflict, by its very nature, is a contest between armed groups. Accordingly, the efficacy of the law that regulates armed conflict is contingent on the recognition of this reality. Treating nonstate operatives as a conglomeration of civilians taking a direct part in hostilities might, in the abstract, seem an appealing response to addressing legal authorities related to such conflicts, but it fails to account for the collective/corporate nature of armed conflict. The defining distinguishing factor between a true civilian and a member of a belligerent group is subordination to the will of belligerent leadership. Whether the result of a voluntary choice, or obligatory service, belligerent operatives do not act as autonomous agents in armed conflict. Instead, they execute the will of belligerent leadership.

As noted above, the traditional LOAC targeting framework acknowledges this reality by permitting presumptive-based uses of deadly force. Once an individual is identified as a member of an enemy belligerent group, friendly forces are entitled to presume the individual represents a hostile threat. However, perhaps even more important, friendly forces act in order to disable the ability of the enemy belligerent operative from executing the will of enemy leadership. In so doing, not only is the belligerent threat eliminated but the corporate capacity of the enemy is degraded.

It has become too simple to assume that the direct participation in hostilities equation proposed by the DPH Study is an adequate substitute for this traditional targeting authority. That equation, while certainly an important component in addressing the threat posed by true civilians who engage in a continuous combat function, does not provide as coextensive a targeting authority as the belligerent status-based targeting equation. First, it is based on a requirement to assess each targeting decision through the lens of individualized actual threat.[137] While the CCF concept may expand the range of conduct that qualifies as such a threat, it is nonetheless a conduct-based targeting equation. This fails to account for the need for presumptive-based targeting. Second, its almost unitary focus on individualized threat assessment fails to account for the legitimate use of attacking belligerent operatives in order to influence belligerent leadership, not merely to eliminate an individual threat.[138]

These flaws are eliminated by applying status-based targeting authority to NIAC. That targeting authority is derived from the determination that individuals qualify as members of opposition belligerent groups. This determination is unquestionably complex. Furthermore, in the NIAC context it will often be based on an assessment of conduct, for the simple reason that more objective criteria (like uniforms or specialized military equipment) will be unavailable. However, even when status threat recognition is based on similar, if not identical, factors as those at the core of the CCF test, the sum of the analysis is not a determination of conduct-based targeting authority. In short, conduct may be the test for status, but conduct-derived status does not trigger conduct-based targeting authority.

Why, if the equation used for CCF and status determinations is analogous, is this distinction even necessary? This is a logical question, but one that reveals a disconnect between legal theory and operational reality. Treating nonstate belligerent operatives as civilians who lose protection from attack for such time as they take a direct part in hostilities, even under the expanded CCF test, grants them an unjustified windfall. This is because it inverts the traditional targeting presumptions associated with armed conflict. This approach places the burden of validating individualized actual threat on the opposing belligerent operative. This burden allocation is justified when the friendly operative encounters a civilian with no subordinate relationship to enemy belligerent leadership. However, it is unjustified when encountering enemy belligerent operatives. As noted above, imposing validation of actual threat as the condition precedent to targeting members of enemy belligerent groups degrades combat initiative, produces dangerous hesitation, and ultimately provides greater protection for the nonstate operative than for the lawful belligerent. Indeed, this latter consequence seems especially oxymoronic. Why should a fighter without the privilege of operating on behalf of a state benefit from greater protection than his legitimate counterpart?

The only plausible answer to this question is that such an outcome is a necessary consequence of the inherent uncertainty as to who is a civilian and who is a nonstate belligerent operative. The risk of erroneous targeting resulting from this uncertainty is mitigated by applying a unitary conduct-based targeting equation to all nonregular armed forces. Accordingly, any additional protection afforded the nonstate actor is an incidental outcome of a rule intended to maximize protection for true civilians. The more rational response to this uncertainty—at least from an operational perspective—is to develop credible and effective status recognition criteria. Ultimately, if conduct-based status determinations are based on criteria relatively analogous to the CCF test, it should be equally effective in mitigating the risk of targeting errors. However, once that status is established, it would trigger targeting authority traditionally relied upon to bring enemy armed forces into submission as efficiently as possible. It will also protect friendly forces from the risk associated with treating enemy operatives as presumptive civilians.

Conclusion

The concept of Continuous Combat Function (CCF) in the ICRC's Direct Participation in Hostilities Study represents an important step in clarifying when civilians lose their protection from deliberate attack. However, this test was not and should not be considered the controlling methodology for determining when nonstate belligerent operatives may be lawfully attacked. Instead, that determination must be made based on a determination of belligerent status; a determination that triggers a broader targeting authority than CCF and properly allocates risk between competing belligerent forces.

Unfortunately, it is increasingly common to assert that nonstate belligerents are merely civilians directly participating in hostilities. CCF is then relied upon as a logical solution to targeting decisions when engaged in hostilities against nonstate groups. This is a dangerous trend. CCF is linked to the direct participation in hostilities qualifier to LOAC civilian protection from deliberate attack. Because of this, it—like the protection it qualifies—is not a method for determining enemy belligerent targetability. Instead, it must be confined to assessing when a true civilian, an individual acting autonomously and not executing the will of belligerent leadership, loses protection from attack.

If CCF is confined to this context, it is a logical step forward in the LOAC. While not a perfect solution to the challenge of dealing with civilians who engage in conduct inconsistent with their LOAC-derived protections, it at least acknowledges the reality that such conduct is not restricted to raising a rifle and pulling the trigger. Even the Section IX concept—that armed forces may use only minimum necessary force when encountering a civilian who has engaged in CCF and therefore must detain when feasible—is acceptable if limited to the autonomous civilian actor. But if extended to define when members of nonstate belligerent groups may be attacked, the rule loses credibility.

Conflicts between organized nonstate groups and state armed forces are unlikely to abate in the foreseeable future. Indeed, what is far more likely is that the definition of noninternational armed conflict, and the security challenges to the state that compel policy makers to invoke LOAC authority in response to nonstate threats, will continue to become more complex. Ultimately, however, armed conflict must be conceived as a contest between organized groups. Deviating from this conception undermines the entire authority/obligation framework of the law. The direct participation in hostilities cannot be a substitute for the reality that armed conflict is a contest between organizations, not a contest between one organization and a conglomerate of like-minded autonomous actors. Armed conflicts—even against the most evasive and loosely organized enemy—are therefore not a one-sided combatant COIN.

3

Protecting Civilians and Risking the Force in COIN

5

VALOR'S VICES: AGAINST A STATE DUTY TO RISK FORCES IN ARMED CONFLICT

Peter Margulies*

Introduction

Suppose that state forces have entered an urban area within the state's nominal political control, but effectively controlled by a terrorist group with a history of civilian attacks and cross-border shelling. State forces have entered for the purpose of stopping these attacks. During this incursion, members of the nonstate armed group have shot at state soldiers from within an apartment dwelling overlooking a key thoroughfare. The state has previously warned civilian residents of the dwelling to evacuate. Receiving new fire from inside the dwelling and not wishing to be pinned down, the commander calls in air support. That stops the hostile fire, but also kills civilians who remain inside. Has the state violated legal and ethical norms that govern the conduct of hostilities?

The traditional answer under international humanitarian law (IHL) would be, "It depends." States bind themselves to follow core principles, including the rules of distinction and proportionality. In the example above, the state has complied with the principle of distinction, which holds that a state may only target combatants.[1] The answer is less certain regarding proportionality, which requires that civilian harm be proportionate to the military advantage obtained in the attack.[2] If the military advantage is significant and the losses are small, the attack would meet the proportionality test. If the military advantage is small and the losses large, the opposite result would apply.

Some prominent scholars, including Michael Walzer, Avishai Margalit, and David Luban, have recently argued that this context-specific analysis fails to adequately protect civilians. For these scholars, military ethics should subject the state to further duties when it uses air power to minimize the risk to its own troops. Under what I call the duty to risk, the state has a categorical duty to risk its own forces.[3] In the example that opened this chapter, a state would have an affirmative obligation to mount a ground assault on the apartment dwelling, even when IHL would not require this decision.

I argue that advocates of the duty to risk misconceive both IHL and military ethics. Principles of IHL and military ethics encourage sound temporal judgment and efficient signaling among parties to armed conflict. Temporal judgment recognizes that human beings fail to make decisions that are consistent over time. Those decisions have two primary flaws. One is myopia, which entails the failure to consider long-term costs.[4] The other is hindsight bias, which narrows harm's perceived causes and naively attributes to one party the absolute power to prevent harm.[5] Violations of the principles of distinction and proportionality reflect myopia, since they erode discipline within armed forces and stoke resentments which prolong conflicts.[6] Duties exceeding IHL requirements may reflect hindsight bias, which blunts the decisiveness commanders need. The effects of unconstrained hindsight bias are particularly problematic in the counterinsurgency (COIN) context, where those effects usually limit only the state, and not violent nonstate actors.[7]

IHL rules also promote efficient signaling among parties to an armed conflict.[8] Current rules check aggressors, including violent nonstate actors, because an aggressor knows that the principles of distinction and proportionality do not undermine a victim's right of self-defense. In contrast, recognition of a duty to risk would encourage aggression by imposing significant new operational restrictions on self-defense rights. The consequence would be more aggressive moves, particularly by nonstate actors, with increased risk of harm to civilians.

In sum, the duty to risk maximizes hindsight bias, imposing what amounts to strict liability on states for civilian losses. The duty to risk also skews incentives for both states and nonstate actors, discouraging technological innovations that can minimize harm to civilians and promoting aggression by nonstate actors that compounds violence. Neither consequence harmonizes with IHL or military ethics.

As an alternative, I suggest a structural approach to COIN. This approach pivots around the relationship between a violent nonstate actor and at-risk civilians. It affords greater protection beyond IHL to civilians located within the shifting sphere of operations of terrorist networks and less protection to civilians in territory, such as Gaza, that terrorist groups effectively control. The approach also promotes the development of new technology that spares civilians, heightened legal and senior official deliberation about options for avoiding civilian casualties, and cooperation with international investigations. Such measures can reduce civilian losses, without the perverse incentives that the duty to risk would establish.

I. CIVILIAN HARM AND THE RATIONALE OF THE DUTY TO RISK

An armed conflict "necessarily places civilians in danger."[9] Some states and violent nonstate actors target civilians. In addition, the use of air power against military targets may harm civilians, particularly when a party to a conflict locates those targets in civilian areas.[10]

The law of armed conflict deals with these concerns in two important ways. The principle of distinction refers to a party's obligation to target combatants and refrain from targeting civilians not participating in hostilities. The proportionality principle requires that attackers avoid excessive collateral harm in achieving a military advantage.[11]

While the law of armed conflict bars the targeting of civilians and limits the collateral harm that civilians experience, some commentators have said that more is required. Focusing their complaints on the use of air power that limits risk to an attacking state's own forces, commentators have criticized military campaigns such as NATO's efforts to stop killing in Kosovo in 1999 and Israel's efforts in 2008–2009 to stop Hamas's launching of rockets from Gaza into Israeli towns. These scholars, including the great political philosopher Michael Walzer, have argued that, above and beyond the law of armed conflict, an attacker has a duty to risk its own forces to spare civilians on the other side.

The duty to risk argument had its genesis in an example provided by Walzer in his classic book, *Just and Unjust Wars*.[12] Walzer discussed a story from World War I in which American soldiers advanced to a French town after taking fire from German soldiers concealed within some of the town's dwellings. The American soldiers feared that German troops had hidden in cellars. For each house, the entering soldiers had to decide whether to provide a warning before throwing a grenade down the cellar stairs. If the cellar contained German troops, a warning would give the enemy troops time to attack. Tossing in a grenade without a warning might, however, kill French civilians who had used the cellar as a hiding place. Walzer's answer, with some modest caveats, was that the soldier must take the risk and issue the warning.

Extrapolating from this anecdote, Walzer and other commentators whom I call protective theorists—so named because of their concern for protecting civilians—have claimed that states must display an "active intention . . . *not* to kill civilians . . . through the risks the soldiers themselves accept."[13] According to the protective theorists, states that do not target civilians and only cause collateral harm that is proportionate to a military goal have nonetheless violated their ethical duties. To satisfy their obligations, states must go one step further and affirmatively risk their own forces.[14] I call this the duty to risk.

The duty to risk approach owes much to theories of strict liability. Luban has acknowledged this, analogizing an attack on a military target to "ultra-hazardous activities" from civilian life such as blasting or transporting dangerous chemicals.[15]

Under this theory, an entity engaging in such hazardous activities is strictly liable for the harms produced, even if the entity was otherwise without fault. Similarly, while Walzer has denied that there is a categorical bar on harming civilians,[16] the standard he fashioned echoes strict liability. According to Walzer, "due care" in the avoidance of harm to civilians entails more than ensuring that such harm is proportionate to a military objective.[17] Due care, under Walzer's formulation, entails more than merely *reasonable* care. In tort law, strict liability is the next available gradation after negligence's requirement of reasonable care.[18] For protective theorists, therefore, strict liability is the rule of choice when assessing collateral harm to civilians.

Unfortunately, the strict liability tenor of the duty to risk ushers in hindsight bias. Because of this inattention to the problem of hindsight bias, the duty to risk upsets IHL's precious equilibrium between hostile parties. The duty to risk also discourages reforms and palliative measures, including more advanced technology and the use of warnings. Furthermore, the duty to risk concept skews signaling about legal norms, and therefore compounds asymmetries between state and nonstate actors. Champions of the duty to risk might have alleviated these concerns if they had accurately assessed compliance with current IHL norms. I address these concerns in the next section of this chapter.

II. THE DUTY TO RISK AS A THREAT TO IHL'S DELICATE BALANCE

The protective theorists claim that IHL provides insufficient protection for civilians. However, this complaint fails to recognize IHL's careful balance of divergent temporal perspectives and organizational incentives. Viewed in this light, IHL protects civilians more effectively than protective theorists assert. Indeed, the protective theorists' formulation of the duty to risk stems from an incomplete understanding of both time and incentives. IHL and military ethics seek to avoid these vices; the protective view embraces them.

The law of armed conflict and military ethics are each studies in equipoise: they promote both decisive military decisions and the exercise of foresight that saves civilian lives. Decisiveness is vital, since hesitation may needlessly prolong an armed conflict with all the suffering that entails.[19] In the fluid environment of war, delays generate opportunity costs, as a resourceful adversary seizes the initiative. Commanders who relinquish their advantage will not be commanders for long, and an IHL that requires such sacrifice will not elicit compliance. At the same time, foresight is crucial for reinforcing enduring values such as prevention of harm to civilians. Civilian lives are valuable in and of themselves; their protection is also an important measure of a force's discipline. An undisciplined force cannot prevail in an armed conflict, or enforce a lasting peace.[20]

In seeking to balance decisiveness and foresight, IHL confronts the same obstacle faced by any legal framework: pervasive distortions in temporal judgment and ambiguities in signaling. By temporal judgment, I refer to the ability to accurately gauge the short- and long-term consequences of decisions. The absence of temporal judgment leaves the field to biases that impair decision making. For example, myopia stresses the present, weighting short-term risks and benefits over long-term effects.[21] Hindsight bias stresses the recent past, overstating the probability that a party could have prevented a particular harm.[22] In demonstrating the truth of maxims such as, "hindsight is 20/20," people attribute harm to the most recent and vivid causative factor.[23] By the same token, people discount data showing that less vivid, more systemic factors play a central part.

In addition to temporal judgment, IHL promotes efficient signaling about a party's intentions and general dispositions.[24] In a world of imperfect information, signaling allows a party to send messages about its compliance with norms.[25] Normative systems also signal about what conduct will create a fair and stable order. Legal systems that send the wrong messages can deepen distrust; in armed conflict, the wrong messages usually lead to needless death.[26]

Signals come in two varieties: operational or holistic. An operational signal seeks to provide information to another party about a particular situation. In contrast, a holistic signal conveys data about the underlying habits and dispositions of the sender. As an example of an operational signal, consider the white flag of surrender. A fighter who flies the flag signals that he will forego hostilities in exchange for being captured, not killed. The receiver, who in the heat of battle cannot conduct a more extended analysis of the sender's sincerity, accepts the white flag as a proxy.

Operational signaling turns on reciprocity: each party to a conflict lives up to the rules because it knows that at some point surrender may serve its tactical or strategic interests. In contrast, holistic signaling does not hinge on reciprocity in a narrow sense, but rather on a party's recognition that it participates in a system of shared understandings.[27] A party may not benefit from holistic signaling in the short term, if its adversary in an armed conflict does not reciprocate. For example, IHL requires that any party to an armed conflict prevent harm to captured adversaries. Al Qaeda would probably harm captives, even if the United States at all times refrained from doing so. Nevertheless, as the United States discovered when it sought to retreat from this IHL norm in the immediate aftermath of the September 11 attacks, failure to comply with IHL also lowers a party's global reputation, impeding multilateral counterterrorism efforts.[28] Holistic signaling situates a party more favorably in this wider arena.

The core IHL doctrines of distinction and proportionality refine temporal judgment. Distinction bars the intentional killing of civilians; without this bar, nations would act myopically and descend into a Hobbesian war of all against all. However, to stave off the countervailing evil of hindsight bias, distinction rejects strict liability. A party making an attack must show due care in ensuring that a target is a combatant,[29] but distinction does not require absolute certitude. Setting the bar too high would encourage perpetual second-guessing about targeting decisions, allowing hindsight bias to run rampant.

Proportionality shows the same careful balance. On the one hand, a law that failed to require proportionality would encourage myopia by giving free rein to the worst impulses of attacking states. Without some limit on collateral damage, a state would lose the ability to discipline its own forces. Proportionality also hedges against hindsight bias. The principle of proportionality does *not* impose an absolute ceiling on collateral damage. Rather, it creates a sliding scale. As expert commentary attests, "proportionality is not an exact science and it is impossible to draw in advance hard and fast rules."[30] Eschewing the protective theorists' drift toward strict liability, proportionality requires only that attackers avoid *excessive* collateral harm in achieving a military advantage.[31] IHL does not fault an attacker when military success is less resounding than planned or harm surpasses diligent estimates. This reading of proportionality recognizes that the "fog of war" forces a gap between reasonable calculations and unpredictable outcomes.

Under this flexible standard, the increased importance of a military objective will justify greater collateral damage. Suppose, for example, that a drone operator reasonably believed that a strike will kill Dr. Ayman al-Zawahiri, who became the leader of al Qaeda after the death of Osama bin Laden. Even if a strike would also kill ten civilians in Zawahiri's immediate vicinity, it would be proportionate, because of Zawahiri's centrality to al Qaeda. In contrast, a strike that killed an al Qaeda foot soldier and cost ten civilian lives would be disproportionate. A drone operator who reasonably believed that his target was Zawahiri and that only ten civilians would be killed would not face legal sanctions if the target turned out to be someone who closely resembled Zawahiri, or if a substantially larger number of civilians were killed because of factors that the operator could not control.[32]

Military ethics similarly seeks to curb both myopia and hindsight bias. The value of restraint is a recurring theme in military ethics.[33] Military ethics rightly rejects the indiscriminate or disproportionate killing of civilians as an approach that erodes the military as an institution, just as coercive interrogation of captives does. Moreover, as Walzer's cellar story shows, the military has a tradition of valor which contemplates risks to self to protect innocents.[34] However, military ethics does not require unbounded valor. Indeed, military ethics requires a commander to limit the risks imposed on those who serve.[35]

This last norm has often been more honored in the breach. The European leaders who sacrificed thousands in World War I's trench warfare to gain a few feet of Belgian real estate were unethical in the most profound sense, issuing commands without regard to the human cost entailed. Today's COIN operations thankfully do not involve slaughter on this scale. In any COIN operation, however, leaders must ask whether strategic objectives justify the risk they impose on military personnel. In many situations, these goals will warrant the risks involved. Winning over the hearts and minds of the civilian population is a worthy strategic goal.[36] However, imposing risk on one's own forces in a patently futile pursuit of this goal is unethical.

III. THE DUTY TO RISK AND THE PERILS OF HINDSIGHT: EXAMPLES FROM THEORY AND PRACTICE

Because the protective theorists have pressed their case for a duty to risk based on examples such as NATO's campaign in Kosovo and the Israel Defense Forces' (IDF's) Operation Cast Lead in Gaza, a closer look at these examples may furnish some clues regarding the soundness to the protective theorists' vision. Advocates of the duty to risk have argued that these examples show that existing concepts like proportionality and distinction fail to adequately protect civilians. Unfortunately, that understanding rests on a misapplication of both principles.

The misunderstanding starts at the level of theory, and continues on to practice. Protective theorists mistakenly depict opposing commentators as suggesting that a state may protect its own combatants by imposing greater-than-proportionate civilian casualties.[37] In fact, these scholars *agree* that IHL requires that collateral damage be proportionate to military advantage.[38] These scholars have merely restated the law: as long as a state complies with proportionality, it is under no *additional duty* to risk its own forces to safeguard civilians. Rather than acknowledge this, champions of the duty to risk create a straw man.

Once the straw man has made his appearance, mischaracterizing facts on the ground is inevitable. Advocates of a duty to risk underestimated the compliance with IHL of state forces in situations such as Israel's Gaza intervention and NATO's Kosovo campaign. These errors start with the protective theorists' poor grasp of the principle of distinction. Luban underestimates the number of terrorist fighters killed in the Gaza intervention. Discussing an incident in which Israel bombed a police station, Luban asserts that the police officers killed were merely civilians.[39] However, reporting has suggested that so-called police officers were in fact fighters for Hamas,[40] otherwise known as the Islamic Resistance Movement, which the United States and most of Europe have designated as a terrorist organization.[41] Luban's hasty conclusion paralleled the United Nations–sponsored Goldstone Report's flawed analysis of the Gaza police force.[42] The chair of the U.N. team subsequently repudiated assertions in the report that Israeli forces had violated the principle of distinction by targeting civilians.[43] While this acknowledgment of error does not necessarily undercut all of the Goldstone Report's conclusions, it suggests a lack of initial diligence that should prompt skepticism about the report's methodology and findings.[44]

Similarly, in the Kosovo campaign, while questions arose about particular targeting decisions, most experts agreed that NATO forces had targeted Serbian fighters and thereby complied with the principle of distinction.[45] Moreover, even critics of the Kosovo campaign acknowledged that NATO forces had exercised caution in order to reduce civilian casualties.[46]

The protective theorists' errors in application of the principle of distinction undermine their case that existing duties fail to protect civilians. Here, the inner workings of the laws of armed conflict are less important than simple arithmetic. Posit a fixed total of casualties of an attack. Suppose that, out of that total, the attackers killed more combatants than their critics claim. As a corollary, the attackers also killed fewer civilians. Elementary math alone thus weakens the protective theorists' case.

In addition, an operation that kills a higher percentage of combatants reaps a larger military advantage. Under proportionality, a greater military advantage justifies increased risk to civilians. So underestimates of compliance with the duty of distinction inevitably yield underestimates of compliance with proportionality.[47]

The protective theorists' estimates of civilian casualties bear out this concern about inflated numbers. Luban, for example, estimated that more than 1,000 civilians were dead in Gaza, and perhaps thousands in Kosovo.[48] These numbers, offered without empirical support, are substantially greater than other estimates.[49]

Inflated numbers do not necessarily undermine the protective theorists' argument that IHL offers insufficient protection to civilians. After all, even smaller numbers of civilian deaths surely merit concern. However, fewer civilian losses weaken the case for supplementing IHL with a new strict liability rule.

IV. UNINTENDED CONSEQUENCES: THE DUTY TO RISK AS AN IMPEDIMENT TO PRECAUTIONS

The protective theorists' misapprehension of both distinction and proportionality accompanies myopia about attackers' precautions. Under Article 57 of the First Additional Protocol to the Geneva Conventions, states have a duty to take feasible precautions to minimize civilian harm.[50] Many of the military campaigns criticized by protective theorists utilized such precautions, sometimes beyond the "rule of reason" demanded by IHL.[51] However, the duty to risk provides no incentives for such precautions, thus making it less likely that attacking states will implement them.

A. Impeding Precision Weaponry

A cardinal precaution used in modern military campaigns such as in Libya and Kosovo is precision weaponry. Using precision weaponry, forces can target combatants while reducing both risk to their own personnel and collateral harm to civilians. For example, drones have the ability to loiter for long periods, collecting information about life patterns that will refine targeting.[52] Bombers today use precision targeting that allows aircraft to fly at high altitudes.[53] At those heights, military personnel are safer, while precision targeting systems have more time to engage. However, protective theorists have not given credit to states for developing precision targeting systems that zero in on combatants and prevent mistaken targeting of civilians or inadvertent collateral damage. Imposing a duty to risk on states would impede the development of precision targeting, since states would have to risk their own personnel in any case. However, the fine adjustment of incentives is off the protective theorists' radar screen, since they urge strict liability for *any* civilian harm.

B. Discounting Human Intelligence

The protective theorists would also discourage an attacking state's development of human intelligence assets. Intelligence from informants can help an attacking state zero in on combatants, promoting compliance with the principle of distinction and making harm to civilians less likely.[54] American forces benefited from such intelligence after the Sunni awakening in Iraq. The IDF has relied heavily on intelligence from informants in the West Bank and Gaza.[55] The exploitation of human intelligence in precision targeting can reduce risk to one's own forces. However, if attackers were under an affirmative duty to risk their own forces in any case, incentives would shift. A state with a duty to commit ground troops would soon encounter the urge to escalate those forces, if an initial commitment was unproductive. Cultivating human intelligence assets requires patience that commanders often lack when they must commit massive numbers of ground troops. Since more ground troops make for more intense fighting, the result may be more civilians killed.

C. Hindering Warnings

The protective theorists also discourage warnings by attackers. This effect seems particularly perverse, both because warnings save lives and because the duty to risk emerged from Walzer's retelling of the World War I cellar story, which turned on a warning. However, the protective theorists' strict liability approach acts as a disincentive. Further, protective theorists fail to recognize the trade-offs that warnings entail, even though IHL already incorporates these trade-offs.

IHL recognizes that warnings are both a necessary precaution and an element in proportionality analysis. Like other elements of IHL, warnings balance foresight and decisiveness. Consider the question of a warning's specificity. Suppose that a violent nonstate actor has located a mobile rocket-launcher on land connected to a civilian dwelling. A state with reliable information about the rocket-launcher could provide a specific warning, cautioning that within twenty-four hours it will attack the rocket-launcher. An attack would be appropriate under the principle of distinction, since the rocket-launcher is an instrumentality of violence used by an opposing force—that is, a military target.[56] However, once the state supplies this warning, the force controlling the launcher will surely respond by either moving the launcher even closer to the civilian dwelling, or moving it to another location altogether. Either way, the warning will impair the state's ability to achieve its legitimate military objectives.[57]

For this reason, IHL has a more limited requirement for warnings. A state may provide general warnings in advance of an attack[58] and allow civilians to take precautions based on their knowledge of the situation on the ground. In the rocket-launcher scenario, a civilian resident in the building connected to the launcher site can see the launcher from her window. A general warning advises the civilian that her dwelling will not be a safe place in the near future and that leaving for a safer location is prudent. Residents in other locations can in turn provide information to the fleeing resident about the safety level in those areas. While some civilians, such as the aged and infirm, may have greater difficulty in moving, a general warning provided sufficiently far in advance can facilitate efforts at relocation for these individuals, as well. As long as the attacker targets only legitimate military objectives such as the rocket-launcher, residents will still have many safe places available. Residents who remain after such warnings are no different from workers in munitions factories: they have notice that their decision will increase their risk.[59] Collateral harm to this group, while regrettable, will not be a violation of the proportionality principle.

Protective theorists, caught up in hindsight bias, have not even acknowledged the warnings the IDF provided in the Gaza campaign. Warnings are irrelevant if, like the protective theorists, one imposes a strict liability standard on civilian harm. Under a strict liability view, the reasonableness of precautions does not matter. Any harm is by definition the responsibility of the attacker. Consistent with this rationale, the United Nations' Goldstone Report faulted the IDF's extensive warnings to civilians in Gaza[60] because many civilians did not comply.[61] However, an attacker's warnings should be judged by whether their specificity is adequate in light of legitimate military objectives. Assessing warnings based on civilian compliance is unfair, since an attacker has no ability to control the civilian population's response.

Moreover, assessing warnings from the perspective of hindsight is also unwise. Warnings of appropriate specificity and comprehensiveness that do not trigger full compliance are nevertheless far superior to warnings that are amorphous and inconsistent. However, because the duty to risk requires risk to a state's own forces even when the state provides effective warnings, the duty discourages this vital precaution.

To see how the duty of risk discourages warnings, one must consider the trade-offs that warnings engender. As noted above, warnings provide information not only to civilians but also to the enemy. Even under current law, therefore, an attacking state has some incentive to "skate by" with warnings as minimalist as possible. In this respect, military organizations subject to the law of war mimic the conduct of other regulated organizations. When regulated entities feel that a legal regime injures their competitiveness without providing corresponding benefits, those entities drag their feet, sabotaging regulation through dilatory tactics and defiance.[62] Imposing a duty to risk further lessens a state's stake in warnings, encouraging them to do the legal minimum or even take a risk that champions of the duty to risk would not embrace: sinking beneath the floor of required humanitarian protection.[63] Few messages are more counterproductive.

In an additional layer of irony, the duty to risk's essential indifference to warnings in contexts like the Gaza intervention points up an inconsistency with its embrace of warnings in the World War I cellar scenario. Walzer's evaluation of the latter hinges on the soldier's issuing a warning before throwing a grenade down the cellar stairs. The inescapable corollary of Walzer's view is that a soldier who issued a warning but received no response would be free under both IHL and military ethics to use his grenade. Unfortunately, neither Walzer nor Luban explain why warnings are inadequate in the larger context of Gaza. In both contexts, warnings will not be completely effective in preventing harm to civilians. In the cellar scenario, for example, civilians might be too afraid to reveal themselves, or might be the captives of German soldiers who are also hiding in the cellar. Yet Walzer and Luban do not argue that the silence that would greet a warning in these situations required *additional steps*, such as the soldier's blind descent down the cellar stairs. To be consistent, the protective theorists would have to impose even more exacting duties on the World War I soldier. They reject this step, apparently conceding that both law and ethics cannot require unbounded valor. However, they failed to provide a uniform rationale for rejecting this further expansion of duties on the cellar stairs, while requiring it in contexts like Gaza.

D. The Incoherence and Unmanageability of the Duty to Risk

This failure of consistency points up the protective theorists' lack of a coherent prescription for the amount of valor they would require. Luban, perhaps recognizing this problem, suggests a metric that he calls the risk transfer ratio.[64] Under one reading of this ratio, military ethics would bar tactics that result in heightened risk to civilians but produce smaller increments of safety to soldiers.

In easy situations, this prescription is sensible. For example, it would fit the cellar scenario, where a soldier could issue a warning that would appreciably lower risks to civilians with only a modest diminution in his own safety.[65] Arguably, Luban's ratio would also permit bombing of a classic military target such as a munitions factory, where another alternative such as a covert ground assault behind enemy lines would subject soldiers to far greater risk while producing at best equal reductions in risk to civilians. However, Luban's ratio runs out of steam when dealing with more difficult situations. On occasion, Luban notes, soldiers might have to endure greater risks to spare civilians from harm.[66] In some situations, Luban may well be correct. However, the devil is in the details. Luban acknowledges that "how much extra risk soldiers must shoulder is... not a question susceptible to precise... or... general answers."[67] Despite this vagueness, Luban is not reticent about attaching blame to soldiers who fail to meet his standards. Commanders who fail to calibrate the risk-transfer ratio in particular operations to Luban's fine-grained specifications will presumably violate ethical norms, IHL, or both. The shifting implementation of the risk-transfer ratio assures that commanders will be wrong far more often than they are right. Luban fails to address how commanders could cope with these uncertain conditions. His failure to offer a consistent standard is a recipe for hindsight bias.

V. THE DUTY TO RISK AND THE INTERACTION OF IHL AND INTERNATIONAL HUMAN RIGHTS LAW

Broken down into its component parts, the case for the duty to risk thus turns on incoherent tests and stories selectively retold. This section takes up one of the latter. Using an idiosyncratic hypothetical, protective theorists argue that their opponents would improperly provide less protection to enemy civilians than to a state's own nationals.

However, this argument ignores nuances in the interaction of IHL and international human rights law (IHRL).

To contend that the law requires protection for enemy civilians that matches the protection a state affords its own civilian citizens, Margalit and Walzer posit Lebanese villagers who join Hezbollah's efforts to take an Israeli town. They argue that these villagers are entitled to the same regard as Israeli civilians taken hostage by Hezbollah.[68] This argument ignores the collaboration in the town scenario between Lebanese villagers and Hezbollah. Margalit and Walzer acknowledge that the Lebanese villagers might wish to take over Israeli land. If the villagers' acquisitive motivation drives their role in a cross-border attack, they lose their immunity from attack. The villagers are not involuntary human shields, but direct participants in hostilities,[69] helping to clear out Israelis from the town to make it "safe" for the invaders. The principle of distinction might well permit targeting of such individuals, who had knowingly crossed the border of another state in operational support of an invading force. At the very least, proportionality would analogize the acquisitive civilians to munitions workers, who do not count as civilians in proportionality estimates involving attacks on weapons factories.

The protective theorists' insistence that a state owes the same minimum duties to both its own nationals and those acting at the direction of a hostile party also suggests an unduly rigid view of the separation of human rights law and IHL. When IHL applies, the best view is that it preempts the field—it is *lex specialis*, as commentators have suggested, displacing other fields of law. When conflict reaches a level of frequency and intensity that moves beyond the capability of ordinary law enforcement, the law of armed conflict should govern.[70] Moreover, IHL should govern whether the conflict is international, that is, between states, or is noninternational in character. An internal rebellion may be sufficiently intense to qualify as an armed conflict. The International Criminal Tribunal for the Former Yugoslavia (ICTY) decisions take this as a given,[71] and countless nations' experiences with civil wars (of which America's was a particularly bloody example) confirms it. If one party to a conflict not of an international character attacks, the party attacked must have a right to self-defense under international law governing the use of force. The U.N. Charter, which mentions armed attack but says nothing about the source of that attack,[72] echoes this proposition.

A categorical choice of a law enforcement paradigm over IHL in such contexts would result in confusion for both attackers and defenders, producing a crazy quilt of conflicting obligations. For example, while IHL ties proportionality to military advantage, human rights law requires what commentators call "strict proportionality,"[73] which often entails heightened duties to spare civilians. Human rights law also conditions the use of lethal force on a concrete need for self-defense or defense of others.[74] In contrast, under IHL, one party can target an adversary's combatants, as well as civilians taking a direct part in hostilities, without a specific showing of the need to use lethal force.[75] Under IHL, once a combatant has identified appropriate targets, he may use lethal force, whether or not the target has initiated fire.

While both human rights law and IHL bar wanton killing, the two bodies of law have very different default stances that the protective theorists often ignore.[76] The default stance of human rights law is often inappropriate for situations of armed conflict. Consider the Lebanese nationals in Walzer and Margalit's hypothetical who aid Hezbollah's efforts to capture an Israeli town. Under a law enforcement paradigm, individuals who assume control of property owned by another without the owner's consent would be considered trespassers. Law enforcement officials would have the power to evict the trespassers and the discretion to arrest them, but would be permitted to use deadly force only in self-defense.[77] If one of the trespassers fired on law enforcement officers, the officers would be required to exercise due care to avoid harm to others in the group of trespassers who had not initiated fire. These limited state measures suffice in ordinary times. During armed conflict, however, such curbs provide an adversary with incentives to use its own civilians in combat and impede a state's efforts to protect its citizenry.

Rejecting a categorical preference for the law enforcement over the IHL paradigm does not preclude resort to human rights principles in appropriate contexts. Suppose, for example, that terrorist violence is episodic in nature, a state has complete control over territory, and the number of suspected terrorists is small. Under those conditions, a state should only use lethal force when such force is necessary to protect lives. In *McCann v. United Kingdom*, British forces tailed three IRA members in the British territory of Gibraltar whom the government correctly suspected were engaged to explode a car bomb within approximately 24 to 48 hours.[78] Security forces used lethal force when the suspects became aware of the surveillance and took action that the officers believed showed the intent to detonate the bomb immediately. The European Court of Human Rights found that the government failed to pursue alternatives short of deadly force, such as wounding the suspects, and that therefore the government had violated the suspects' right to life.[79] Nine judges disagreed, contending that the government had taken adequate precautions in dealing with three concededly dangerous individuals who were executing a terrorist plot and that the majority's contrary view was marred by hindsight bias.[80] However, the parties conceded that IHRL was the appropriate legal framework. Therefore, government personnel using lethal force had to show some evidence of an imminent threat to life. No party argued that IHL governed, which with today's technology would have permitted the government to use a drone to kill the terrorists.

IHL, in contrast, would not have required following the *McCann* suspects, at least where the conflict had become more intense than episodic and the suspects were operating in an area not wholly under the government's control. IHL recognizes that in those situations, reliance on ordinary law enforcement processes poses unreasonable risks. In an armed conflict, viewing individuals accompanying an invading force as mere trespassers subject to arrest would put the invaded state at a grave disadvantage, given the difficulty of using ordinary law enforcement procedures during a period of intense violence.[81] This result would merely encourage violence against the state, raising the risks of harm to all civilians.

VI. THE DUTY TO RISK AS FLAWED SIGNALING

If protective theorists allow hindsight bias to distort their assessment, they also fundamentally misunderstand how rules shape the incentives for each party. Historically, the law of armed conflict has promoted signaling between parties to a conflict that reinforce rules of the road.[82] These rules are not random conventions but norms that stop needless suffering. Unfortunately, some proposed innovations in the law of armed conflict in the last forty years have undermined rules of the road, promoting signaling that instead breeds mistrust and extends violence. The duty to risk, despite its origins in the urge to protect civilians, is yet another installment in this trend.

To appreciate the signaling that the law of armed conflict has historically encouraged, consider the age-old war crime of perfidy. A party that engages in perfidy, or "false flag," signals surrender while planning to lure in opponents for the kill. Allowing perfidy would send a perverse message, encouraging parties to deceive each other instead of cooperating on ground rules.[83] A legal system that condoned perfidy would also promote needless slaughter, since a risk-averse receiving party would doubt the sincerity of any surrender plea and therefore in all cases intensify its attack.

Some proposed innovations in the law of armed conflict have retreated from this promotion of straightforward signaling. For example, Additional Protocol I to the Geneva Conventions for the first time allowed fighters to pursue hostilities dressed in civilian garb, as long as the fighters revealed themselves at the last possible moment before launching an attack.[84] Billed as a measure to help indigenous people fight off imperial domination,[85] the Additional Protocol I changes have nurtured greater uncertainty about the intent and capacity of civilians. While traditional rules of armed conflict required wearing a uniform or some distinguishing emblem, current norms leave states to puzzle out less reliable distinctions between fighters and those hors de combat.[86] When states are stumped, they resort to the same deadly risk-aversion that drives the rule against perfidy: they shoot first and ask questions later. The result is more peril for genuine civilians.

The duty to risk sends perverse signals in a different respect. Both ethical and legal rules on the conduct of war must be consistent with rules governing the use of force. Those rules generally limit the use of force to cases involving self-defense to address an imminent threat.[87] However, *jus in bello* rules that hamper a party's resort to self-defense shift the equilibrium, giving the advantage to an aggressor. Consider a party's consideration of an unprovoked attack on another. *Jus in bello* restrictions like distinction and proportionality preserve self-defense as an option for the party attacked, although they constrain the operational contours of that response. Without this balance, self-defense would be ineffective, giving attackers who violate the *jus ad bellum* an undeserved edge. An effective self-defense option preserves equilibrium because it obliges a prospective aggressor to think twice.[88]

A legal norm that sends this signal about thinking twice before an unprovoked attack will influence any party that fears another party's capacity for self-defense. Any party—state and nonstate actor alike—that controls territory has such a stake. States with a monopoly on force within their territory are also in some fashion accountable to their civilian citizens.[89] Civilians will protest, legally or outside the law, if a state's rash decisions expose them to harm through another party's exercise of the right to self-defense. Some terrorist groups, such as Hamas or the now-defunct Tamil Tigers of Sri Lanka, also control territory and are accountable to constituencies on that territory.[90] The accountability of both states and nonstate actors will vary. However, try as they might, no state or nonstate leader can wholly suppress popular indignation.[91]

Because territorially-rooted parties, state and nonstate actor alike, are accountable to people within that territory, they have an incentive to avoid aggressive conduct that will trigger another's right of self-defense. Legal norms' cautionary signal to would-be aggressors thus reduces future violence. Indeed, this is exactly what seems to have happened in Gaza, where the vigorous Israeli response to Hamas's repeated rocket attacks on Israeli towns yielded a lower level of Hamas-initiated violence.[92] In contrast, unduly burdensome restrictions on the right to self-defense encourage aggression.[93]

Champions of a duty to risk ignore the importance of signaling and thus fail to understand how undue operational restrictions on a state's right of self-defense encourage aggression. Nonstate actors will risk their own civilians in violation of their duties under the law of war, if they derive a tactical advantage from doing so. Luban notes this phenomenon,[94] but does not recognize the connection between operational restrictions on self-defense and the incentives of aggressors.[95] Addressing this nexus is crucial to both IHL and military ethics.

The duty to risk misapprehends the incentives of violent nonstate actors in another sense that is inherent in asymmetric warfare. For nonstate actors like Hamas or al Qaeda with a long track record of disregarding IHL, military victory means something very different than what it does to a member of a state force. A state force wins only if it stops violence directed at the state by defeating an opponent. However, a nonstate actor whose principle mode of conducting hostilities entails targeting civilians—either those of a state opponent or moderates from its own community[96]—wins when it kills opponents. For the terrorist, such killing is a symbolic act which expresses solidarity with other terrorist groups.[97]

These asymmetrical incentives affect both proportionality and military ethics. Even protective theorists concede that force protection is at least relevant to determination of military advantage. Exposing forces to targeting by terrorist groups thus presents an exponentially greater risk than the risk taken in combat with an actor that complies with IHL and diminishes military advantage to a greater degree. Under IHL, a state would be free to take this difference into account. Moreover, from the standpoint of military ethics, a decision by a commander to reject a legal course of action and thereby risk exponentially greater numbers of casualties should have a clear strategic justification. Any other decision fails to attach sufficient importance to the lives of state combat personnel. In some situations, a long-term strategic advantage in combat against a nonstate actor could justify such a decision, as we shall see. However, the protective theorists' approach fails to identify such grounds, which in any event would be permissive, not mandatory: states would be *allowed* to make such decisions, but would not be required to do so, because a mandate would skew incentives and favor nonstate actors.

Nevertheless, states may well have a duty to risk their forces in particular situations. Compliance with IHL norms like distinction and proportionality may require that a state risk its forces in an engagement with its adversaries, or rule out an attack. In Kosovo, NATO commanders refused to attack a bridge used by Serbian forces, because an attack would have killed a large number of civilians.[98] Israel has acknowledged that the attack on a U.N. facility during the Gaza campaign failed to comply with IHL.[99] Today's technology, including use of unmanned aerial vehicles (UAVs, or drones) with superior reconnaissance and imaging capabilities, might have prevented such harm. Refinements in targeting processes might have also led to a better result. These are the issues that a worthwhile code of military ethics should address, as I suggest in the next section.

VII. A STRUCTURAL APPROACH

In earlier work I have recommended a structural approach to IHL; this approach, with some modest changes, also seems appropriate here. A structural approach focuses on the distinct roles played by players, including nonstate actors and the varied constituencies that speak for states.[100] By promoting dialogue among state constituencies and between certain nonstate actors and the civilians those actors purport to champion, a structural approach aims to realign the incentives that the duty to risk distorts.

A. The Reciprocity Ratio

In determining when military ethics permits the risk of forces to minimize civilian harm beyond what IHL requires, senior COIN commanders should consider the nature of the nonstate actor and that actor's relationship to civilians. When a commander knows that a nonstate actor values the killing of state forces as much or more than attaining a strategic victory, military ethics would preclude putting state forces in harm's way.[101] Moreover, requiring such a sacrifice would skew incentives for nonstate actors, encouraging decisions that enhanced the danger to civilians. However, a state may impose rules of engagement entailing a superproportionality obligation when the relationship between a nonstate actor and civilians suggests that over time civilians may reciprocate for this regard by shifting their allegiance to the state.[102] I call this calculation the reciprocity ratio. If Re equals reciprocity, Ri is risk, S is a short-term calculation, and T is a temporal calculation in the intermediate or longer term, than a short-term risk is appropriate when $ReT \geq RiT$.

In other words, when there is a likelihood that forces or civilians in time may reciprocate, and that likelihood equals or exceeds the near-term risk to one's own forces, risk is appropriate, since today's risk may result in lowered risk tomorrow. When a probability of reciprocity over time falls below the short-term risk, a state should not undertake super-IHL duties. In the latter case, moreover, maintaining ordinary proportionality signals to the nonstate actor that a state will engage in robust self-defense, thus shaping the nonstate actor's ex ante calculation on aggressive behavior.

Generally, the reciprocity ratio will vary with the relationship between the nonstate actor and the civilians it controls. The more civilians the nonstate actor controls and the greater the analogy between that nonstate player and a state, the harder it will be to turn a civilian population against the nonstate actor through mere forbearance. The futility of forbearance in turn makes the harm to one's own forces more difficult to justify. Therefore, in fashioning rules of engagement (ROE),[103] commanders should distinguish between what I call territory-based terrorist groups and transnational terrorist networks.

A territory-based nonstate actor controls a significant chunk of territory within a state and commands the allegiance of a substantial number of civilians. While exact parameters are difficult to pinpoint without some measure of arbitrariness, such a group would generally feature at least: (1) fifty square miles of contiguous territory; (2) ten thousand civilians under control within that territory; and (3) two-thirds of civilians within the location either tacitly or actively supporting the group. Hamas and the now-defunct Sri Lankan Tamil Tigers (Liberation Tigers of Tamil Eelam, or LTTE) are examples of a territory-based group. The territory-based group has more ability to enlist the population it controls in its various activities, through the provision of social services that inspire community loyalty.[104] A state confronting a territory-based nonstate actor should decide that, consistent with proportionality, it need not lower civilian deaths below the proportionality threshold. Eschewing a lower ROE standard treats the group the way it wishes to be treated—as a state. States starting a war know the result is putting their own civilians at risk. That is one aspect of the "right authority" of states[105]—they act as a clearinghouse for citizen inputs.[106] The "clearinghouse" nature of states makes states an appropriate party to wars, which could dissolve into a struggle of all against all if groups of civilians could too readily engage in warfare. So a state fighting a territory-based terrorist group may seek to extract concessions from that group just as it would from a state: through ROE in which proportionality is both ceiling and floor.

With network-based groups such as al Qaeda, the calculus for ROE is different. There, the group may have only an instrumental and short-term need for the community where it is currently located. Community anger at the group may not be a disincentive to future violence, since the group quickly exits, moving on to greener pastures if attacks on a state elicit robust self-defense measures at its present location. Here, state attackers should be careful to avoid civilian casualties and set ROE at superproportionality levels. Indeed, a showing of restraint may win hearts and minds in a way that is far more difficult with a community that has deep bonds of identity with a territory-based group.

B. Most Precise Available Weaponry

Military ethics should require the use of weapons of the maximum precision practically available to that state. A state like the United States should endeavor to use such weapons, even when estimated collateral harm from an attack without such sophisticated systems would satisfy the proportionality principle.[107] This does not mean that a state must rush on to the battlefield weapons that have not been fully tested, or must commit to across-the-board use of weapons that are expensive, because the newness of the weapon limits economies of scale. However, a nation should use its best efforts to bring new technology online, when that technology will improve targeting and reduce collateral damage. While one might be concerned that such a requirement of ethics might actually discourage the development of precision weapons, since states that developed such weapons would then be subject to heightened ethical requirements, that fear seems misplaced. Enough strategic incentives exist to develop such weapons even if their dissemination results in the imposition of new ethical burdens. Moreover, the precision imperative is consistent with the reciprocity ratio discussed above. Nationals whose hearts and minds are up for grabs will doubtless have a rough idea of the technological capacities of a country seeking to suppress an insurgency. Particularly when that country is the United States, nationals may well have an exaggerated idea of the counterinsurgency's technological strengths. Failure to exploit the technological edge available would result in a loss of credibility.

C. Tailoring Force for Use against Dual-Use Objectives

Modern warfare has often seen disputes about attack on dual-use objectives, such as power plants and communications centers. On the one hand, dual-use facilities that assist one party's ability to use force are clearly military objectives.[108] On the other hand, complete destruction of these facilities can cost the lives of civilians who work at these sites, and can also impede a civilian population's ability to recover after conclusion of a conflict. To solve this dilemma, military ethics, if not IHL, should require the least amount of force necessary to achieve an immediate military advantage, when the attacking state can readily repeat this level of force because of its own superior technology and military might.

With power plants or similar targets, tailoring the use of force may entail an attack that will temporarily halt the target's functioning, without total destruction of the facility. For example, an attacking force may disable broadcast transmissions through cyberattacks or attacks on wiring.[109] Attacks may still have short-term costs for civilians who depend on goods that the target provides, such as electricity. However, the tailoring of force to function will spare civilians working at or near the targets. An attacking state will assume the burden of repeating this attack if a conflict lasts longer than anticipated. But this burden seems reasonable as a matter of ethics, if not IHL, when the attacking state has a clear technological edge and greater military resources.

D. Legal Deliberation

Similarly, COIN should include a process imperative. While proportionality has traditionally been a case-by-case inquiry, carefully designed systems can squeeze out much of the human error that often afflicts such determinations. The comprehensive integration of military lawyers into the targeting process will assist with this task. For example, during the Kosovo campaign, the U.S. Joint Staff developed a matrix that rated a target's military value and classified projected collateral harm in gradations of high, medium, and low, as well as other variables, such as the probability that a bomb or missile would miss its target, and the effect on civilians if this event occurred.[110] Since a miss would obviously have a greater effect on civilians in a densely populated area, the U.S. military took this risk into account as well, requiring a correspondingly greater military justification for an attack in that context. Moreover, any particularly sensitive targets, including dual-use targets such as power plants, should require approval from senior military or civilian leadership.[111] The requirements of proportionality clearly informed these judgments, but the methodical and analytical approach reflected an ethical and strategic commitment beyond proportionality's dictates. Moreover, an attacker should strive for a balance between planned and unplanned targets.[112] Whenever possible, a state should develop a list of targets in advance, to permit the careful review that eliminates mistakes. Human intelligence and other resources can be collected and synthesized at this stage.

E. Duty of Investigation

A nation should undertake as a matter of both IHL and ethics to promptly and comprehensively investigate plausible reports of avoidable civilian losses. Even where a nation has few or no ground troops and does not systemically violate IHL, it may commit IHL violations in particular situations. Prompt investigation will address such problems. Investigation has both ex ante and ex post uses. First, ex ante, the prospect of investigation will ensure that commanders make decisions that comply with IHL. Second, ex post, investigation will uncover mistakes that have occurred and supply a database for future improvements. It will also flag commanders who have violated IHL and encourage commanders who comply with the law of armed conflict.

F. An Ethic of Engagement with International Entities

States should also engage with international tribunals and other entities. These tribunals are flawed, yet the failure to engage is often a self-fulfilling prophecy. Only engagement improves the tribunals and allows them to gather other perspectives. A policy of engagement also recognizes that such tribunals are a pervasive aspect of modern international law and help adjudicate the boundaries between IHL and human rights. As a pragmatic matter, moreover, like most structural rules that comprise a set of self-constraint,[113] such rules encouraging engagement redound to a state's benefit in the intermediate and long term.

This ethical commitment to engagement makes sense as a matter of temporal judgment and holistic signaling. A state faces short-term pressure *not* to cooperate with transnational tribunals or entities: those entities may cramp the style of a country with far-flung diplomatic, political, and military responsibilities, such as the United States. These concerns helped persuade the United States to limit its cooperation with the International Court of Justice (ICJ) in the Nicaragua case.[114] Alternatively, a nation like Israel, long a favorite target of other countries for political and ideological reasons, may feel that it will not receive a fair hearing. This concern led to the Israel's failure to cooperate in the ICJ's "Wall" case.[115]

While these short-term moves were expedient, in the longer term they were unwise. Lack of cooperation has helped reinforce a strand in international discourse that sees the United States and Israel as outliers in the international legal system.[116] This is a caricatured and unfair view. However, a lack of cooperation invites this result.

Current models of U.S. engagement with international entities provide a contrasting vision. Initially, the United States flatly opposed the International Criminal Court (ICC), viewing it as an entity with only a limited upside and a significant potential for second-guessing American forces deployed around the world.[117] While the United States has not ratified the Statute of Rome, which created the ICC, it has participated in a recent round of talks addressing whether and how to include the crime of aggression within the ICC's jurisdiction.[118] Another model from the human rights realm is the United States' engagement with the Inter-American Commission on Human Rights (IACHR), where the United States responds to complaints even though it does not recognize the commission's decisions as binding.[119]

An ethic of engagement promotes sound signaling and temporal judgment, despite its occasional amenability to short-term bumps in the road. First, consider the relatively low opportunity costs. A state that participates does not give up the right to criticize a result as unfair. The appropriate analogy is to a litigant before a court—the litigant does not waive the right to appeal because she has appeared in a case before a trial court. An ethic of engagement also reinforces that a state has nothing to fear from a transparent investigation; this message promotes holistic signaling, proving a state's bona fides in the international legal system. Willingness to participate can also have a useful ex ante effect; it will oblige the state to use methods that may minimize casualties and other damage, including the use of the most sophisticated technology available and the use of the least possible force to disable dual-use facilities such as communications hubs. States that rightly point to the distorted incentives wrought by the duty to risk should not ignore their own responsibility to set norms and live by them. That is the heart of self-governance, domestically and internationally, and an essential element of effective counterterrorism.

Conclusion

A structural approach that considers the likelihood of reciprocity and pushes states to calibrate force when possible, use precautions, and deliberate with a greater range of actors about lethal tactics will lack the sentimental appeal to valor of the duty to risk. However, this approach will also be free of the hindsight bias that plagues the duty to risk's strict liability. Moreover, it will send cleaner signals to both states and nonstate actors, pushing territorial nonstate actors to consider blowback from their own constituencies instead of (as in the duty to risk) permitting them to strike but withholding from states an adequate right of self-defense.

This approach also avoids the more obvious errors of the duty to risk's extrapolation from Walzer's cellar scenario. Building a theory from a vivid example can obscure the facts in the wide run of cases, including the actual facts—which champions of the duty to risk get wrong—on civilian casualties in campaigns like Kosovo and Gaza. Indeed, to the extent that the cellar scenario turned on the soldier's warning, a structural approach which incentivizes warnings is actually closer to the spirit of the story itself. The approach recommended here may not produce more heroes, but it may save the lives of more civilians.

6

AGENCY OF RISK: THE COMPETING BALANCE BETWEEN PROTECTING MILITARY FORCES AND THE CIVILIAN POPULATION DURING COUNTERINSURGENCY OPERATIONS IN AFGHANISTAN[1]

Chris Jenks[2]

Introduction

> Leaders prepare to indirectly inflict suffering on their Soldiers and Marines by sending them into harm's way to accomplish the mission. At the same time, leaders attempt to avoid, at great length, injury and death to innocents. This requirement gets to the very essence of what some describe as "the burden of command." ... *Ultimate success in [Counterinsurgency Operations] is gained by protecting the populace, not the [military] force.... [yet] combatants are not required to take so much risk that they fail in their mission or forfeit their lives.*[3]

In 2010, nine years into Operation Enduring Freedom in Afghanistan, casualties among both Afghan civilians and members of the U.S. military were at their highest levels to date. Taliban[4] insurgents employing suicide and improvised explosive devices directly caused the vast majority of those casualties. Yet attitudes and perceptions among both the Afghan population and the U.S. military reflect differing conceptions of blame and responsibility. Afghans blame the mere presence of the United States as the underlying cause for Taliban attacks and resulting civilian casualties, while members of the U.S. military question whether self-imposed limitations on employing force has led to increasing numbers of U.S. service members wounded and killed.

One of the origins of this angst is the degree and manner by which risk is borne by the two groups, Afghan civilians and U.S. military, as a result of the ongoing counterinsurgency operations. At the strategic level, military doctrine, including the language in the quotation above from the U.S. military's 2006 counterinsurgency manual, provides fundamental principles that govern the conduct of military operations and, in so doing, shapes the parameters of this risk. Moving from the strategic level through the chain of command to the tactical level, the soldier and marine on the ground, the contours of risk are operationalized through the war-fighting command promulgating rules of engagement and, in Afghanistan, the tactical directive.

The tactical directive "provides guidance and intent for the use of force by" U.S. and coalition military forces operating in Afghanistan.[5] Previously classified to protect the force and still classified in parts, beginning in 2008 three successive commanders of the International Security Assistance Force (ISAF) in Afghanistan have made portions of the directive public.

The 2008 tactical directive stressed minimizing death or injury of innocent civilians and reinforced the idea of proportionality, "requisite restraint, and the utmost discrimination in our application of firepower."[6] This iteration, however, did not place any specific limitations on the certain types of force, relying instead on "[g]ood tactical judgment" to minimize civilian casualties.[7] In contrast, the 2009 directive dictated that "use of air-ground munitions and indirect fires against residential compounds is only authorized under very limited and prescribed conditions..."[8]

The 2009 tactical directive "de-emphasized airstrikes, artillery and mortars. This transferred some of the risk in skirmishes from Afghan civilians to Western combatants. In the past, American patrols in contact often quickly called for and received fire support. Not anymore. Many firefights are strictly rifle and machine gun fights."[9] In accordance with this perspective, in not providing the fire support that otherwise may be available, the tactical directive increased engagement times with the enemy, which in turn heightened the risk to troops on the ground. The resulting concern of some U.S. service members wasn't so much that the tactical directive transferred risks away from the civilians to the U.S. military, but that it transferred risk away from the enemy.[10] Yet, a 2010 review of the tactical directive "found no evidence that the rules restricted the use of lifesaving firepower" or even "a single situation where a soldier has lost his life because he was not allowed to protect himself."[11]

At first blush, the current iteration of the tactical directive, which General David Petraeus issued in 2010, differs only slightly from the 2009 version. And those differences seem little more than an alternatively worded means to the same conceptual end—the importance of protecting the Afghan population. But the 2010 tactical directive seemingly alters the risk relationship and balance between Afghan civilians and the U.S. military. The 2009 directive expressly acknowledged that "the carefully controlled and disciplined use of force entails risk to our troops," recognizing that protecting the force must, on some level and at some times, be subordinate to protecting the civilian population. In partial yet profound contrast, the 2010 tactical directive lists protecting Afghan civilians and the men and women in uniform as coequal moral imperatives.

Utilizing both the tactical directive and doctrinal concepts from the counterinsurgency manual, this chapter will explore the allocation of risk between the military force and Afghan civilian population. The chapter first reviews civilian and military casualty figures and then uses those numbers as a touchstone against which to consider each group's perception of the risk they face.

To set the conditions for that comparison, the chapter discusses the allocation of risk outlined in recent counterinsurgency doctrine and how that allocation translates from the conceptual or strategic level to the operational reality of soldiers and marines in harm's way at the tactical level. This chapter examines whether that translation is conceptually consistent and tactically viable.

While the concept of the U.S. military accepting increased risk in order to protect the civilian population is codified as doctrine, how well is the military translating, and training, that doctrine? As one commentator stated, "[n]o one wants to advocate loosening rules that might see more civilians killed. But no one wants to explain whether the restrictions are increasing the number of coffins arriving at Dover Air Force Base, and seeding disillusionment among those sent to fight."[12] This chapter seeks not to provide that explanation but to prompt a discussion on whether there is consistency in risk tolerance between U.S. military counterinsurgency doctrine and the execution of that doctrine at the tactical level in Afghanistan.

I. CIVILIAN AND MILITARY CASUALTIES AND THEIR CAUSES[13]

A. Civilian Casualties

According to the U.N. Assistance Mission in Afghanistan (UNAMA), 2,777 civilians were killed in 2010 as a result of the ongoing conflict in Afghanistan, a 15 percent increase from 2009.[14] This follows a four-year trend in which each year more civilians were killed in Afghanistan than the year prior.[15] In terms of wounded civilians, UNAMA documented 4,343 conflict-related injuries in 2010, a 22 percent increase from 2009.[16]

While the total number of civilian casualties has been increasing, in both 2009 and 2010 the percentage of civilian casualties caused by pro-government forces (including the Afghan, U.S., and coalition militaries) *decreased*.[17] In 2010, UNAMA claimed that pro-government forces were responsible for 440 civilian deaths or 16 percent of the total, a 26 percent decrease from 2009.[18] Pro-government forces were also purportedly responsible for 400 civilian injuries or 9 percent of the total, a 13 percent decrease from

2009.[19] In terms of the manner by which Afghans are wounded or killed by coalition forces, UNAMA stated that "[a]erial attacks claimed the largest percentage of civilian deaths caused by Pro-Government Forces in 2010, causing 171 deaths (39 percent of the total number of civilian deaths attributed to Pro-Government Forces)."[20] However, that figure represents a 52 percent decrease in Afghan civilian fatalities stemming from coalition air strikes from 2009.[21]

Conversely, anti-government elements, such as the Taliban, were responsible for 2,080 deaths in 2010 or 75 percent of the total civilian deaths, a 28 percent increase from 2009.[22] This continues, and widens, the trend recognized by UNAMA in 2009, that "more civilians are being killed by AGEs than by PGF."[23] Anti-government elements injured some 3,366 civilians or 78 percent of the total, a 21 percent increase from 2009. Anti-government element suicide and improvised explosive device attacks caused the greatest overall number of killed and wounded Afghan civilians.[24]

B. U.S. Military Casualties

Not until 2008 did U.S. troop levels in Afghanistan exceed 30,000.[25] But by November 2009, there were 68,000 troops in Afghanistan and roughly 100,000 by mid-2010. Such variance either skews quantitative comparison, or at a minimum renders statistical analysis of U.S. casualty rates over time beyond the scope of this chapter. Accordingly, this section will refer to U.S. casualty data from only 2010 to provide a frame of reference and comparison and not as the basis for empirical analysis.

In 2010, 499 U.S. troops were killed in Afghanistan.[26] Improvised explosive devices were responsible for 268 of those fatalities, or roughly 54 percent.[27] In terms of injuries, 5,173 U.S. service members were wounded in Afghanistan in 2010.[28]

C. Casualty Comparison

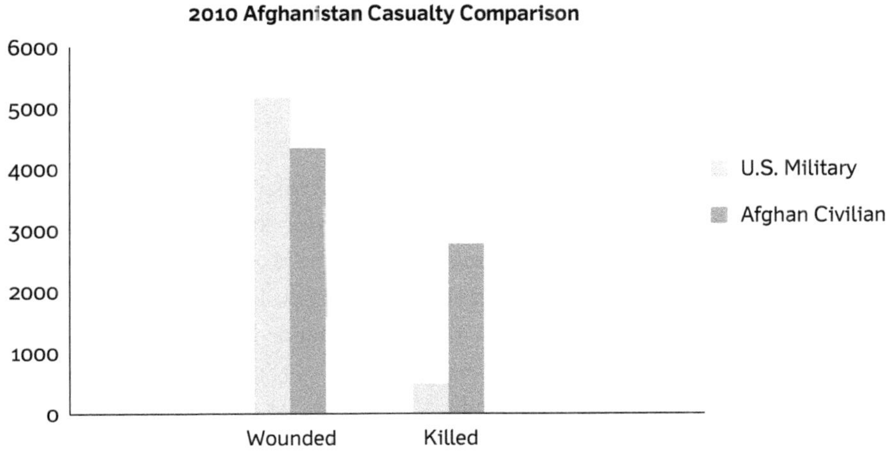

FIGURE 6.1

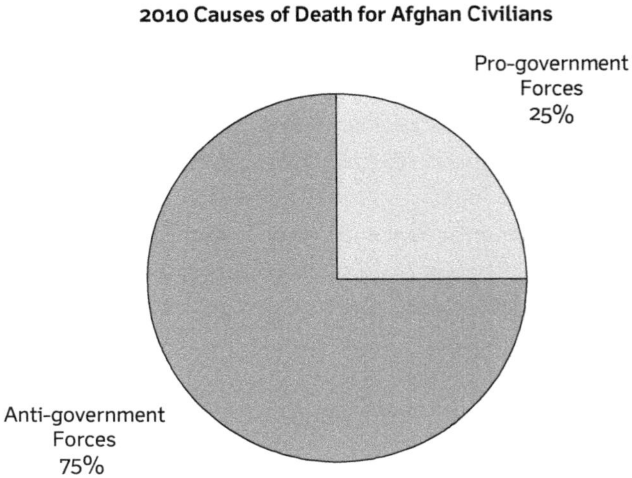

FIGURE 6.2

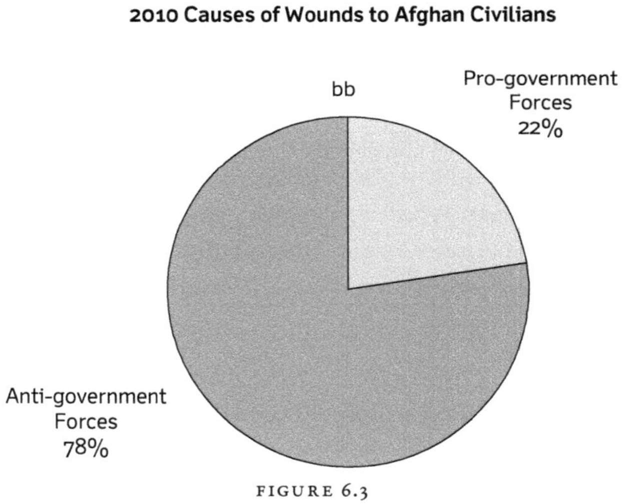

FIGURE 6.3

Using 2010 for comparison purposes, and speaking only in total numbers and ratios, more U.S. forces were wounded, 5,173, than the total number of Afghan civilians wounded, 4,343. Within that total number, the United Nations claims that U.S. and coalition forces were responsible for wounding 400 Afghan civilians. Thus, U.S. forces were wounded almost eleven times for every one instance when they wounded an Afghan civilian.

But the fatalities discussion is almost flipped. In 2010, 2,412 Afghan civilians were killed compared to 499 U.S. service members.[29] Within that total number, the United Nations claims that U.S. and coalition forces were responsible for the deaths of 440 Afghan civilians. Thus, U.S. forces were killed at roughly the same ratio by which they killed Afghan civilians.[30]

That comparison provides a reference point for a normative discussion on how the two sides, Afghan civilian and U.S. military, perceive the risk that the casualty rates depict, the extent to which those perceptions are consistent with the numerical indicia of risk, and ultimately how that impacts the overall counterinsurgency effort. From the perspective of the U.S. military, examining U.S. service member perceptions of risk requires first briefly reviewing the modern doctrine by which the United States purports to conduct counterinsurgency campaigns, filtered through the ISAF and actualized through the tactical directive.

D. U.S. Military Counterinsurgency Doctrine

In 2006, the U.S. Army and Marine Corps filled a doctrinal gap that spanned two decades or more by issuing a publication exclusively devoted to counterinsurgency operations, the type of armed conflict soldiers and marines were—and are—fighting in Iraq and Afghanistan.[31] The publication provides principles and guidelines for counterinsurgency (COIN) operations, including lessons learned thus far in Iraq and Afghanistan and those learned and to some extent forgotten, or at least neglected, from Vietnam.

The doctrine details a variety of guidance on the risk relationship between military forces and the civilian population, including the following:

- "The military forces' primary function in COIN is protecting the populace."[32]
- "The importance of protecting the populace, gaining people's support by assisting them, and using measured force when fighting insurgents should be reinforced and understood."[33]
- "In conventional conflicts, balancing competing responsibilities of mission accomplishment with protection of noncombatants is difficult enough. Complex COIN operations place the toughest of ethical demands on Soldiers, Marines, and their leaders."[34]
- "Limiting the misery caused by war requires combatants to consider certain rules, principles, and consequences that restrain the amount of force they may apply. At the same time, combatants are not required to take so much risk that they fail in their mission or forfeit their lives. As long as their use of force is proportional to the gain to be achieved and discriminates in distinguishing between combatants and noncombatants. Soldiers and Marines may take actions where they knowingly risk, but do not intend, harm to noncombatants."[35]

The doctrine also identifies paradoxes of counterinsurgency operations, including:

- "Sometimes, the more you protect your force, the less secure you may be."[36]
- "Sometimes, the more force is used, the less effective it is."[37]
- "The more successful the counterinsurgency is, the less force can be used and the more risk must be accepted."[38]
- "Sometimes doing nothing is the best reaction."[39]

The doctrinal guidance on the use of force in a counterinsurgency is markedly different than in conventional armed conflicts where the focus is to "concentrate the effects of combat power at the decisive place and time."[40]

The counterinsurgency doctrinal guidance is then filtered through the International Security Assistance Force mission, which is

> [i]n support of the Government of the Islamic Republic of Afghanistan, ISAF conducts operations in Afghanistan to reduce the capability and will of the insurgency, support the growth in capacity of the Afghan National Security Forces, and facilitate improvements in governance and socio-economic development in order to provide a secure environment for sustainable stability that is observable to the population.[41]

The question then becomes: how should the United States reduce the capability and will of the insurgency? And the answer is both through offensive or kinetic operations as well as through denying the insurgency the ability to operate by improving governance and socioeconomic development. Protecting the civilian population is inextricably linked to both. The tactical directive provides the means, or parameters on force as a means, to accomplishing the mission, shaped by the counterinsurgency doctrine.

II. TACTICAL DIRECTIVE

After revising the tactical directive in the summer of 2009, General Stanley McChrystal made portions public "to ensure a broader awareness of the intent and scope" of his guidance to the force.[42] While there had always been limitations on the use of force by U.S. military forces in Afghanistan and at least one prior version of the tactical directive released to the media, General McChrystal's modifications sparked discussion and controversy, much of which has continued and remained not just unresolved but unaddressed by the U.S. military.[43]

As discussed in the introduction of this chapter, contrary to his predecessor, General McChrystal allowed the use of air-to-ground munitions and indirect fires under "very limited and prescribed conditions." In his iteration of the directive, General McChrystal acknowledged the implicit trade-off inherent in limiting the use of force at the tactical level in support of the broader strategic goal of Afghan civilian support.

> We must fight the insurgents, and will use the tools at our disposal to both defeat the enemy and protect our forces. But we will not win based on the number of Taliban we kill, but instead on our ability to separate insurgents from the center of gravity—the people. That means we must respect and protect the population from coercion and violence—and operate in a manner which will win their support.

This is different from conventional combat, and how we operate will determine the outcome more than traditional measures, like capture of terrain or attrition of enemy forces. We must avoid the trap of winning tactical victories—but suffering strategic defeats—by causing civilian casualties or excessive damage and thus alienating the people.

While this is also a legal and a moral issue, it is an overarching operational issue—clear-eyed recognition that loss of popular support will be decisive to either side in this struggle. The Taliban cannot militarily defeat us—but we can defeat ourselves.

I recognize that the carefully controlled and disciplined employment of force entails risk to our troops—and we must work to mitigate that risk wherever possible. But excessive use of force resulting in an alienated population will produce far greater risks. We must understand this reality at every level in our force.[44]

Some U.S. service members were critical of the 2009 tactical directive. As a junior U.S. Army officer in Afghanistan queried:

> [m]inimizing civilian casualties is a fine goal, but should it be the be-all and end-all of the policy? If we allow soldiers to die in Afghanistan at the hands of a leader who says, "We're going to protect civilians rather than soldiers' what's going to happen on the ground? The soldiers are not going to execute the mission to the best of their ability. They won't put their hearts into the mission. That's the kind of atmosphere we're building."[45]

One noncommissioned officer sent an e-mail to a member of the U.S. Congress complaining that the rules of engagement (ROE) which flowed from the tactical directive were "too prohibitive for coalition forces to achieve sustained tactical success."[46] Another noncommissioned officer commented to a reporter that he "wish[ed] we had generals who remembered what it was like when they were down in a platoon.... Either they have never been in real fighting, or they forgot what it's like."[47] One news story quoted U.S. service members in Afghanistan as complaining that the tactical directive "handcuffed" them.

The directive's limitations on the employment of indirect fire and close air support received much of the frontline criticism. One marine infantry officer said he had stopped requesting air support during ground engagements as the approval process was too time consuming and tethered him to a radio. Moreover, the officer claimed that air support didn't arrive, was late when it did arrive, or that pilots were hesitant to conduct the requested air strike of ground targets. Alternatively, some units describe "decisions by patrol leaders to have fellow soldiers move briefly out into the open to draw fire once aircraft arrive, so the pilots might be cleared to participate in the fight."[48] While those are perhaps anecdotal and isolated examples, they occurred within a time frame in which both U.S. military and Afghan civilian casualty rates increased. And while at the same time the percentage of Afghan casualties caused by the U.S. military significantly decreased, there does not appear to have been a corresponding increase in Afghan civilian perceptions of safety, security, and the legitimacy and utility of the U.S. military presence in their country. From the Afghan civilian perspective, that the casualty-producing entity is predominantly the Taliban is little comfort and does not constitute protection. From the U.S. service member perspective, if the Afghan civilians refuse to place blame for civilian casualties on the entity actually causing them, and the only real beneficiaries of U.S. restraint are the enemy, then why should those service members accept more risk?

The criticism of the tactical directive seemed to reach its zenith in the spring and summer of 2010. In the spring came word of an ISAF proposal to award a medal to U.S. service members for "courageous restraint for holding fire to save civilian lives."[49] According to an ISAF statement: "[w]e routinely and systematically recognize valor, courage and effectiveness during kinetic combat operations.... In a COIN campaign, however, it is critical to also recognize that sometimes the most effective bullet is the bullet not fired."[50] As one story noted, "[a] combat medal to recognize a conscious effort to avoid a combat action would be unique."[51] The courageous restraint effort, which would seemingly recognize the tactical application of some of the counterinsurgency doctrinal points listed above, was short lived.

By the summer, one U.S. Army colonel claimed the troops "hated" the tactical directive and that "right now we're losing the tactical-level fight in the chase for a strategic victory."[52] At this point and unrelated to the tactical directive, General McChrystal resigned as the ISAF commander.[53] But the tactical directive loomed large in the interim while General Petraeus was awaiting confirmation to succeed General McChrystal. At his confirmation hearing, General Petraeus submitted an opening statement acknowledging that some U.S. service members were concerned over the tactical directive:

> Our efforts in Afghanistan have appropriately focused on protecting the population. This is, needless to say, of considerable importance, for in counterinsurgency operations, the human terrain is the decisive terrain. The results in recent months have been notable. Indeed, over the last 12 weeks, the number of innocent civilians killed in the course of military operations has been substantially lower than it was during the same period last year. And I will continue the emphasis on reducing the loss of innocent civilian life to an absolute minimum in the course of military operations.
>
> Focusing on securing the people does not, however, mean that we don't go after the enemy; in fact, protecting the population inevitably requires killing, capturing, or turning the insurgents. Our forces have been doing that, and we will continue to do that. In fact, our troopers and our Afghan partners have been very much taking the fight to the enemy. Since the beginning of April alone, more than 130 middle and upper-level Taliban and other extremist element leaders have been killed or

captured, and thousands of their rank and file have been taken off the battlefield. Together with our Afghan counterparts, we will continue to pursue relentlessly the enemies of the new Afghanistan in the months and years ahead.

On a related note, I want to assure the mothers and fathers of those fighting in Afghanistan that I see it as a moral imperative to bring all assets to bear to protect our men and women in uniform and the Afghan security forces with whom ISAF troopers are fighting shoulder-to-shoulder. Those on the ground must have all the support they need when they are in a tough situation. This is so important that I have discussed it with President Karzai, Afghan Defense Minister Wardak, and Afghan Interior Minister Bismullah Kahn since my nomination to be COMISAF, and they are in full agreement with me on it. I mention this because I am keenly aware of concerns by some of our troopers on the ground about the application of our rules of engagement and the tactical directive. They should know that I will look very hard at this issue.[54]

In early August 2010, General Petraeus issued an updated tactical directive. In the first paragraph of the revised directive, General Petraeus cautioned that "[s]ubordinate commanders are not authorized to further restrict this guidance without my approval." This requirement sought to address concerns that the issue with the tactical directive under General McChrystal was not so much the limitations he imposed, but that those limitations were a floor, not a ceiling, and that several layers of command between ISAF and a U.S. Army or Marine Corps unit in contact were adding additional restrictions or requirements.

But contrary to General McChrystal's express acknowledgment of the limitations on the use of force equating to increased risk for U.S. service members, General Petraeus's version placed protecting Afghan civilians and the force on the same level.

We must balance our pursuit of the enemy with our efforts to minimize loss of innocent civilian life, and with our obligation to protect our troops. Our forces have been striving to do that, and we will continue to do so.

In so doing, however, we must remember that it is a moral imperative both to protect Afghan civilians and to bring all assets to bear to protect our men and women in uniform and the Afghan security forces with whom we are fighting shoulder-to-shoulder when they are in a tough spot.[55]

This language seems, if not inconsistent with counterinsurgency doctrine, an avoidance or obfuscation of the subordinate relationship between protecting the force and the civilian population which the doctrine emphasizes and the harsh consequences that flow from that subordination during use-of-force situations.

Yet, the current tactical directive is not receiving as much open criticism as its predecessor, at least from U.S. service members. Its impact on the "hearts and minds" of the population is unclear though. Despite the report that in the third quarter of calendar year 2010 anti-government elements caused 90 percent of Afghan civilian deaths and injuries, in the minds of many Afghans the true cause of the casualties was ISAF's presence in Afghanistan.[56]

Even with the changed wording the tactical directive emphasizing protecting civilians, those civilians feel, or think, otherwise. This is apparently due in part to continuing challenges ISAF faces in strategic communications—a 2010 poll revealed that 40 percent of those interviewed believed ISAF was in Afghanistan "to destroy Islam or to occupy or destroy the country." Additionally, "only 8 percent of interviewees in the south knew the story of the 9/11 attacks and as a result had no understanding of the justification for the conflict with the Taliban and al Qaeda."[57] Further supporting an argument that Afghan perceptions of their safety and future in Afghanistan are not positive, beginning in 2008 and continuing through 2010, more Afghans were seeking asylum in foreign countries than at any point since the 2001 U.S. invasion.

As one commentator aptly noted, "[a]n American counterinsurgency campaign seeks support from at least two publics—the Afghan and the American. Efforts to satisfy one can undermine support in the other."[58] Overt, or at least publicized, U.S. military criticism of the current directive seems to have abated. But is that indicative of greater acceptance of the implicit risk trade-offs the directive represents? A lesser dislike of the current directive's requirements than those in prior versions? Or is the absence of overt comment simply masking continued divergence between doctrinal counterinsurgency guidance and tactical realities on the ground? General McChrystal subordinated the military to the civilian population in terms of risk allocation. General Petraeus seemed to consider them coequal. What then of the U.S. Army platoon leader who, prior to leading his soldiers on patrol in Afghanistan, told them, "[w]e are going to go up there and take care of each other. That is going to be our number one priority."[59] Thus, at the same point in time the operational commander considered protecting the military and civilian population equally important while the tactical leader on the ground considered force protection more important. The first needed step is to recognize this cognitive dissonance. From there, the task becomes reassessing how the U.S. military operationalizes counterinsurgency doctrine.

III. TRANSLATING AND TRAINING DOCTRINE

The counterinsurgency doctrine lays out competing interests, protection of the force and of the civilian population. On the one hand, the doctrine claims that "ultimate success in [Counterinsurgency Operations] is gained by protecting the populace, not the [military] force"[60] while "[a]t the same time, combatants are not required to take so much risk that they fail in their mission or forfeit their lives."[61]

The U.S. military acknowledges that "[e]thically speaking, COIN environments can be much more complex than conventional ones" and that "[t]he fortitude to see Soldiers and Marines closing with the enemy and sustaining casualties day in and day out requires resolve and mental toughness in commanders and units. Leaders must develop these characteristics in peacetime through study and hard training. They must maintain them in combat."[62]

There have now been three iterations of how doctrinal concepts of risk allocation in counterinsurgency are implemented through the tactical directive and a wide range of service member responses. There are a few lingering questions regarding the evolution of this doctrine, primarily: Are the implications of this tactical directive conceptually consistent with doctrine, and has the military discussed the ethics of risk allocation necessary to conduct counterinsurgency operations?

The service member complaints about the tactical directive and the short-lived nature of the courageous restraint medal suggest a disconnect between doctrine and practice. What kind of discussions is the military fostering through its professional military education and other training? The U.S. Army's Center for the Army Profession and Ethic (CAPE) provides one example.[63] One of the CAPE training vignettes involves a new platoon leader who overhears one of his noncommissioned officers telling the soldiers in the platoon that protecting each other, that ensuring every member of the platoon returns home uninjured, is their mission. While the vignette sets the conditions for the very discussions this chapter suggests, CAPE training is relatively new, is not widespread, and, more important, not required. One commentator, a U.S. Army officer who served as a platoon leader in Afghanistan, recently wrote in *Army Magazine* that the lack of operationally contextualized ethics training amounted to a structural weakness.[64] The officer claimed that "we need to begin shaping a different future with a community conversation about ethics, anger and how best to prepare ourselves for war."[65]

That U.S. service members are in harm's way in Afghanistan and are also trying to protect the civilian population is implicitly understood—within the military. Outside that group, there is a perception that the military has shifted the risks of war to contractors and that Americans on the whole are indifferent to foreign national civilian casualties.[66] But is there an explicit training component in place that will ensure all service members in the U.S. military have a shared understanding when doctrine espouses the dichotomy that "[u]ltimate success in [Counterinsurgency Operations] is gained by protecting the populace, not the [military] force,... [yet] combatants are not required to take so much risk that they fail in their mission or forfeit their lives?"[67]

The U.S. military inculcates its service members with the concepts of selfless service and duty, among other values.[68] The military ethos at its core focuses on service members looking for the safety and welfare of their fellow service member—the proverbial, but in combat literal, soldier or marine on your left and right. While that focus is understandable and even commendable, without frank and candid discussions on where and how risk is apportioned and accepted in counterinsurgency operations, is the military unintentionally sowing the seeds of dissonance?

APPENDIX 1

2008 Tactical Directive—General McKiernan

1. We are here to win and that victory is an Afghan victory; a victory which creates a secure population which enjoys freedom of movement, effective governance, viable institutions, and economic progress. We must always keep in mind that what we do and how we do it must support the Afghan people and the Government of the Islamic Republic of Afghanistan (GIRoA). We must continue to take the fight to the enemy in partnership with the ANSF to defeat the insurgency and provide security to the population. The way we act, the techniques we use, and the means we employ must serve to protect and defend the Afghan public and reinforce their confidence in GIRoA and the forces fighting on their behalf. We will take a comprehensive approach wherein ISAF operates in a complementary way with GIRoA and the ANSF. With that as background, I direct as follows:

2. The support of the Afghan people for the GIRoA and their collective support for ISAF are critical to defeating the insurgency we are fighting. We have that public support at the national level. We cannot take it for granted and must strive to deepen and broaden it.

3. We must partner and conduct combined operations with Afghan National Security Forces (ANSF) in support of Afghan objectives to the maximum extent possible. ISAF independent operations must be the exception.

4. Our actions both on and off the battlefield are important to our success. We must maintain our professionalism at all times, and always keep in mind the consequences of our actions. Respect for the Afghan people, their culture, their religion, and their customs is essential:

a. Unless there is a clear and identified danger emanating from a building and to do otherwise would threaten our ANSF partners and ourselves, all searches and entries of Afghan homes, mosques, religious sites or places of cultural significance will be led by ANSF. All responses must be proportionate and the utmost of care should be taken to minimize any damage.

b. All personnel will demonstrate respect for Afghans, Afghan culture, Afghan customs, and Islam in their actions and words. On the road and in vehicles, ISAF personnel will demonstrate respect and consideration for Afghan traffic and pedestrians.

c. In order to minimize death or injury of innocent civilians in escalation of force engagements, Commanders are to set conditions through the employment of techniques and procedures and, most importantly, the **training** of forces to minimize the need to resort to deadly force. Signals, signs, general and specific warnings (visual and audible) must be unambiguous and repeated to ensure the safety of innocent civilians.

5. We are engaged in a counterinsurgency in an extremely demanding environment. We are fighting an enemy that often cannot be identified before he has struck and then once he has, he hides among the civilian population. The battle is often waged among civilians and their property. We must clearly apply and demonstrate proportionality, requisite restraint, and the utmost discrimination in our application of firepower. No one seeks or intends to constrain the inherent right of self defense of every member of the ISAF force. However, Commanders must focus upon the principles which attach to every use of force—be that self defense or offensive fires. Good tactical judgment, necessity, and proportionality are to drive every action and engagement; minimizing civilian casualties is of paramount importance.

6. Whenever we believe we may have caused civilian casualties or civilian property damage we will immediately investigate the incident. If it is determined ISAF caused those casualties or that damage, ISAF will immediately acknowledge that fact. Acknowledgement by media, key leader engagement, by shura or other means, must happen at each level of command as appropriate. There must be a battle drill in place at each tactical level of the organization, and all investigations will be in cooperation with our Afghan partners.

7. We presently have the momentum on the battlefield and should endeavor to maintain it. In equal measure we must maintain the support of the Afghan people. We must remember that ultimately the solution in this war will be political, not military action. As such, we must always be cognizant of the consequences of our actions and public perceptions. I have every confidence in the dedication and competence of the members of our force to operate effectively within this challenging environment. Do not hesitate to pursue the enemy, but stay true to the values of integrity and respect for human life. Living these values distinguishes us from our enemies. There is no tougher endeavor than the one in which we are engaged. I direct this guidance to be briefed and explained to every Soldier, Sailor, Airman, Marine, and Civilian (including contractors) of the force as soon as practical.[69]

APPENDIX 2

2009 Tactical Directive—General McChrystal

Our strategic goal is to defeat the insurgency threatening the stability of Afghanistan. Like any insurgency, there is a struggle for the support and will of the population. Gaining and maintaining that support must be our overriding operational imperative—and the ultimate objective of every action we take.

We must fight the insurgents, and will use the tools at our disposal to both defeat the enemy and protect our forces. But we will not win based on the number of Taliban we kill, but instead on our ability to separate insurgents from the center of gravity—the people. That means we must respect and protect the population from coercion and violence—and operate in a manner which will win their support.

This is different from conventional combat, and how we operate will determine the outcome more than traditional measures, like capture of terrain or attrition of enemy forces. We must avoid the trap of winning tactical victories—but suffering strategic defeats—by causing civilian casualties or excessive damage and thus alienating the people.

While this is also a legal and a moral issue, it is an overarching operational issue—clear-eyed recognition that loss of popular support will be decisive to either side in this struggle. The Taliban cannot militarily defeat us—but we can defeat ourselves.

I recognize that the carefully controlled and disciplined employment of force entails risk to our troops—and we must work to mitigate that risk wherever possible. But excessive use of force resulting in an alienated population will produce far greater risks. We must understand this reality at every level in our force.

I expect leaders at all levels to scrutinize and limit the use of force like close air support (CAS) against residential compounds and other locations likely to produce civilian casualties in accordance with this guidance. Commanders must weigh the gain of using

CAS against the cost of civilian casualties, which in the long run make mission success more difficult and turn the Afghan people against us.

I cannot prescribe the appropriate use of force for every condition that a more complex battlefield will produce, so I expect our force to internalize and operate in accordance with my intent. Following this intent requires a cultural shift within our forces—and complete understanding at every level—down to the most junior soldiers. I expect leaders to ensure this is clearly communicated and continually reinforced.

The use of air-to-ground munitions and indirect fires against residential compounds is only authorized under very limited and prescribed conditions (specific conditions deleted due to operational security).

(NOTE) This directive does not prevent commanders from protecting the lives of their men and women as a matter of self-defense where it is determined no other options (specific options deleted due to operational security) are available to effectively counter the threat.

We will not isolate the population from us through our daily conduct or execution of combat operations. Therefore:

Any entry into an Afghan house should always be accompanied by Afghan National Security Forces (ANSF), with the support of local authorities, and account for the unique cultural sensitivities toward local women.

No ISAF forces will enter or fire upon, or fire into a mosque or any religious or historical site except in self-defense. All searches and entries for any other reason will be conducted by ANSF.

The challenges in Afghanistan are complex and interrelated, and counterinsurgencies are difficult to win. Nevertheless, we will win this war. I have every confidence in the dedication and competence of the members of our force to operate effectively within this challenging environment. Working together with our Afghan partners, we can overcome the enemy's influence and give the Afghan people what they deserve: a country at peace for the first time in three decades, foundations of good governance, and economic development.[70]

APPENDIX 3

2010 Tactical Directive—General Petraeus

This directive applies to all ISAF and US Forces-Afghanistan (USFOR-A) forces operating under operational or tactical control... Subordinate commanders are not authorized to further restrict this guidance without my approval.

Our counterinsurgency strategy is achieving progress in the face of tough enemies and a number of other challenges. Concentrating our efforts on protecting the population is having a significant effect. We have increased security in some key areas, and we have reduced the number of civilian casualties caused by coalition forces.

The Afghan population is, in a number of areas, increasingly supportive of the Government of the Islamic Republic of Afghanistan and of coalition forces. We have also seen support for the insurgency decrease in various areas as the number of insurgent-caused civilian casualties has risen dramatically. We must build on this momentum.

This effort is a contest of wills. Our enemies will do all that they can to shake our confidence and the confidence of the Afghan people. In turn, we must continue to demonstrate our resolve to the enemy. We will do so through our relentless pursuit of the Taliban and others who mean Afghanistan harm, through our compassion for the Afghan people, and through the example we provide to our Afghan partners.

We must continue—indeed, redouble—our efforts to reduce the loss of innocent civilian life to an absolute minimum. *Every Afghan civilian death diminishes our cause.* If we use excessive force or operate contrary to our counterinsurgency principles, tactical victories may prove to be strategic setbacks.

We must never forget that the center of gravity in this struggle is the Afghan people; it is they who will ultimately determine the future of Afghanistan... Prior to the use of fires, the commander approving the strike must determine that no civilians are present. If unable to assess the risk of civilian presence, fires are prohibited, except under of the following two conditions (specific conditions deleted due to operational security; however, they have to do with the risk to ISAF and Afghan forces).

(NOTE) This directive, as with the previous version, does not prevent commanders from protecting the lives of their men and women as a matter of self-defense where it is determined no other options are available to effectively counter the threat.

...Protecting the Afghan people does require killing, capturing, or turning the insurgents. Indeed, as I noted earlier, we must pursue the Taliban tenaciously. But we must fight with great discipline and tactical patience.

We must balance our pursuit of the enemy with our efforts to minimize loss of innocent civilian life, and with our obligation to protect our troops. Our forces have been striving to do that, and we will continue to do so.

In so doing, however, we must remember that it is a moral imperative both to protect Afghan civilians and to bring all assets to bear to protect our men and women in uniform and the Afghan security forces with whom we are fighting shoulder-to-shoulder when they are in a tough spot.

We must be consistent throughout the force in our application of this directive and our rules of engagement. All commanders must reinforce the right and obligation of self-defense of coalition forces, of our Afghan partners, and of others as authorized by the rules of engagement.

We must train our forces to know and understand the rules of engagement and the intent of the tactical directive. We must give our troopers the confidence to take all necessary actions when it matters most, while understanding the strategic consequences of civilian casualties. Indeed, I expect our troopers to exert their best judgment according to the situation on the ground. Beyond that, every Soldier, Sailor, Airman, and Marine has my full support as we take the fight to the enemy.

...Partnering is how we operate. Some civilian casualties result from a misunderstanding or ignorance of local customs and behaviors. No individuals are more attuned to the Afghan culture than our Afghan partners. Accordingly, it is essential that all operations be partnered with an ANSF unit and that our Afghan partners be part of the planning and execution phases. Their presence will ensure greater situational awareness. It will also serve to alleviate anxiety on the part of the local population and build confidence in Afghan security forces.

I expect every operation and patrol to be partnered. If there are operational reasons why partnership is not possible for a particular operation, the CONOP approval authority must be informed...

Partnership is an essential aspect of our counterinsurgency strategy. It is also an indispensible element of the transition of security responsibility to ANSF.

Again, we need to build on the momentum we are achieving. I expect every trooper and commander to use force judiciously, especially in situations where civilians may be present. At the same time, we must employ all assets to ensure our troopers' safety, keeping in mind the importance of protecting the Afghan people as we do.

This is a critical challenge at a critical time; but we must and will succeed. I expect that everyone under my command, operational and tactical, will not only adhere to the letter of this directive, but—more importantly—to its intent.

Strategic and operational commanders cannot anticipate every engagement. We have no desire to undermine the judgment of tactical commanders. However, that judgment should always be guided by my intent. Take the fight to the enemy. And protect the Afghan people and help our Afghan partners defeat the insurgency.[71]

"Civilian casualties are an enduring challenge of war that we must strive to reduce to an absolute minimum."
GENERAL DAVID H. PETRAEUS,
August 31, 2010

7

NEW APPROACHES TO REDUCING AND MITIGATING HARM TO CIVILIANS

Gregory S. McNeal*

Introduction

The wars in Iraq and Afghanistan highlighted the strategic importance of the U.S. commitment to civilian protection. Both wars were eventually fought as a counterinsurgency (COIN) and both revealed how protecting civilians is a central feature of COIN. This chapter will examine the importance of civilian casualty mitigation in U.S. counterinsurgency operations, and it will describe theoretical and practical lessons learned regarding civilian casualties and situate them in a broader strategic context. The chapter concludes by describing the U.S. military's newest doctrinal publication aimed specifically at preventing and mitigating harm to civilians.

Minimizing civilian casualties has become an increasingly important component of mission success for U.S. forces, regardless of the type of operation in which U.S. forces have been engaged (COIN, counterterrorism, major combat operations, etc.). Moreover, civilian casualties will likely remain a concern in almost all conceivable future military scenarios. Whether it is a conflict on the Korean Peninsula with the attendant postconflict political ramifications that flow from the infliction or prevention of civilian casualties, or counterterrorism or even counternarcotics operations, each type of operation would be largely shaped by civilian casualty concerns. Even in lawless, largely ungoverned territories like Somalia, the ability to conduct sustained operations in theater requires accounting for the negative consequences of civilian harm.

Three geopolitical developments have brought about a greater emphasis on the strategic consequences of civilian casualties. First, war is increasingly transparent and evaluated by external actors with technology and global telecommunications systems transforming both the conduct and the assessment of war.[1] Insurgents, journalists, nongovernmental organizations (NGOs), and private citizens utilize global telecommunications systems to transmit near real-time battlefield information to international bodies, which assess the humanitarian impact and conduct of war.[2] Second, U.S. forces are held to a higher standard because of their superior capabilities and precision weaponry.[3] Moreover, U.S. information operations claiming that the U.S. use of force is "the most discriminate in history [heightens] expectations of minimal civilian harm in war."[4] Third, insurgents and enemy states recognize the strategic advantages that flow from falsely accusing or purposefully causing civilian casualties.[5] These adversaries know that charges of U.S.-caused civilian casualties can shape international opinion, serve as a recruiting tool, and diminish U.S. operational flexibility.[6] Accordingly, these adversaries routinely violate legal and ethical rules "by collocating military and civilian objects, using human shields, and refusing to identify their combatants."[7] These three geopolitical developments have made avoiding civilian casualties not only a legal obligation, but also a mission-essential strategic consideration for U.S. forces.

Some may contend that the law of armed conflict—the body of international law that regulates the conduct of armed hostilities—is adequate to address the challenge of reducing and mitigating harm to civilians. The lesson of counterinsurgency though, is that complying with legal restraints merely establishes the baseline for acceptable conduct.[8] In fact, mission success may require meeting expectations that civilians will be safe from a variety of harms, not just the collateral effects of U.S. use of armed force. Harm to civilians, even if lawful, can undermine strategic policy objectives and can assist the enemy by inciting increased opposition to U.S. forces.

The implications of these insights are that winning hearts and minds through civilian protection may at the same time put one's own forces at greater risk.[9] These judgments are fraught with political considerations. Such political concerns about preventing civilian casualties are one of the most significant drivers of the conduct of military operations. While the military may believe that collateral damage sensitivity sometimes inappropriately relegates mission accomplishment to a secondary objective, the human rights community, in contrast, may fear that force protection is the paramount concern in Western military operations. Lost in the debate between the military and human rights groups is the important viewpoint of the civilian population, who in the context of a counterinsurgency campaign may be choosing whether to support counterinsurgents or armed groups.

While the view of the civilian population is critical in a counterinsurgency campaign, so too is the view of allies and regional partners. Thus, while it is possible that the U.S. public may be willing to support a strategy that elevates force protection over protection of civilians, the international community is far less forgiving. For example, during the 1990s the U.S. public did not significantly object to collateral damage resulting from numerous air attacks against Iraqi military targets after the 1993 Gulf War (admittedly not a counterinsurgency). At the same time though, Arab states did object, and the perception that the Iraqi people were suffering as a result of U.S. military actions damaged international support for U.S. operations. "Even Britain and France, which generally supported U.S. efforts and took part in the attacks on air defenses in the no-fly zones, seemed to distance themselves from the cruise missile strike. Similar protests, especially among the Arab states…followed every instance of Iraqi collateral damage."[10] More recently, a 2003 Pew Global Attitudes study noted that "more than 90 percent of Jordanian, Moroccan, Palestinian, and Turkish respondents and more than 80 percent of Indonesian and Pakistani respondents felt that the United States 'didn't try very hard' to avoid Iraqi civilian casualties."[11] While the 1993 Gulf War, the air strikes following it, and the Iraq War as of 2003 were not counterinsurgencies; the strategic imperative of preventing harm to civilians was nevertheless an important consideration. This is especially the case when one considers that even outside the Muslim world more than "70 percent of Brazilians, French, Russians and South Koreans" believed that the United States did not try very hard to avoid civilian casualties in Iraq in 2003.[12] Such viewpoints, be they in counterinsurgency or other operations, can undermine coalition support and can paint a narrative that works its way into the in-theater narrative, undermining local support for the work of counterinsurgents.

I. COIN DOCTRINE

The U.S. Army's *Counterinsurgency Field Manual* (FM 3-24)[13] serves as the bedrock of COIN doctrine for modern-day conflicts. It is a helpful starting point for noting the imperative of preventing and mitigating harm to civilians. The *Counterinsurgency Field Manual* explains that when forces "fail to provide security or threaten the security of civilians, the population is likely to seek security guarantees from insurgents, militias, or other armed groups. This situation can feed support for an insurgency."[14] Note the manual's emphasis on the harm to civilians and its impact, irrespective of the legality of the harm, which may have otherwise satisfied international law principles of military necessity, distinction, proportionality, and humanity. In fact, COIN doctrine specifically instructs that "…in COIN environments, the number of civilian lives lost and property destroyed needs to be measured against how much harm the targeted insurgent could do if allowed to escape. If the target in question is relatively inconsequential, then proportionality requires combatants to forego severe action, or seek non-combative means of engagement."[15] This form of proportionality is keyed to the operational circumstances a unit finds itself engaged in, and is thus a prudential concern, not a legal consideration. Thus while the law of armed conflict may allow for the killing of a relatively inconsequential combatant directly participating in hostilities, prudence may dictate a noncombative means of engagement if that can better serve the operational imperatives of the counterinsurgency mission.[16]

Such prudential considerations require an understanding of first-order (tactical), second-order (operational), and even third-order (strategic) fallout from the use of armed force. As the *Counterinsurgency Field Manual* notes:

> The COIN environment requires counterinsurgents to not only determine the kinds of weapons to use and how to employ them but also establish whether lethal means are desired—or even permitted...Soldiers and Marines require an innate understanding of the effects of their actions and weapons on all aspects of the operational environment. Leaders must consider not only the first-order, desired effects of a munition or action but also possible second- and third-order effects—including undesired ones...Achieving the desired effects requires employing tactics and weapons appropriate to the situation.[17]

In a similar vein, while the legal principles of discrimination and proportionality set the minimum acceptable limits for combatants, practical considerations will force counterinsurgents to weigh the ethical and moral impact of their use of force. "Frequent and swift reliance upon force, or routine application of maximum allowable (versus minimum required) firepower can cause unnecessary civilian harm and thereby antagonize the local population. Such actions can crucially affect the attitudes and motivations of sympathetic or neutral civilians."[18] The counterinsurgent who only engages in self-assessment, merely viewing his actions from the viewpoint of what is lawful, may deem these considerations unimportant. But, the counterinsurgent's self-assessment is not the critical determinant of success in a counterinsurgency, rather the important viewpoint is that of the populace. From this viewpoint, actions on the part of counterinsurgents "that cause unnecessary harm or death to noncombatants may create more resistance and increase the insurgency's appeal—especially if the populace perceives a lack of discrimination in their use."[19] These perceptions may create more enemies than they eliminate.

II. CIVILIAN HARM AND ASYMMETRIC CONFLICT

While preventing and mitigating harm to civilians is a central consideration for counterinsurgents, asymmetric opponents seek to exploit civilian harm. Their goal, when faced with the reality that they cannot defeat occupying forces, is to generate enough opposition to cause them to leave. "One way of doing so is by increasing segments of the population that come to hate the occupiers enough to move toward violent opposition to them."[20] Insurgents "hide among the civilian population, emerge to ambush the occupiers in the city streets, and blend back into the civilian population from whom they cannot be distinguished. This pattern makes it difficult for the occupying forces to avoid civilian casualties."[21] The insurgent is counting on harm to civilians as a means to radicalize those who were previously not sympathetic to the insurgency.

There are three mechanisms by which civilian harm can be linked to support for an insurgency or at least increased violence against counterinsurgents. First, civilians may be driven to participate in violence out of a "desire to avenge specific harm to family, friends or neighbors."[22] Under this theory, civilian harm will lead to "increasing long-run trends in violence as individuals take up the fight in response to specific slights...."[23] Second, civilian harm may increase anger, making individuals more susceptible to recruitment by insurgent groups.[24] This is distinguishable from the revenge motivation because one need not be directly exposed to or personally affected by civilian harm.[25] Rather, the anger may be driven by a feeling of antipathy toward counterinsurgents. Third, civilian harm may hamper the ability to gather intelligence about the operations of insurgents. Civilian harm incidents may anger others in the civilian population or may compel them to believe that they are more vulnerable to harm, and they may consequently be less willing to cooperate with the force they see as responsible for inflicting harm.[26] "To the extent that counterinsurgent forces engage in unpopular and aggressive operations that generate specific local grievances, they are likely to facilitate increased recruitment and support for insurgent groups."[27] The task for the counterinsurgent is thus to minimize harm to civilians and thereby reap the benefit of decreased recruiting potential for insurgent forces. Thus, contrary to the perceived conflict between force protection and minimizing civilian harm, a long-term view of a counterinsurgency campaign demonstrates that reducing civilian casualties is not in conflict with the objective of force protection.

In light of the reality that asymmetric opponents will intermingle with civilians, blurring their identities and seeking to engage in combat in areas heavily populated by civilians, the task of minimizing civilian harm is easier said than done. "Rules of engagement designed to protect civilians tend to place soldiers at greater risk; conversely, rules that stress force protection usually come at the expense of civilians. A sliding scale of violence means that there is no natural level of civilian casualties. Nor are particular tactics always necessary or inevitable. The civilian toll varies depending on how risk is allocated and how rules of engagement are defined."[28] While there is no doubt that the restrained use of force can increase physical risk in the short run, strategic success in counterinsurgency demands increased physical risk for counterinsurgent forces. Unfortunately, these risks may seem to come without a coordinate reward because success in COIN is elusive and difficult to measure, oftentimes lying in avoiding or mitigating harms rather than achieving measurable victories.

III. THE DEVELOPMENT OF CIVILIAN CASUALTY MITIGATION DOCTRINE

Given the likelihood of greater risks to U.S. forces in counterinsurgency, the challenge for the military is to expand beyond a strict law of armed conflict mind-set, to an approach that integrates civilian casualty mitigation principles across the spectrum of routine and predeployment training, doctrine, and education. To that end, the U.S. military created the first manual in the history of warfare solely focused on mitigating harm to civilians. The manual, technically termed *Army Tactics, Techniques, and Procedures 3-37.31 Civilian Casualty (CIVCAS) Mitigation* (*Army CIVCAS Manual*)[29] is the U.S. Army's doctrinal publication for mitigating harm to civilians. The manual is intended to provide the doctrinal guidance and direction necessary for army units to minimize harm to civilians, and when such harm occurs, to provide guidance about how to manage the consequences of that harm. While the manual is intended for use by the U.S. Army, it was drafted with input from other branches of the armed forces, members of foreign militaries, and civilian experts from NGOs and the legal academy.[30] The manual builds on lessons learned from the U.S. military's experience in all contemporary conflicts, especially those in Afghanistan and Iraq, as detailed in the Joint CIVCAS Study.[31] The manual is intended to remedy some of the serious deficiencies noted in the Joint CIVCAS Study, which examined the causes of civilian harm and techniques to prevent civilian casualties during combat. That inquiry was a research effort that combined external academic rigor with professional military expertise to create the first comprehensive assessment of civilian protection. Writing the Foreword to the Joint CIVCAS Study, then-General David H. Petraeus noted that avoiding civilian casualties "is a central operational challenge in Afghanistan and Iraq and it will be a challenge in any future conflict as well."[32] One of the key recommendations of the Joint CIVCAS Study was that the military must develop a handbook for civilian casualty response, and the CIVCAS manual aims to fill that doctrinal gap.

The heart of the *Army CIVCAS Manual* is a six-step mitigation cycle, intended to go beyond merely adhering to the law of armed conflict's principles for targeting and use of force. Effective CIVCAS mitigation requires more than ensuring soldiers who make decisions about the use of force have been trained on the law of armed conflict, and it is not restricted to the moment a decision is made to engage a target. The six steps in the CIVCAS mitigation cycle are: prepare, plan, employ, assess, respond, and learn.

Preparation includes training to ensure members of the military are prepared to minimize the likelihood of harm to civilians, and if it occurs, preparation is meant to ensure they are ready to act to mitigate the effect of such harm. The doctrinal aim is to ensure that soldiers are not solely focused on fighting against a hostile enemy but that they also recognize that they may encounter civilians who are not hostile. With this aim in mind, "Preparation both helps forces begin to understand the operational environment and provides a base set of planning approaches and tactics, techniques, and procedures (TTP) appropriate to that environment."[33] During the preparation stage, it is important to bear in mind, "…there is no single 'silver bullet' means of solving CIVCAS. The approach lies in specifically understanding patterns of life and the operating environment that forces are entering, ensuring subordinate organizations understand the implications of civilian casualties and the long term effect on operations, ensuring that they have the technical ability and resources for them to understand the operating environment, principally from an intelligence perspective, and then plan operations accordingly."[34]

The *Army CIVCAS Manual* highlights the importance of preparing for potential incidents of civilian harm by encouraging leaders to establish a permanent CIVCAS Tracking, Assessment, and Response Cell to assist in effective CIVCAS mitigation. This recommendation is clearly an outgrowth of the Joint CIVCAS Study, which noted that no U.S. military entity exists with sole responsibility for tracking civilian harm incidents in the United States' wars. While some may criticize the publication of the *Army CIVCAS Manual* as a concession to the critiques of human rights groups, some of whom argue that the United States does not exercise adequate care in its operations, the *Army CIVCAS Manual* instead recognizes that despite the best efforts of armed forces, harm to civilians will nevertheless occur, as it is an unavoidable feature of armed conflict. Thus, the *Army CIVCAS Manual's* prepare step emphasizes that "[s]ystems should be established in advance to respond to CIVCAS incidents to include reporting, tracking, investigation, public response, and making amends to families and communities through the recognition of harm, appropriate compensation, and apologies and dignifying gestures."[35]

Planning, the second step in the CIVCAS mitigation cycle, focuses on the importance of intelligence and decision-making processes. Armed forces should employ intelligence to not only identify potential threats but also to understand the civilian population in the area of operations. In particular, armed forces should take note of the fact that "enemies may be co-located with their family members and other civilians."[36] Relevant intelligence concerns for preventing harm to civilians include a focus on social variables, such as cultural considerations and other questions such as "the degree to which the population accepts the mission's legitimacy; the degree to which the government is accountable to its people; and the degree to which regional neighbors and the broader international community support the civilians."[37] Counterinsurgents should employ intelligence assets and analysis to understand how civilian casualties may retard the fostering of local host nation ownership and capacity development. "Host nation ownership is dependent on inclusive partnerships that include private sector actors and large segments of civil society that may be adversely affected by civilian casualties."[38] Armed forces may want to partner with others to better facilitate intelligence collection as host nation government officials, leaders in host nation security forces, local leaders, and NGO personnel can be valuable sources of information regarding the civilian population. However, as the controversy surrounding drone strikes in the Afghanistan theater has made clear, information from sources must be cross-checked, as some individuals may attempt to manipulate armed forces into taking actions that support ulterior motives.[39] The planning stage in the CIVCAS mitigation cycle, taken together with preparation, ensures that armed forces are aware of the potential for civilian harm and take early steps to minimize its likelihood.

In the employment step, counterinsurgents conduct operations with an eye toward preventing or mitigating harm to civilians. As such, they will likely need to exercise levels of restraint that exceed the baseline requirements of the law of armed conflict.[40] Combatants are governed in their use of force by the principles of discrimination and proportionality. "Discrimination refers to the ability to direct action solely against an intended legitimate target, and proportionality accounts for the suitability of an action when considering other options and the potential impact on other actors besides the intended target."[41] As discussed earlier, these law of armed conflict considerations may be inadequate for the purposes of a counterinsurgency, where armed forces may also need to consider how the potential for civilian casualties affects mission success.[42] This is not to say that engagements cannot be conducted if civilians will be harmed—military necessity may still justify neutralizing a target despite the risk of incidental civilian harm—rather, a focus on civilian casualty mitigation merely counsels that armed forces should look beyond the immediate lawfulness of an action to the broader second- and third-order effects.[43] Furthermore, counterinsurgents can employ information operations techniques to aid in civilian casualty avoidance, and if such incidents occur, to mitigate the fallout from such incidents or address associated rumors that may be circulating among the population. "Many conflicts are as much information wars as they are military contests, and civilian casualties in particular are an issue that can be exploited."[44] For example:

> [a] best case scenario is coupling a key leader engagement with a media engagement, where leaders can meet privately first and come to an agreement about what happened and what is to be done from that point forward. Then, marry this up with an aggressive media campaign to have them then go public and speak about what they've just discussed, which is helpful in quelling the follow-on, second and third order effects of CIVCAS.[45]

Regardless of the specific steps counterinsurgents take, in most circumstances their cooperation with other stakeholders will be outcome determinative. Thus, counterinsurgents who work with other actors such as host nation security forces, governmental agencies and local leaders are likely to employ their forces in a way that is sensitive to the needs and risks in the local area of operations. Counterinsurgents should also seek to work with "multinational partners, intergovernmental and international organizations, USG agencies, NGOs, the private sector, and the media."[46] Many of these actors may be unwilling to work with the military, but if they understand the counterinsurgent's goal as preventing civilian harm, their cooperation may be forthcoming and may prove to be essential.

Despite efforts aimed at training, preparing, and employing forces with an eye toward avoiding civilian harm, such incidents are often unavoidable and it is unreasonable to believe they can be completely eliminated. Thus, "[w]hen civilian casualties occur, the most important step in consequence management is to determine the facts of the incident, including the numbers and severity of civilian casualties,"[47] this task is known as the assessment step. Success in this step sets the stage for effective response (the next step in the cycle) and for the development of lessons learned to prevent future harm to civilians (the final step in the cycle). For instance, given the fact that many civilian casualties in Afghanistan occur in escalation of force (EOF) incidents, Coalition forces can "ask Afghans what [EOF] measures they believe to be most intuitive and effective... [and] hold forums to aid communication with local nationals regarding culturally effective measures for stopping traffic and individuals. At the same time, the Coalition should undertake to educate the Afghan people on EOF procedures and the steps they can take in EOF situations to minimize misunderstandings and avoid civilian casualties."[48] Tasks for counterinsurgents include gathering accurate information regarding civilian casualties, developing a uniform database to track information on civilian casualty incidents at all levels of command, and constantly analyzing data to develop mitigation measures to take in the future.

The goals of response efforts must address to some extent the expectations of the population. In Afghanistan, "[t]he population generally sought two things following Coalition-caused civilian casualty incidents: some form of redress or compensation for losses incurred, and a form of apology and explanation of what happened, which illustrates ISAF remorse and acceptance of responsibility for the incident."[49] How counterinsurgents respond to unavoidable civilian harm can, if not carefully conducted, compound the harm that flows from the initial civilian casualty incident.

> For example, monetary compensation may be appropriate in some situations; in others, it would be perceived as an insult. Depending on cultural norms, it may be appropriate to contact the victim's family directly, while in other situations, this should never be done. Sometimes, an intermediary such as a tribal leader should be used instead.[50]

These cultural considerations are not innate to counterinsurgents from different backgrounds with different training and experiences; thus, preparation beforehand will largely determine the counterinsurgent's success in the response step.

To effectively respond to civilian casualty incidents, counterinsurgents must be aware of those incidents. Importantly, the counterinsurgent's perspective regarding the credibility of claims of civilian harm is largely irrelevant. Rather, local assessments and beliefs regarding the credibility of the claims are what matters. As such, counterinsurgents must ensure that civilians can "be confident that they can report civilian casualty incidents without fear of retribution or harassment."[51] While allegations may be legitimate, mistaken, or purposely false, the counterinsurgent should nevertheless treat the civilians with respect and appear to take their claims seriously, as negative perceptions can harm the counterinsurgency mission.

> If an allegation is clearly well-founded, Army units should immediately proceed to the last stage (making amends). Conversely, if an allegation is clearly incorrect,

> leaders should explain this in as much detail as possible. When in any doubt, leaders should opt for an investigation of an incident. Often, the process itself can play a major part in addressing local anger and concerns. The worst thing that can be done is to ignore a grievance, and local civilians will not necessarily accept Soldiers' credibility without question. If an investigation is required, Soldiers should explain the relevant procedures, clearly establish the expected timeframes, and identify the assistance and cooperation required. The entire process of consequence management works best when it is done in a transparent fashion and in frequent contact with victims, survivors, or their community representatives.[52]

The enemy will use civilian harm, and the response of the counterinsurgent to that harm to their advantage.[53] They will likely "use local folklore, religious and historical allusions, and language to great effect"[54] and will use relations with local media to co-opt the counterinsurgent's message of goodwill and remorse. As such, counterinsurgents must anticipate these measures and attempt to disrupt them while implementing their own communication techniques. Partnering with local forces and tribal leaders can also be of great value in both mitigating CIVCAS and responding to them when they inevitably occur. As a baseline consideration, "[d]ata suggests that Coalition forces have a reduced rate of civilian casualties during operations where they are partnered with Afghan National Security Forces (ANSF)."[55] Some of the benefits of partnering noted by Coalition forces include "... increased awareness of cultural cues that help [ISAF] discriminate true threats and better communication with locals that facilitates intelligence gathering."[56] A number of ISAF units also claimed that partnering with ANSF and local tribal leaders allowed ISAF to better gauge the veracity of CIVCAS claims when conducting Battle Damage Assessments (BDAs).[57] Moreover, in terms of planning and response coordination:

> Local leaders and security forces had a better understanding of the local environment and culture, including Pashtunwali code whereby harming a member of an Afghans family could create enmity within that family for generations. Familiarity with acceptable ways to respond to civilian casualty incidents according to Pashtunwali was invaluable. Afghans could help bridge the culture gap in redress, avoiding a backlash from Coalition forces to make amends.[58]

Counterinsurgents must become learning organizations, which is why the final step in civilian casualty mitigation is "learning"; mandating that counterinsurgents be able to learn and adapt quickly and continuously.

> Effective force preparation and the ability to learn and adapt are complementary and synergistic. A force that is effectively prepared is better positioned to adapt in precise and measured ways, instead of inventing entirely new ways to operate. At the same time, a force that is well prepared for a specific environment but

is unable to learn and adapt will become less capable as that operational environment changes.[59]

Several factors have been identified that hinder the ability of Coalition forces to learn and adapt to civilian casualty issues. Chiefly, "the status of data and reporting on civilian casualties" combined with the fact that "no organization was tasked to provide operational analysis of lessons learned"[60] slowed the progress of learning and adaptation. Furthermore, "[e]ven when lessons were identified, they were often not shared across the force because of largely ineffective means of integrating and disseminating lessons, both in-theater and in US institutions."[61] Thus, a critical task for counterinsurgents is to collect, analyze, and disseminate information about civilian casualty mitigation throughout the fighting force. Importantly, "[l]essons on civilian casualties do not come solely from actual civilian casualty incidents, but also from incidents where there was a high risk of civilian casualties but they were avoided. Such incidents are referred to as 'near misses' and can offer valuable lessons and best practices."[62]

Following up on the Joint CIVCAS Study mentioned above, *Adaptive Learning for Afghanistan* (ALA) assessed the state and degree to which lessons learned in regard to civilian casualty mitigation had been institutionalized for U.S. forces both in-theater and in predeployment training. While ALA reported a number of improvements, it stressed the need to further disseminate and institutionalize lessons learned. Some of the improvements noted in-theater included:

> [1]... use of alternate tactics, including the employment of snipers or maneuver elements instead of airstrikes when there were civilian casualty concerns.... [2] The use of tactical patience to verify positive identification (PID) and collateral damage concerns... [3] Modifications in air-ground communication and procedures... [4] Modified [tactics, techniques, and procedures (TTP)] in counter-terrorism operations... [5] Increased partnering between Coalition and Afghan forces.[63]

Although these in-theater adjustments were promising indications that lessons were in fact being learned, there were still shortcomings in certain areas. The primary obstacle for in-theater units was in knowledge sharing. "For example, due to leader emphasis on the need to reduce and mitigate civilian casualties, units adjusted their TTP accordingly. While many of these practices and lessons were captured by a variety of methods, they were not always shared with other units in theater."[64]

In terms of predeployment training, the ALA study found that many positive changes had been made "to integrate CIVCAS reduction and mitigation into the 'road to deployment' to Afghanistan."[65] While there remain some barriers in the way of knowledge sharing, such as limited access to shared networks, the overall improvements have been promising. As the ALA study explains, "The desired end state is a force that understands the operational environment, has the basic tool set to meet operational requirements, and can use those tools as a baseline to learn and adapt as needed to achieve the mission."[66] Thus, such an end state will never be perfected, but instead, requires an ongoing effort to implement adaptations drawn from lessons learned.

As the preceding discussion makes clear, preventing and mitigating harm to civilians is best understood as a recursive process. Success in protecting the civilian population in counterinsurgency (or in any operation for that matter) requires steps aimed at training for the reality that harm to civilians will be unavoidable. It further requires planning, information sharing, and integration of lessons learned on an ongoing basis to ensure that incidents may be avoided in the future. Perhaps most important, avoiding civilian casualties in a counterinsurgency may require the exercise of courageous restraint by counterinsurgents who may have a legal right to undertake a military action, but who must recognize the strategic consequences of such actions.

Conclusion

Protecting the population and winning hearts and minds are well-known central planks in counterinsurgency theory. However, achieving these goals is oftentimes easier said than done, especially when the reality of modern operations is a transparent conflict environment in which enemy forces will seek to purposefully cause harm to civilians and exploit such harm for their own ends. Based on the United States' experiences in Iraq and Afghanistan, the U.S. military has created the world's first manual directed at preventing and mitigating harm to civilians in combat. As the discussion in this chapter highlights, the *Army CIVCAS Manual* is merely the first step in cataloging and sharing lessons learned. Effective civilian casualty mitigation in counterinsurgencies and other operations will require sustained efforts on the part of the military to act in a way that may exceed the baseline requirements of the law of armed conflict. Ample evidence—ranging from restrictive rules of engagement to a consistent focus on "lessons learned"—suggests the U.S. military is already exceeding the requirements of the law of armed conflict. Such actions may not placate critics of U.S. operations, but it may go a long way toward laying the foundation for success in counterinsurgency operations.

4

New Problems for Unconventional Battlefields

8

TERRORISM AND THE LAWS OF MULTIDIMENSIONAL WARFARE

Dr. Boaz Ganor[1]

Introduction

The laws of war, as codified in The Hague and Geneva Conventions and in related international treaties and protocols, are largely accepted by most nation-states, and have consequently become a central component of accepted international law. These laws, which have evolved and been transformed over the years, were designed to minimize, as much as possible, the level of human suffering that occurs in the context of armed conflict.[2] Central to this objective is the premise that civilians are to be protected from intentional killing and arbitrary injury, and that military commanders are committed to upholding the principles of military necessity and proportionality in their activities; in other words, a military objective should be achieved with as little harm as possible to noninvolved civilians. Historically, these principles have formed the cornerstone of international humanitarian law governing the conduct of war.

However, despite their previous adoption and application, today these principles are largely anachronistic in the framework of warfare between a nation and a nonstate actor (organization, group, network, unit, or cell). International humanitarian law (IHL), or the laws of war, was developed to govern traditional warfare and international armed conflict between sovereign states and their conventional armies. In fact, the term "international armed conflict" refers to conflicts between two or more states and does not account for the role of nonstate actors in international conflicts and war.

As such, the role, rights, and obligations of nonstate actors in the current legal framework are either neglected or disputed.[3] Even the categories of "combatant" and "civilian," which serve as a foundation for the rules of war (through the principle of distinction), do not adequately reflect the realities of warfare on the ground today, warfare in which soldiers do not necessarily wear uniforms, in which rockets are launched from kindergartens, and in which humanitarian protections can be exploited as weapons.[4] Indeed, in many cases, nonstate actors perceive the humanitarian rules that make up IHL as constituting the "soft underbelly" of the state they oppose. While focusing their activities on causing intentional harm to civilians, nonstate actors will simultaneously take advantage of the protections afforded to civilians in order to hide among them, blend in with them, and effectively use them as "human shields."

The principle of proportionality proves especially problematic, since it relies on these fundamentally flawed categories and a vague and essentially unenforceable balance between military necessity and collateral damage to civilian life. In addition to the problems of defining the principle of proportionality (e.g., "excessive" and "military advantage" are ill-defined in Additional Protocol I of the Geneva Conventions),[5] multiple variables influence the principle of proportionality and collateral damage—the difficulty in comparing civilian harm and military gain, tactical intelligence failure, the precision of the weapons used, etc.—as Michael Schmitt argues.[6] Thus, this new kind of warfare clearly requires a reexamination of the laws of war and the introduction of a new legal approach on three fundamental levels: a new term for and definition of modern warfare, which will reflect current realities and cover nonstate actors; new categories to define involved actors (civilians, combatants, etc.) within this paradigm; and, most important, an objective framework for calculating proportionality, based on these new categories and incorporating a new proportionality rationale.

I. THE CURRENT LEGAL FRAMEWORK: INTERNATIONAL HUMANITARIAN LAW (IHL)

Several basic humanitarian principles lie at the core of IHL, governing the conduct of military action in international and noninternational armed conflict. Namely, military commanders are required to differentiate between civilians and combatants in all of their operations, as well as between civilian infrastructure and legitimate military objectives, and to only intentionally target the latter. They must also ensure the humane treatment of all prisoners of war (POWs). In seeking to achieve specific military goals, the actual target of a military operation must be considered a legitimate "military objective." While accidental harm caused to civilians by such an operation would not, in and of itself, be considered illegal, several precautionary principles are designed to limit collateral damage and even prohibit military action when the cost to civilians is too high. These basic principles have been codified in the Geneva Conventions and other conventions and protocols, and form four basic rules that govern the conduct of war:

(1) The principle of distinction and civilian immunity: Parties to a conflict are at all times required to "distinguish between the civilian population and combatants and between civilian objects and military objectives," and should accordingly "direct their operations only against military objectives."[7] Civilian populations enjoy general protections against any dangers that arise out of military operations and should not be the object of attack.[8]

(2) The principle of proportionality: Even when targeting legitimate military objectives, military commanders must ensure that the losses to the civilian population and the damage to civilian property are not "excessive in relation to the concrete and direct military advantage anticipated."[9] This principle requires that a reasonable balance be struck when military commanders launch attacks that put civilians or civilian objects at risk, in which the military advantages of an attack are weighed against any undesirable collateral effects. Essentially, harmful effects must not be disproportionate to the military advantages of targeting the objective.[10]

(3) Respect for the basic rights of POWs: Although combatants and others taking a direct part in hostilities are legitimate targets of attack, once they have been captured or have surrendered, they become entitled to POW status, which affords them certain rights and imposes a number of obligations on the forces holding them.[11] As long as the combatant has abided by the provisions of IHL, POW status ensures that he will not face criminal prosecution for his acts on the battlefield. Other rights are designed to secure the welfare of captured combatants and their return home after the cessation of hostilities.[12]

(4) Use of legal weapons only: Combatants are forbidden to use weapons that are indiscriminate in nature, or that would cause greater suffering than is required to eliminate enemy combatants. Weapons that cause widespread, long-term damage to the natural environment are also prohibited. Specific treaties also restrict and prohibit the use of biological, chemical, and other unconventional types of weapon.[13]

Defining the status of those involved in war—distinguishing between who can be designated a "civilian" and who a "combatant"—fundamentally determines what obligations and rights apply to those involved in war: who can be targeted, who protected, who captured, and who prosecuted. Under IHL, the term "combatant" refers to those who have a "right to participate directly in hostilities," and who consequently are lawful targets of military action during wartime.[14]

In general, a combatant may be a soldier acting within the framework of a regular army, a militia member on active duty, or a reserve soldier.[15] The conditions for being considered a combatant/POW are further detailed in the Third Geneva Convention, Article 4, as well as in Articles 43 and 44 of Additional Protocol I, and include the following:[16]

- those who wear identifying symbols (that distinguish them from the civilian population);
- those who carry weapons openly; and
- members of a hierarchical organization, with a clear chain of command, that adheres to the laws of war.[17]

Regarding those individuals who are not members of regular armed forces, Article 4 of the Third Geneva Convention stipulates that members of other militias and volunteer corps (including organized resistance movements), are also considered combatants, as long as they "belong to a Party to the conflict" and fulfill similar conditions:

(a) they are commanded by a person responsible for his subordinates;
(b) they wear a fixed, distinctive emblem recognizable at a distance;
(c) they carry arms openly; and
(d) they conduct their operations in accordance with the laws and customs of war.[18]

Civilians, who may spontaneously organize themselves to take violent action in resistance to a foreign invader, are afforded the same status of combatants and POWs, as long as they clearly display their weapons.[19] When civilians are in no way involved in combat, they are immune from intentional attack and cannot be held captive for any reason.

Civilians, or noncombatants, are those who do not participate in hostilities and are thus protected persons, immune from being intentionally targeted. Although a certain amount of collateral damage is allowed in operations that target lawful military objectives, a violation of IHL would occur if civilians or civilian objects were intentionally targeted—if the planning, preparation for, and conduct of an operation were specifically designed to harm civilians.

Humanitarian protections for civilians are suspended, however, when such individuals take a "direct part in hostilities," at which point they lose their privileges as civilians, becoming legitimate military targets during armed conflict. It is permissible to attack them only when they violate the prohibition against being involved in combat, and when their involvement in the combat is direct.[20]

However, it is important to recognize that in a war situation, injury to civilians during combat does not, in and of itself, constitute a violation of the principles of civilian protection under IHL. Such a violation is dependent on both the *intent* of the military action—whether it targeted a legitimate military objective, such that harm to civilians was incidental and unintentional (collateral damage)—and on the *extent* of injury caused to these civilians—whether the unintentional harm was proportional to the military advantage anticipated when the military objective was targeted.

In the current legal framework, in times of armed conflict, harm to civilians is thus lawful if their death or injury is the unintentional consequence of an attack on a lawful military objective.[21] However, the fact that civilian loss can be justified when a combat operation targets legitimate military objectives, increases the risk of unnecessary causalities and disregard for civilian life on the part of planners and perpetrators of battle.

It is in seeking to prevent exactly this tendency that certain precautionary principles and obligations are included in the framework of IHL. The principle of proportionality is one such obligation; it attempts to limit the collateral damage caused by military operations that involve civilian life or objects, and provides criteria—albeit vague ones—for determining how much unintentional damage to civilian life and infrastructure can be justified under international law. It requires that a reasonable balance be struck, which weighs the military advantages of an attack against its undesirable collateral effects. Essentially, such harmful effects must not be disproportionate to the "concrete and direct military advantage anticipated."[22]

A. Old Laws and New Threats: Multidimensional Warfare

While these basic principles are certainly crucial to protecting human life, it is clear that they and the ensuing laws governing international armed conflict (IAC) are not fully applicable to a situation of modern warfare in which an armed conflict, international in character, is conducted between a state and a nonstate actor.

Indeed, modern warfare between states and nonstate actors has been termed and defined variously over the years. The most prominent of these definitions—"asymmetric warfare"—does not adequately reflect the characteristics of the phenomenon, and can even be misleading. For example, a campaign defined as "asymmetric" in nature is widely understood to include a struggle between two actors of different strengths (military, economic). As such, the immediate association arising from the term "asymmetric warfare" is that of a struggle between the young shepherd David and the giant Goliath. The nation state is "Goliath": the strong party armed with a regular military and an enormous capability to generate firepower in all arenas—air, land, and sea. Standing before Goliath is "David": a pseudo-military organization with few and reduced military capabilities. This David and Goliath framework may appear applicable in the context of modern warfare. However, contrary to the popular understanding of "asymmetric warfare," the balance of power between the two actors does not necessarily favor "Goliath"—that is, the state.

Even though, prima facie, a state has stronger military, intelligence, and economic resources and capabilities than does a nonstate actor or pseudo-military organization, a modern liberal-democratic state is subject to the rules of war, which, in effect, limit the resources it can utilize and which restrict, at least in part, its ability to operate and maneuver freely. A nonstate actor, on the other hand, does not adopt these values and is not committed to their application. In fact, such actors actually make cynical use of IHL. As a result, a form of "reverse asymmetric warfare" is established: in a conflict portrayed as a war between David and Goliath, Goliath (the state) is bound by its hands and feet by IHL and by its own liberal values, while David (the nonstate actor) is exempt from all moral or legal restraints.[23]

An alternative term used to describe modern military campaigns is "low intensity warfare." However, yet again, in the modern age this term simply does not fit: terrorist[24] organizations may use nonconventional means of combat such as biological, chemical, radiological, or even nuclear weapons in their attacks, means that are far from "low intensity" in nature. Additionally, "conventional" attacks may result in the deaths of thousands (e.g., the attacks of 9/11). In effect, referring to this type of military activity as "low intensity" downgrades the severity of the threat posed by nonstate actors.

In reality, the modern military campaign between a state and a nonstate actor or actors traverses multiple dimensions, and includes terrorist activity, guerrilla warfare, and insurgency. It is conducted on military, political, criminal, media, psychological, and economic planes. The modern campaign is waged at varying levels of intensity, from low intensity (e.g., through the use of "cold" weapons) to high intensity (through unconventional attacks[25]). The effect of such a campaign is wide-ranging, and can cause anywhere from a few to hundreds of thousands of causalities. Given these circumstances, it appears that the most suitable definition of the modern campaign is actually "*multidimensional warfare*": that is, warfare conducted simultaneously in different dimensions between states and nonstate actors.

Since the very nature of such a conflict involves different actors, tactics, and risks, it is not suitable to apply outdated models to multidimensional warfare.

1. Problems of Distinction and Definition

In particular, it appears that some of the criteria that serve as the basis for distinguishing between civilians and combatants may no longer be relevant in the framework of multidimensional warfare. In applying the current legal conditions required to be considered a combatant to the facts "on the ground," for example, it is unrealistic to expect all combatants to be affiliated with an army or militia that is organized in a hierarchical manner with a clear system of command and control. It is in the very nature of the nonstate actor to sometimes have an amorphous framework. It may even have a network-type construction, organized as a clique, matrix, or chain of activists.[26] Sometimes a group of activists may combine to form a cell, in order to promote political goals. This makes the requirement of an organizational, hierarchical framework problematic.

2. The Problematic Proportionality Principle

The next normative principle of IHL that must be adjusted to fit the realities of multidimensional warfare is the proportionality principle. As discussed above, the principle of proportionality is designed to introduce an early obligation—at the planning, preparation, and implementation stages of a combat operation—whereby military commanders must predict the extent to which the objective of their combat operation may endanger noninvolved civilians, taking into consideration that operation's design, location, execution, and the resources to be used in its implementation. All of these considerations must be balanced against the military necessity and importance of conducting the operation as planned; at the same time, alternatives must also be examined that may lead to the same end goal, but with a lesser degree of anticipated civilian injury.

Despite a lack of clear criteria for determining proportionality in advance, it is clear that as the operational need to perform a military action increases, and as long as there are no reasonable alternatives to it, so too will the range of damage considered proportional increase. It is also reasonable to assume that it is possible to retrospectively identify events in which military commanders have severely deviated from accepted standards of proportionality, effectively crossing a red line. Consider, for example, the military act of targeted killing, designed to eliminate a senior terrorist designated as a "ticking bomb," who is hiding in a civilian environment.[27] The targeted killing is achieved by a focused aerial attack, the use of accurate munitions, and a small amount of explosives. Even if the operation causes harm to a small number of uninvolved civilians, it will not necessarily be considered disproportionate, since it was designed to limit civilian causalities and prevent a concrete terrorist attack from occurring imminently. If the same aerial operation had been executed using inaccurate munitions and a very large quantity of explosives, however, which might cause severe large-scale harm to uninvolved civilians, it would then be considered a violation of the proportionality principle.

The principle of proportionality reflects one of the most central humanitarian values embodied in the laws of war. It is accepted as a primary means of judging the legality—and thus the equivalent moral value of an attacking party's conduct in a war situation.[28] Despite this, nation-states, philosophers, strategists, and legal experts have yet to decide on an internationally accepted scale by which to determine the proportionality of various military actions. That is, they have yet to arrive at normative consensus on where exactly a red line can be drawn between a proportional and a disproportional action. In the absence of such a scale, all judgment on the proportionality—and thus the legality—of military action is highly subjective.[29] The lack of a concrete definition and a related equation for calculating proportionality makes this concept difficult to apply. After all, how much loss—death, injury, or destruction—can be unintentionally inflicted before it is considered excessive, relative to the aims of a military operation?

On one hand, a subjective assessment may overstate the importance of military necessity and consequently result in disregard for civilian lives. Alternatively, it may lead to a disregard of military necessity and its many challenges, and of the complex circumstances under which an attacking party functions during warfare with nonstate actors. Subjective assessment also often leads to unjustified accusation that attacking forces have used allegedly disproportionate force in their operations.

The controversial calculation of proportionality is further complicated when the lives lost belong to those who, in the current framework, would be considered civilians, but who potentially serve as passive or active supporters of combat; who are members of nonstate, pseudo-military organizations; or who were forced onto the battlefield as human shields.[30]

In fact, in some cases the principle of proportionality can actually be exploited. By taking cover within civilian areas and using civilian infrastructure and transport to conduct their operations, nonstate actors and other parties to war may intentionally use civilians to prevent their law-abiding enemies from attacking, knowing that the latter are fully aware that the potential for civilian causalities would be too high—too disproportionate—to permit an attack. In such a situation, the ability of law-abiding actors to effectively target legitimate military objectives is essentially paralyzed or neutralized, and civilians are more likely to be exploited.[31]

Therefore, as it now stands, the principle of proportionality is essentially a problematic, unenforceable norm laced through with subjectivity. Thus, the need for a new legal framework arises, one that will include expanded categories of actors, clearer distinctions among groups, and a more concrete framework for determining proportionality.[32]

In reexamining IHL and in introducing new models for determining proportionality, however, we must preserve the humanitarian values that serve as the foundation of the laws of war, such as the desire to safeguard the lives and property of civilians not participating in a military campaign (i.e., uninvolved persons). While it is clear that the proportionality principle and the other humanitarian values discussed thus far are largely anachronistic, as expressed in current IHL and in the context of multidimensional warfare between a state and a nonstate actor, that does not mean they lack value entirely.

In aiming to minimize human suffering and prevent civilian casualties, the fundamental humanitarian principles these rules embody undoubtedly reflect an important moral achievement of the human race. They are the product of wide-ranging agreement on the basic principles that should govern the conduct of military campaigns. From this standpoint, any attempt to update or modify IHL and adapt it to modern warfare requires great caution so as not to damage what has been achieved thus far. The baby must not be thrown out with the bathwater.

B. Multidimensional Warfare: New Categories of Involved Actors

The protections afforded to civilians during war under international law are based on the fundamental distinctions drawn between the civilian population and those considered to be involved in hostilities. As discussed, the current legal framework distinguishes between two groups—civilians and combatants—in affording these rights and defining state obligations. Within the multidimensional warfare model, however, the legal framework would be expanded to account for new categories of involved actors—specifically, the following four distinct categories: (1) combatants, (2) militia combatants, (3) combat-supportive civilians, and (4) civilians (*see* Diagram A).

(1) *Combatants* include all regular military personnel (conscripted and career personnel), soldiers, noncommissioned officers, and officers affiliated with either an organized, hierarchical military framework or a nonhierarchical framework, representing either a state or a nonstate actor that is actively engaged in hostilities.

(2) *Militia combatants* include those affiliated with either a state actor or a nonstate actor (a terrorist organization, group, network, unit, or cell), who usually undergo some basic training and then return to civilian life until called upon to participate in a military operation or attack. When affiliated with a nonstate actor, such combatants are, in effect, members of a militia. When such combatants fight on behalf of a state, they are most likely reserve soldiers or contractors. In any case, militia combatants remain immersed in civilian life until they are called to active duty.

(3) *Combat-supportive civilians* are civilians who assist in the activities of a state or nonstate military organization. It is possible to classify the activities of combat-supportive civilians into two subgroups, according to the characteristics of the assistance they provide: (a) those who offer operational assistance, and (b) those who offer nonoperational assistance:

   a. *Operational assistance* includes involvement in the operational, combat-related activity of an operation. This category can include two types of activity:

      i. *Ongoing operational activity*—This can include the production or purchase of combat resources, as well as the provision of basic military instruction.

      ii. *Concrete (combative) activity*—This can include participation in the operational preparation of concrete attacks (e.g., the transfer of combat resources or combatants to the location of an operation, the collection of tactical intelligence during the combat event, and the like).

   b. *Nonoperational assistance* includes both active and passive assistance to the military organization, including involvement in the organization's administrative activity, financial donations, and the supply of nonoperational equipment.

(4) *Civilians* can be further divided into two subgroups:

   a. *Noninvolved civilians*—Those who are in no way involved in combat activity and who do not assist the organization or its combatants. These civilians are entitled to all the protections currently afforded to civilians in a wartime setting, as prescribed by international law.

   b. *Civilians involved by force*—Those who are involved against their will, under threat or by force from the military organization (human shields).

The accurate definition of the many actors involved in war is important to the subsequent delineation of rights and protections each actor will be afforded under the multidimensional warfare model. The range of permissible activity that can be conducted in regard to these actors will depend on how they are defined and what status they are assigned, such as whether they can be deliberately targeted, whether they are eligible for POW status, and what restraints on collateral damage an attacking force is obligated to maintain. The multidimensional model retains the essence of the principles of distinction and immunity, but rests on these new, much more specific and inclusive categories of actors, as well as on different levels of obligation of the attacking force.

Diagram A presents these associated obligations and rights. Green shading indicates a situation in which the intention to deliberately target the actor in a combat situation is permitted.[33] Red shading indicates a situation in which intentionally targeting the actor should be prohibited.

Furthermore, once these actors have become collateral damage—unintended causalities of an attack on a lawful military objective—each subcategory is also associated with a certain level of required obligation, liability, and legitimacy. According to the multidimensional model, different degrees of obligation are assigned to the unintentional harm caused to different actors—a concept represented by the "level of precautionary obligation."

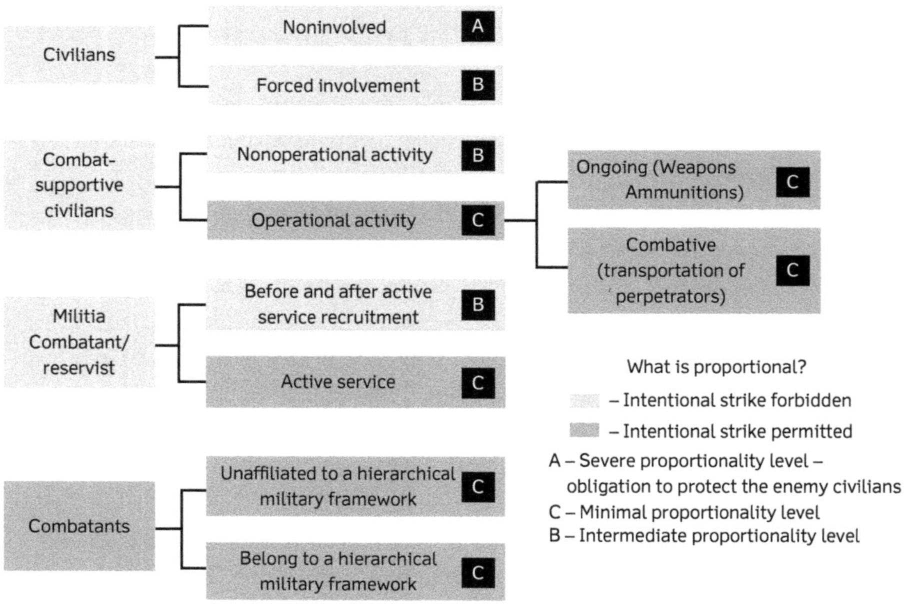

DIAGRAM A. Classifications of Actors in a Multidimensional Warfare: Civilians—Combatants

1. Levels of "Precautionary Obligation"

A wide variety of actors are involved in multidimensional warfare, all of whom—both the attacking and attacked parties—have complex characteristics. In order to navigate this complex reality, and assign different degrees of legitimacy to the targeting of, or unintentional harm caused to, different actors, it is necessary to distinguish among three different levels of "precautionary obligation," as stipulated by the multidimensional warfare model, defined below:

Level A—A High Level of Precautionary Obligation: This level includes the group of noninvolved civilians (not including civilians involved by force). Intentionally targeting elements in this category is absolutely forbidden. In the case of a military operation that may cause unintentional harm to these elements (collateral damage), the highest level of precautionary obligation is required. Even when unintentional, the loss of noninvolved civilians is of the greatest consequence in the context of war, and Level A status is designed to reflect this. Attacking forces have the greatest obligation to protect these actors.

Level B—An Intermediate Level of Precautionary Obligation: The intermediate level includes militia combatants not on active duty (including reservists not on active duty), as well as nonoperational combat-supportive civilians[34] and civilians who find themselves involved under threat or force (human shields). While intentionally targeting elements in this category is forbidden, if a military operation may cause unintentional harm to them, an intermediate level of precautionary obligation is required. The level of obligation toward these individuals, although high, is nevertheless less stringent than that of Level A. An attacking party cannot reasonably be expected to have an obligation to safeguard the lives of these actors equal to its obligation to safeguard the lives of completely noninvolved civilians. It is for this reason that a different level of precautionary obligation is assigned to noninvolved civilians, versus civilians involved by force. In both cases, intentional harm to civilians is forbidden,[35] but by assigning civilians involved by force a lower degree of precautionary obligation, the restraints on collateral damage, while clearly not abandoned, are justifiably *relaxed*.[36]

Level C—A Low Level of Precautionary Obligation: This level includes combatants, those affiliated and unaffiliated with a hierarchical military framework, and those representing both state and nonstate actors; militia combatants and reservists on active duty; and combat-supportive civilians involved in ongoing or concrete operational activity. Intentionally targeting elements in this category in multidimensional warfare is permitted, since they are active in hostilities and therefore represent legitimate military objectives. However, even in these cases it is necessary to strive, during combat, to minimize loss of life as much as possible.

According to this rationale, intentional harm to a combat-supportive civilian involved in ongoing or concrete operational assistance to enemy combatants is permitted. Such allowance is based on the understanding that operational combat support to an enemy constitutes an important contribution to the war effort, and is initiated by willing actors who knowingly put themselves in harm's way and thereby become legitimate military objectives in modern warfare. This actually reflects the current standard of IHL, which specifies that the protection of civilians during wartime is suspended when such individuals take direct part in hostilities, at which point they effectively earn combatant status and become legitimate military targets.

II. LAWFUL AND UNLAWFUL COMBATANTS

In the current IHL framework, those individuals engaged in hostilities who are considered terrorists do not fit neatly into the "civilian" or "combatant" category of actor. They actively participate in hostilities, so they are not civilians, but they do not meet the conditions for being considered lawful combatants either. Faced with this ambiguous situation, states such as the United States and Israel have referred to such individuals as falling into a third category of actor—that of "unlawful/illegal combatants."[37] Although the term "unlawful combatant"[38] is outside the scope of the Geneva Conventions and other treaties of IHL, it has been used frequently in military manuals, legal literature, and case law to refer to those who take direct part in hostilities without being entitled to do so and in violation of IHL. Among the many implications of this situation, those deemed "unlawful combatants" do not qualify for prisoner of war status if they fall into enemy hands, nor do they enjoy the privileges associated with protected persons.[39]

The use of the "unlawful combatant" category can be seen as an attempt to fine-tune and adjust IHL to accommodate the challenges of terrorism and multidimensional

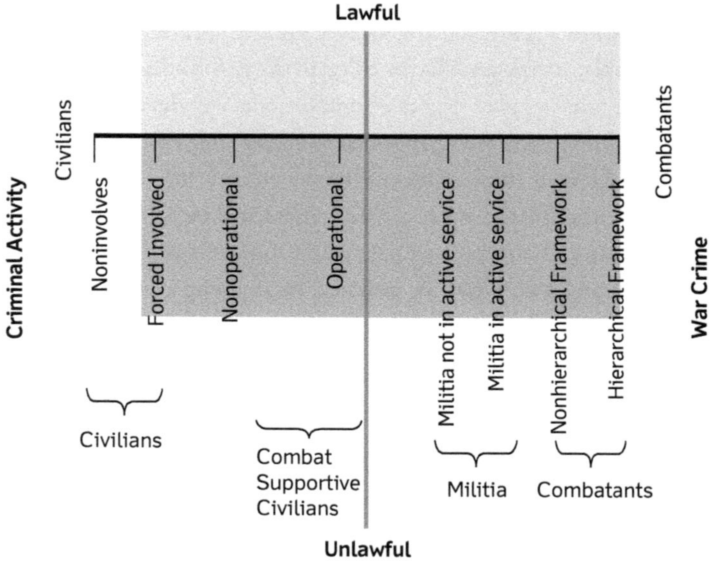

DIAGRAM B. A Third Category—Unlawful Combatants?

warfare. It seems, however, that the adoption of this third category, which is riddled with legal and ethical problems, has not proven the best approach.

In contrast to the attempt to use the "unlawful combatant" category to both define and deal with the illegal activities of terrorists, each category of actor in the multidimensional warfare model can itself be considered lawful or unlawful, based on the legitimacy of the actors' actions and their adherence to the laws of multidimensional warfare. In this way, the categories themselves do not denote the legality or illegality of their actions.

Rather, as demonstrated in Diagram B, civilians and combatants represent the two extremes of a spectrum of involvement in a combat situation. Between these two extremes lie the other levels of involvement: militia combatants and combat-supportive civilians (and their subgroups).

Another spectrum, represented by the vertical axis in Diagram B, reflects the legality of the activities of each category and subgroup of actor involved in multidimensional warfare. As noted, according to the multidimensional warfare model, each category of actor can be considered lawful or unlawful. A lawful combatant would be a combatant that abides by IHL in its military activities, and an unlawful combatant would be a combatant that disregards IHL in conducting its activities. Similarly, lawful militia combatants or lawful combat-supportive civilians would be those who follow the laws of war. Those who do not, regardless of the nature and level of their involvement in combat activities, would be regarded as unlawful.

The aforementioned models are designed to be used first and foremost as guidelines for determining the status of combatants and combat-supportive civilians involved in multidimensional warfare. Furthermore, these models can be used as a scale for assessing the legality and morality of the behavior of those actors involved in multidimensional warfare, thus providing additional tools and guidelines for responding to terrorism.

A. A New Model of Proportionality: Rationale and Equation

In seeking to remedy the apparent ambiguity of the proportionality principle, one primary question emerges: is it feasible to devise a scale and objective criteria for the examination of proportionality of military action in the framework of multidimensional warfare? Such a scale would need to balance military necessity with civilian causalities, transform qualitative values into quantitative factors, and take into account the unique characteristics and complexity of multidimensional warfare. This equation should also refer to the new categories of involved actors proposed by the multidimensional warfare framework, and differentiate among the varying levels of precautionary obligation regarding targeting and collateral damage. While certainly open to further development and fine-tuning, a concrete and quantifiable equation to calculate proportionality in the framework of multidimensional warfare is indeed possible.

In this proposed equation, the proportionality (P) of a given military operation is represented by the quotient of military necessity (X*Y*Z) divided by the expected scope of enemy causalities ($Nc^1+Nb^2+Na^3$), as follows:

$$P = \frac{X * Y * Z}{Nc^1 + Nb^2 + Na^3}$$

As a product of three variables, the value assigned to the military necessity of conducting the operation—the equation's numerator—reflects the immediacy of the operation, the ramifications of refraining from the operation, and the number of causalities that would be incurred in the attacking force if an alternative operation were conducted. Similarly, the number of expected enemy causalities—the denominator of the equation—reflects the scope, scale, and characteristics of these causalities, placing weighted value on causalities incurred among noninvolved civilians. By including these variables in the equation—essentially taking into account the unique challenges facing military commanders and posed by the involvement of nontraditional actors in warfare—this equation of proportionality adheres to the values of IHL by protecting the lives of civilians, while simultaneously more accurately reflecting current realities.

1. Calculating Military Necessity

As noted, the numerator in the proportionality equation (X*Y*Z) quantifies the military necessity of conducting a specific military operation by considering three central components, represented by the product of the variables X, Y, and Z.

X—A value representing the urgency of the need for the military operation. This value is determined using a scale based on a time line of operational phases in the preparation and execution of a concrete terrorist or guerrilla attack by an opponent nonstate party (prior to initiation; initiation of the attack; and planning, preparing, and executing of the attack). A numerical value is assigned to the operation based on a scale of 1–10, described below, with the numerical value increasing as the attack being prevented becomes more concrete and imminent:

- Minimal urgency: There is no concrete plan to execute an attack. The military operation is generally meant to deter the nonstate actor from executing attacks or to disrupt its ongoing activities (2).
- Preliminary stage: The military operation is meant to impede an attack that may occur, once the decision to launch an attack has been initiated (4).
- Early stage: The targeted party is in the early phase of executing an attack; that is, the decision to launch an attack has been made, and the targeted party is in the planning phase of the attack (6).
- Advanced stage: The targeted party is at the advanced stage of executing an attack, that is, in the preparation phase of the attack (8).

- "Tickingbomb" situation: The targeted party has finished finalizing preparations for the attack, and is in the phase immediately before executing it (10).

Y—The degree of military necessity in the proportionality equation is meant to reflect, in part, the advantage of the proposed operation over other alternatives, as measured by reducing potential danger to the attacking force. A military goal can be achieved by executing a variety of operations, using different modus operandi, different tactics, and so forth. The "Y" factor thus reflects the number of fighters who might be killed in the least deadly alternative operation. For example, an alternative to an aerial bombing operation aimed at neutralizing a terrorist activist designated as a "ticking bomb" would be an attack by ground forces, which would endanger many of the combatants on the attacking force. The number of likely combatant casualties in such an alternative ground operation would be represented by "Y" in the equation.

Z—An estimation of the number of causalities that would be suffered by the attacking side were the military operation not conducted. For example, the decision to refrain from carrying out a military operation designed to thwart an attack or eliminate a "ticking bomb" may result in the attack being successfully perpetrated, causing a number of casualties. This number of casualties is represented by "Z" in the equation, and constitutes an integral part of the calculation of the military necessity of the execution of a specific operation. The variable "Z" is estimated based on intelligence information and previous experience of similar attacks.[40]

2. Calculating the Expected Scope of Casualties (Collateral Damage)

The proportionality equation also reflects the collateral damage, or number of causalities, expected on the side being attacked, represented by the variable "N." The total number of expected enemy causalities—as noted, the denominator in the equation ($Nc^1+Nb^2+Na^3$)—is the sum of the amounts of three types of casualty (Na, Nb, Nc), based on the three "levels of precautionary obligation" described above. Each level is ascribed a higher or lower value via its corresponding exponent, thus enabling different categories of casualty to hold different weight in the calculation.

Level A—A High Level of Precautionary Obligation: In the case of a military operation that may cause unintentional harm to groups of noninvolved civilians (not including those who are involved by force), the highest level of precautionary obligation is required. The number of causalities in a given operation that falls at this level is represented by the variable (Na) in the proportionality equation. The value of the damage caused to noninvolved civilian causalities (Na) is expressed by increasing the exponent to 3 (Na^3), to reflect that it is forbidden to intentionally target them, and that their unintentional loss should be prevented to the extent possible.

Level B—An Intermediate Level of Precautionary Obligation: In the case of a military operation that causes unintentional harm to militia combatants not on active duty, nonoperational combat-supportive civilians, and civilians involved by force, an intermediate level of proportionality is required. The number of causalities in a given operation that falls at this level is represented by the variable (Nb) in the proportionality equation, and the value of the damage to these elements is expressed by the exponent of 2 (Nb^2).

Note that, using this equation, the number of those civilians involved by force who are unintentionally killed in battle is raised to the second power, rather than to the third power (the value assigned to noninvolved civilians). This distinction is actually designed to discourage nonstate actors from exploiting humanitarian protections to further their military objectives, including using civilians as human shields in battle. In effect, assigning civilians involved by force a less stringent level of precautionary obligation actually helps discourage the use of human shields. It also more accurately reflects the challenges and complexities facing military commanders who are pitted against enemy forces that, though unconcerned with fulfilling humanitarian obligations themselves, nevertheless and simultaneously take advantage of their enemies' attempts to do so.

Level C—A Low Level of Precautionary Obligation: The loss of combatants, militia combatants, and operational combat-supportive civilians is represented by the variable (Nc) in the proportionality equation, and is raised to the power of 1 (Nc^1) to reflect that such deaths, whether intentional or unintentional, are permitted in the multidimensional war framework. Here, precautionary obligation is the least stringent.

The expected scope of casualties (collateral damage) is therefore the sum of the number of Level C casualties to the first power (Nc^1), Level B causalities squared (Nb^2), and Level A causalities cubed (Na^3). As explained above, the severity of the loss and the legitimacy of the damage caused to each group of causalities is expressed by its corresponding exponent.

3. Determining the Proportionality of a Military Operation

Once the proportionality equation has been calculated, the resulting quotient determines whether, and to what extent, the military operation in question is proportional. For the military operation to be considered proportional, "P" must be greater than one ($P > 1$).[41] Calculations that result in a quotient of less than one would be considered disproportional.

For example, in one possible scenario, military commanders are planning an aerial bombing of terrorist headquarters located in a residential neighborhood. The bombing has been planned based on intelligence information, and in order to thwart an attack considered to be a "ticking bomb" (X), which might result in approximately 30 casualties (Z). An alternative ground operation would likely endanger 10 soldiers from the attacking side (Y). Expected enemy casualties include 5 enemy combatants (Nc), about 10 nonoperational combat-supportive civilians situated at the headquarters (Nb), and possibly up to 10 noninvolved civilians (Na).

$$P = \frac{10 * 10 * 30}{5^1 + 10^2 + 10^3} = \frac{3000}{1105} = 2.7$$

In this scenario, the quotient (2.7) is greater than 1, indicating that a proportional level of force would be used in the operation, based on available intelligence and related estimates. On the other hand, if, in the same operation, an estimated 20 noninvolved civilians causalities (Na) were expected rather than 10, then the equation would provide a different result:

$$P = \frac{10 * 10 * 30}{5^1 + 10^2 + 20^3} = \frac{3000}{8105} = .37$$

In this scenario, the quotient (.37) is less than 1, indicating that a disproportional level of force would be used in such an operation.

4. *Calculating Proportionality on the Battlefield*

The equation proposed above serves as a tool for estimating the proportionality of a concrete, single, proactive, *premeditated* military operation. In a sense, it is relevant to situations in which commanders have a certain amount of time, however limited, and a secure environment in which to plan and calculate potential casualties prior to an operation: that is, targeted killing, aerial operations targeting defined military objectives, and the like, based on available intelligence information. If the resulting calculation reveals that the operation would be disproportional in nature, commanders could potentially cancel plans for the operation prior to its execution.

However, it would seem that the equation as such is not applicable to a battlefield scenario in which a military force on a mission is ambushed and must react in self-defense immediately and almost automatically. In such situations, the immediate necessity is to neutralize the source of fire or ensure the escape of the attacking force. After the situation has been stabilized, the original proportionality equation can once again become relevant.

We must thus also determine whether or not the equation can be applied to operations that are not preplanned or premeditated, operations or attacks that are called for in the heat of battle, as new situations and circumstances present themselves. As part of a broad military operation lasting days, weeks, or months, a military force may find itself dragged into local activities and operations that are dictated by developments in the campaign or by the enemy in the field.

How, then, can the proportionality equation assist in the formation of policy when military commanders are faced with immediate, unanticipated threats on the battlefield? In such cases, the equation can indeed be used, with slight adjustments to the value of each variable. To restate the proportionality equation:

$$P = \frac{X * Y * Z}{Nc^1 + Nb^2 + Na^3}$$

X—In battlefield situations, the value of the variable "X" in this equation, representing the urgency of the operation to be performed, would be calculated using a different scale, as detailed below:

- The need to neutralize an immediate threat from which ongoing fire is directed at civilian territory (20).
- The need to neutralize immediate threat from which ongoing fire is directed at the attacking forces (15).
- The need to disrupt the ongoing military activity of the enemy (10).
- The need to capture or neutralize enemy combatants who have taken shelter in a facility but do not pose any immediate or concrete threat to attacking forces (1).

Y—The variable "Y" is used in the same manner as in the original equation—that is, as representing the advantage of the operation over other alternatives, based on a desire to reduce potential dangers to the attacking force. Thus, the number of likely attacking force combatant causalities in the least deadly alternative operation would be represented by "Y" in the equation.

Z—The variable "Z" reflects disruption of the planned military operation, should the objective not be seized or neutralized. In such a case, "Z" in the proportionality equation can be determined by the following scale:

- Not seizing the objective is not detrimental to the continuation of effective combat activities and the achievement of the original goals of the military operation. The value of "Z" is therefore lowest (1).
- Not seizing the objective would be slightly detrimental to the continuation of combat activities and the original aims of the military operation. The value of "Z" is (10).
- Not seizing the objective would be highly detrimental to the continuation of combat activities and the original aims of the military operation. The value of "Z" is high (15).
- Not seizing the objective would endanger the primary combat effort and may hinder military achievements or result in a complete failure of the entire campaign. The value of "Z" is the highest (20).

For example, during an ongoing military campaign, an attacking force is fired upon by an enemy force entrenched in a mosque ($X = 15$). Based on intelligence information, inside the mosque there are an estimated 10 enemy combatants ($Nc = 10$). There is a possibility that 20 noninvolved civilian worshipers are present ($Na = 20$). The commanders estimate that an alternative operation—a ground attempt to seize the mosque, in lieu of an aerial bombing—may cause the death of 20 combatants, as a result of improvised explosive devices (IEDs) spread along the routes to the mosque's entrance (Y = 20). Refraining from neutralizing or seizing the objective might prevent the continued advance of the forces in this direction, and even endanger the goals of the overall military campaign (Z = 20).

$$P = \frac{15 * 20 * 20}{10^1 + 0^2 + 20^3} = \frac{6000}{8010} = .74$$

Although this is an extreme case in which refraining from neutralizing the objective would endanger the overall campaign, in this scenario, the quotient (.74) is less than 1. This indicates that, based on available intelligence and ensuing estimates, a disproportional level of force would be used in the operation, and therefore the operation should be prohibited.

5. Continuation of the Scenario: Introducing Advanced Warning

However, before launching an attack that targets a mosque or another civilian facility, the attacking force must take concrete action to notify noninvolved civilians of the coming danger and allow them to leave the combat area prior to the attack. Such notification should include a clear warning in the local language by means of loudspeakers, leaflets, text and telephone messages, or any other form of communication that will reach the desired parties. The attacking force must also introduce a humanitarian ceasefire, as soon as it is militarily possible, whose objective is to allow noninvolved civilians to leave the facility unharmed.[42]

According to the multidimensional warfare model, the obligation to provide advanced warning would have further implications. As outlined above, civilians can be attributed Level A status if they are completely uninvolved in hostilities, or Level B status if they are involved unwillingly as human shields. Each status will lend a different value to the appropriate variable in the proportionality calculation, and thus a different expected level of obligation to protect the civilians in question, and thus a different level of legality associated with a military operation. The obvious difficulty here is distinguishing among civilians present in the combat area: are they involved by choice, or against their will?

If an effective warning has been issued by the attacking force and evacuation is possible, then it can be assumed that a civilian remaining in the facility to be attacked no longer retains the status of Level A personnel. It may be assumed that that individual is now in the facility either unwillingly as a human shield (Level B), or willingly, as a nonoperational combat-supportive civilian (Level B, as well).

For example, in the scenario described above, in which attacking forces encounter enemy fire emanating from a mosque, advance warning could change the results of the proportionality equation. The attacking force could provide advance warning over a loudspeaker or by other means, calling on the entrenched combatants and civilian worshippers inside the mosque to exit prior to the pending attack. The force could announce two unilateral ceasefires so as to allow all those inside the mosque to exit prior to it being bombed from the air. If, after such precautions have been taken, no one exits the mosque, then the status of those civilians in the mosque changes from that of noninvolved civilians (Level A) to that of civilians involved by force (Level B). The values of Na and Nb are thus reversed (20 to 0):

$$P = \frac{15 * 20 * 20}{10^1 + 20^2 + 0^3} = \frac{6000}{410} = 14.6$$

In this scenario, the quotient (14.6) is greater than 1, indicating that a proportional level of force would be used in an operation in which advanced warning was provided. This example demonstrates that the use of a civilian facility and human shields would not prevent an attacking force from seizing and neutralizing the threat emanating from the mosque, as long as the necessary precautions had been taken and warnings were given. Such an attack would not violate the proportionality principle. Using this rationale, the incentive for the enemy force to then use human shields would be minimized.

According to the multidimensional warfare model, and similar to the current IHL clause imposing an obligation to advance warning (see footnote 42), an attacking force would not be obligated to give warning in all cases. A force could act according to military necessity, with minimum risk to its own combatants. If providing warning or introducing a ceasefire would put the attacking force's troops or its operation's military goals at risk, then such action would not be obligatory.

Conclusion—A Decision-Making Support Tool

The suggested equation is a conceptual exercise intended to introduce and encourage a new way of thinking about proportionality and the principle of distinction during wartime. The proposed equation is supposed to enhance the debate by addressing key dilemmas related to the issue of proportionality, and the means to quantify these dilemmas. The equation is a draft that requires validation and adaptation through empirical case studies. In any case, the proposed proportionality equation is not meant to serve as the sole tool for enabling decision makers to determine the legitimacy of planned military actions, but rather as a means of supporting such a decision.

In seeking to further develop this model, it is important to note that the primary weakness of the equation is that its variables are derived from intelligence information that is in the hands of the attacking force. It does not reflect the results of an actual operation but rather reflects the estimates and considerations anticipated and embodied before an operation's execution. Such estimates may retrospectively be proven to have been mistaken. The strength of the equation, however, is that it allows for the quantification of qualitative variables and for consideration of the types of calculation and estimate relevant to determining proportionality. It thus allows for an independent, unbiased, and valued judgment of the degree of proportionality of a military operation, prior to a decision on the execution of the operation.

The equation also allows for retrospective examination, enabling analysts to accurately reconstruct the context of the original decision-making process. When an attacking party is accused of having used disproportionate force in its military activities, this equation can be used to reveal the system of considerations that guided decision making prior to an operation's launch (based on the intelligence information available at the time). An accused party can also present the actual proportionality calculation performed at the time of planning, prior to making the decision to conduct the military action in question. Furthermore, the equation can reflect the efforts made to reduce unintentional harm to the attacked party's noninvolved civilians.

The growing challenges posed by multidimensional warfare, paired with an increase in the number of nations being pulled onto this new battlefield, necessitates a new legal approach, one suited to addressing the related issues of proportionality and the use of force. Experts in the fields of law, political science, philosophy, and ethics, as well as decision makers and military actors, must combine their efforts to reconceptualize, further develop, and ultimately construct a contemporary, clear, and normative system—an objective and accepted system—that can be used to examine the issue of proportionality, and to fairly determine the legitimacy of military operations. Adoption of the proposed proportionality equation, or the development and consolidation of a similar equation, represents a first step toward achieving this crucial objective. The suitable adaptation of IHL to the rules of modern multidimensional warfare, whether through the adoption of an additional protocol or through other legal instruments, is the next step required by reality.

9

THE USE OF FORCE REGIME AND UNCONVENTIONAL THREATS: ANWAR AL-AWLAKI AS A CASE STUDY

Robert Chesney*

Introduction

Anwar al-Awlaki was a dual Yemeni-American citizen who emerged in recent years as the leading English-language proponent of violent *jihad*, including explicit calls for the indiscriminate murder of Americans. According to the U.S. government, moreover, he had taken on an operational leadership role within the organization al Qaeda in the Arabian Peninsula (AQAP), recruiting and directing individuals to participate in specific acts of violence. He is now dead, killed in an airstrike carried out by the U.S. government in Yemen.

Was it lawful to kill al-Awlaki? That question can be addressed from many perspectives, including through the lens of international humanitarian law, international human rights law, and—in light of his citizenship—the U.S. Constitution. In this chapter, however, I wish to focus on a distinct consideration: did the killing of al-Awlaki violate the U.N. Charter system's rules governing the use of force in international affairs?

I. BACKGROUND

Before coming to grips with the legal issues, a close review of the underlying fact pattern is in order. I begin below with a sketch of AQAP and its relationship with what we might call "core al Qaeda" or, simply, "al Qaeda," a topic that takes on significance in light of the U.S. government's claim that a state of armed conflict exists between it and al Qaeda. Next, I review Anwar al-Awlaki's background and activities. Last, I survey what is known about the use of force by the United States in Yemen in relation to AQAP in general and al-Awlaki in particular.

A. AQAP in Relation to al Qaeda

What is the relationship of the entity now known as AQAP to the entity we label al Qaeda? This is a difficult question for several reasons. As an initial matter, we lack access to the classified intelligence that would be most useful to answering it. Second, it is in any event difficult to map familiar notions of organizational structure on to al Qaeda. It might best be described as a network blending elements of hierarchy and centralization with elements of disaggregation, fluid individual relationships, and franchise-like connections to separate organizations, all against the backdrop of a larger, multifaceted movement associated with violent Islamist extremism.[1]

For some entities that today brandish the al Qaeda "brand," the relationship is a relatively new phenomenon in which a previously independent organization has for whatever reason decided to at least portray itself as part of the al Qaeda network. This appears to be the case, for example, with al Qaeda in the Islamic Maghreb (AQIM), which emerged in Algeria in the 1990s under the name the Salafist Group for Preaching and Combat and which aimed to overthrow the Algerian government.[2] The Salafist Group had no particular ties to al Qaeda until a few years ago when under pressure from declining membership and having a new leader, it reached out to al Qaeda.[3] An alliance was announced in September 2006, and by January the group had changed its name and reoriented its activities away from just Algeria.[4]

Contrast that with al Qaeda's history of direct involvement in Yemen. According to the 9/11 Commission Report,[5] the key figure in al Qaeda's early relationship to Yemen was Abd al Rahim al Nashiri, a citizen of Saudi Arabia who had fought against the Soviets in Afghanistan and then returned there in the mid-1990s with a group of fighters whom Osama bin Laden attempted to recruit into al Qaeda. Nashiri initially resisted swearing an oath of loyalty to bin Laden, and for a time went to live in Yemen. He later returned to Afghanistan, however, and eventually agreed to join al Qaeda. Sometime in 1998, Nashiri proposed to bin Laden that al Qaeda attack a U.S. Navy vessel in Yemen, and bin Laden agreed. Eventually this resulted in the failed attack on the *USS The Sullivans* in January 2000 and the successful attack on the *USS Cole* in October 2000. Nashiri subsequently became "chief of al Qaeda operations in and around the Arabian Peninsula,"[6] and continued to orchestrate attacks (including the bombing of a French ship) until he was captured in the United Arab Emirates in November 2002.

In the years immediately following Nashiri's capture, al Qaeda's operational activities in Yemen were limited. From roughly 2003 to 2006, al Qaeda focused its efforts on the Arabian Peninsula instead of Saudi Arabia, with Yemenis encouraged to travel to Iraq to fight.[7] Things began to change after twenty-three imprisoned al Qaeda members escaped from a jail in Sanaa.[8] Many were recaptured, but two who were not—Nasser Abdul Karim al-Wuhayshi and Qasim al-Raymi—went on to establish "al Qaeda in Yemen" in order to renew operations there. Wuhayshi had joined al Qaeda in the late 1990s, serving as a "personal assistant" to bin Laden. Under his leadership, al Qaeda in Yemen began a series of attacks, including the murder of Western tourists and an attack on the U.S. embassy in Sanaa in 2008. At the beginning of 2009, moreover, Wuhayshi pronounced that al Qaeda operations in Saudi Arabia and Yemen were merging and henceforth would operate under the collective heading of AQAP. AQAP has, since then, been remarkably active, including but not limited to its attempt to destroy a U.S. passenger jet bound for Detroit on Christmas Day 2009 and its "cargo jet" plot in 2010 involving explosives hidden in packages shipped via overnight delivery services. As a Carnegie Endowment report emphasizes, however, the "raised profile of the current incarnation of the organization should not detract from an awareness of al-Qaeda's enduring presence in Yemen."[9]

The picture that emerges from this brief sketch is complicated. AQAP appears to be merely the latest iteration of al Qaeda's long-standing operational presence in Yemen, contrasting sharply with the lack of historical ties to al Qaeda when it comes to some other current al Qaeda franchises such as AQIM. On the other hand, AQAP appears to operate without direct lines of control running to bin Laden or other senior al Qaeda leaders. Whether it is best perceived as part and parcel of al Qaeda, then, or instead simply an affiliated but independent franchise, would depend on how one defines organizational boundaries in this context in the first place and how one interprets the information available as to this question.

B. Anwar al-Awlaki in Relation to AQAP

At the time of the 9/11 attacks, Anwar al-Awlaki was an imam at a mosque in Northern Virginia. He soon became a public figure of sorts thanks to his public pronouncements condemning the 9/11 attacks from an Islamic perspective.[10] Over time, however, his publicly stated views appeared to change, taking on an increasingly anti-Western tinge.[11] He left the United States, first for the United Kingdom and then later for Yemen. Anwar al-Awlaki was in hiding in Yemen, and, far from denouncing indiscriminate violence, he emerged as a prominent English-language propagandist for violent *jihad*, calling for the indiscriminate murder of Americans and others.[12] According to the U.S. government, moreover, he became part of AQAP,[13] and not just as an ideologue or propagandist. The government asserts that he took on an operational leadership role in connection with specific attacks.[14]

The U.S. government's attention actually had been drawn to al-Awlaki much earlier. The FBI became interested in him in 1999 in light of a position he had held at an Islamic charity suspected of channeling money to extremists and because he had been in at least brief contact with individuals indirectly linked to both bin Laden and Omar Abdel Rahman, the so-called "Blind Sheik" associated with the 1993 World Trade Center bombing and the 1995 landmarks-and-tunnels plot in New York City.[15] Two of the future 9/11 hijackers had attended a mosque in San Diego where al-Awlaki had been the imam, and apparently spent a substantial amount of time in conference with him. The FBI ultimately concluded that these contacts were innocent, but not everyone involved in the investigation agreed.[16]

In any event, al-Awlaki's extremist views—whether preexisting or newly developed—would not become widely known to the general public until media reports in late 2009 began to emphasize that the perpetrator of the Fort Hood massacre, Major Nadal Malik Hasan, had been in touch with al-Awlaki by e-mail.[17] Soon thereafter, al-Awlaki gained still further notoriety when media reports asserted that he had been involved with Umar Farouk Abdulmutallab, the would-be "Christmas Day bomber," who unsuccessfully attempted to ignite an underwear bomb on a flight in 2009.[18]

More recently, the U.S. government has set forth its view that al-Awlaki was not merely a propagandist of *jihad* but an active member of AQAP. In a declaration filed by the government in connection with the aforementioned American Civil Liberties Union (ACLU) lawsuit, the Director of National Intelligence asserted that:

> Anwar al-Aulaqi has pledged an oath of loyalty to AQAP emir Nasir al-Wahishi, and is playing a key role in setting the strategic direction for AQAP. al-Aulaqi has also recruited individuals to join AQAP, facilitated training at camps in Yemen in support of acts of terrorism, and helped focus AQAP's attention on planning attacks on U.S. interests.[19]

The declaration adds that al-Awlaki personally instructed Abdulmutallab "to detonate an explosive device aboard a U.S. airplane," as part of a larger shift toward an operational leadership role with AQAP.[20]

In May 2010, al-Awlaki for the first time in a public setting expressly endorsed the use of violence not just against American military targets but also against American civilians.[21] When asked whether he supported operations "target[ing] what the media calls 'innocent civilians,'" al-Awlaki responded:

> Yes.... The American people in its entirety takes part in the war, because they elected this administration, and they finance this war. In the recent elections, and in the previous ones, the American people had other options, and could have elected people who did not want war. Nevertheless, these candidates got nothing but a handful of votes. We should examine this issue from the perspective of Islamic law, and this settles the issue—is it permitted or forbidden? If the heroic *mujahid* brother Umar Farouk could have targeted hundreds of soldiers, that would have been wonderful. But we are talking about the realities of war....
>
> For 50 years, an entire people—the Muslims in Palestine—has been strangled, with American aid, support, and weapons. Twenty years of siege and then occupation of Iraq, and now, the occupation of Afghanistan. After all this, no one should even ask us about targeting a bunch of Americans who would have been

killed in an airplane. Our unsettled account with America includes, at the very least, one million women and children. I'm not even talking about the men. Our unsettled account with America, in women and children alone, has exceeded one million. Those who would have been killed in the plane are a drop in the ocean.[22]

According to analyst Thomas Hegghammer, al-Awlaki was "not a top leader in AQAP's domestic operations, but he is arguably the single most important individual behind the group's efforts to carry out operations in the West."[23] Hegghammer explains that al-Awlaki was "most likely part of a small AQAP cell—the Foreign Operations Unit—which specializes in international operations and keeps a certain distance to the rest of the organization."[24] Indeed, he believes al-Awlaki may have been the head of that cell, arguing that "intelligence analysts familiar with his e-mail communications" have long suspected as much:

In public, Awlaki cast himself as an ideologue who supports armed struggle against the West, but is not directly involved in operations. In private, however, he has spent the past year actively recruiting prospective terrorists by e-mail and taking part in face-to-face indoctrination of operatives in Yemen.[25]

C. The United States and the Use of Lethal Force in Yemen

It is tempting to begin a discussion of the U.S. government's past uses of lethal force in Yemen with the November 2002 incident in which, it appears, a U.S.-operated drone fired a Hellfire missile into a vehicle traveling through the desert, killing the occupants. Because of the lengthy fallow period that seems to have followed that attack, however, I will confine the discussion in this subsection to events beginning in late 2009.

The week prior to Abdulmutallab's unsuccessful attempt to take down a passenger jet on Christmas Day 2009, the government of Yemen claimed credit for conducting a pair of attacks on AQAP targets, including an airstrike meant to kill Wuhayshi, al-Shihri, and al-Awlaki.[26] The media reported that the United States also was involved (at least in terms of providing intelligence, but possibly more directly), and this prompted al-Awlaki's father to argue that it is illegal for the United States to attack its own citizens and that his son "should face trial if he's done something wrong."[27] Journalists near this time began to focus on whether the United States had orchestrated the attack to kill al-Awlaki and what legal grounds might support such a policy.[28] Then, a few weeks later, Dana Priest of the *Washington Post* wrote an article asserting that:

U.S. military and intelligence agencies are deeply involved in secret joint operations with Yemeni troops who in the past six weeks have killed scores of people, among them six of 15 top leaders of a regional al-Qaeda affiliate, according to senior administration officials.[29]

Priest described President Obama as having signed off on the December 24 attack on al-Awlaki's house, with the caveat that al-Awlaki was "not the focus of the strike." She went on to reveal, however, that al-Awlaki "has since been added to a shortlist of U.S. citizens specifically targeted for killing or capture by the [Joint Special Operations Command]." According to Priest's account:

> After the Sept. 11 attacks, Bush gave the CIA, and later the military, authority to kill U.S. citizens abroad if strong evidence existed that an American was involved in organizing or carrying out terrorist actions against the United States or U.S. interests, military and intelligence officials said. The evidence has to meet a certain, defined threshold. The person, for instance, has to pose "a continuing and imminent threat to U.S. persons and interests," said one former intelligence official. The Obama administration has adopted the same stance. If a U.S. citizen joins al-Qaeda, "it doesn't really change anything from the standpoint of whether we can target them," a senior administration official said. "They are then part of the enemy."[30]

Subsequently, Scott Shane of the *New York Times* reported that al-Awlaki had been added to a similar list maintained by the CIA on the ground that he had become personally involved in operational planning, that "international law permits the use of lethal force against individuals and groups that pose an imminent threat to a country," and that the individuals on the CIA list "are considered to be military enemies of the United States" within the scope of the congressional authorization for the use of military force enacted after 9/11.[31]

In late March, State Department Legal Adviser Harold Koh gave a much-noted speech to the American Society of International Law in which he addressed in more detail the legal argument in favor of using lethal force in circumstances such as this.[32] Koh endorsed the propositions that "the United States is in an armed conflict with al-Qaeda, as well as the Taliban and associated forces" and that the United States "may use force consistent with its inherent right of self-defense under international law." He then addressed the factors that the United States considers in connection with specific targeting decisions, describing this as a case-by-case process turning on such considerations as "the imminence of the threat," "the sovereignty of the other states involved," and "the willingness and ability of those states to suppress the threat the target poses." Koh added that the proposed attack must also conform to "law of war principles," including the principle of "*distinction*, which requires that attacks be limited to military objectives and that civilians or civilian objects shall not be the object of attack," and the principle of "*proportionality*, which prohibits attacks that may be expected to cause incidental loss of civilian life, injury to civilians, damage to civilian objects, or a combination thereof, that would be excessive in relation to the concrete and direct military advantage anticipated."

As Ken Anderson observed, Koh's comments seemed to affirm not just the existence of an armed conflict between the United States and al Qaeda, but also a long-standing U.S. government position regarding the right to use force in self-defense even absent connections to an existing armed conflict.[33] The *Washington Post* editorial page subsequently praised the speech on similar grounds.[34] Others found his analysis unpersuasive.[35] The ACLU, for example, wrote a letter to the President expressing "profound concern about recent reports indicating that you have authorized a program that contemplates the killing of specific terrorists—including U.S. citizens—located far away from zones of actual armed conflict. If accurately described, this program violates international law."[36] Meanwhile, the United Nations' Special Rapporteur on Extrajudicial, Summary, or Arbitrary Executions—Professor Philip Alston of New York University School of Law—produced a report for the U.N. Human Rights Council that advanced legal arguments relating to the use of force that in many ways appeared to conflict with the views of the U.S. government, above all in connection with the use of force in response to terrorism in locations physically removed from conventional battlefields.[37] As Alston summarized things in a separate statement published on the ACLU's website:

> The United States has endorsed "a broad and novel theory that there is a 'law of 9/11' that enables it to legally use force in the territory of other States as part of its inherent right to self-defence on the basis that it is in an armed conflict with al-Qaeda, the Taliban and undefined 'associated forces.'" This expansive and open-ended interpretation of the right to self-defence threatens to destroy the prohibition on the use of armed force contained in the U.N. Charter, which is essential to the rule of law.[38]

For a time in the fall of 2010, it appeared that the legality of killing al-Awlaki might be put to the test in a judicial forum. In late August, the ACLU joined forces with the Center for Constitutional Rights (CCR) to represent al-Awlaki's father in a suit against President Obama, CIA Director Leon Panetta, and Secretary of Defense Robert Gates, requesting, among other things: (1) a declaratory judgment to the effect that international law forbids the use of lethal force outside of armed conflict except insofar as the targets "present concrete, specific, and imminent threats to life or physical safety, and there are no means other than lethal force that could reasonably be employed to neutralize the threats,"[39] and (2) an injunction forbidding the use of lethal force against al-Awlaki except on those terms. The suit proved short-lived, however. Ultimately, the court did not reach the merits. Instead, it granted the government's motion to dismiss on the grounds that al-Awlaki's father lacks standing to invoke al-Awlaki's asserted rights and that the arguments in any event present "political questions" that are not justiciable in the American legal system.[40]

Against this backdrop, the United States eventually carried out a strike, in Yemeni territory, which resulted in al-Awlaki's death. The strike raised a host of legal questions, all of which are important in their own way. The question at hand, however, is whether this action violated the U.N. Charter system's rules restricting the use of force in international affairs.

II. OBJECTIONS FOUNDED IN THE U.N. CHARTER

Did the use of force by the U.S. government against al-Awlaki in Yemen violate the U.N. Charter rules regarding the use of force in international affairs? The better view is that it did not.

Article 2(4) of the U.N. Charter provides that member states "shall refrain in their international relations from the... use of force against the territorial integrity or political independence of any state, or in any other manner inconsistent with the purposes of the United Nations" (which purposes are defined in Article 1 to include, among other things, the goal of "maintain[ing] international peace and security").[41] Article 2(4) is subject to exceptions, however, including Article 51's preservation of the right of self-defense and the Chapter VII mechanism whereby the Security Council may authorize the use of force.[42] Commentators debate the efficacy of the resulting system, but we can at least say that the general aim was to sharply constrict the circumstances in which force lawfully could be employed across borders.

A. Has Yemen Consented to the Use of Force?

The legality of a U.S. strike in Yemen at first blush might seem to turn on the plausibility of an Article 51 self-defense argument, there being no applicable Chapter VII Security Council resolution in this setting. But the need to make such an argument drops out if Yemen has effectively consented to the strike.[43] In that circumstance, there is no infringement of Article 2(4) in the first instance—no offense to Yemen's territorial integrity or its political independence, no threat to international peace and security insofar as the rights of member states are concerned—and hence no need to make exculpatory arguments under Article 51.[44]

The public record provides considerable reason to believe that the government of Yemen has given at least some form of consent to at least some uses of lethal force by the United States—or at least to the use of lethal force by U.S. and Yemeni forces acting in cooperation—on Yemen territory. Whether that consent suffices for purposes of a U.N. Charter analysis and whether in any event it extends to the al-Awlaki scenario are more difficult questions.[45]

Priest reports that "[s]hortly after the Sept. 11, 2001, attacks, [then-CIA Director George] Tenet coaxed [Yemen's President] Saleh into a partnership that would give the CIA and U.S. military units the means to attack terrorist training camps and al-Qaeda targets."[46] Pursuant to this agreement, the United States provided the Yemen security services with equipment and training, and Saleh in turn gave "approval to fly Predator drones armed with Hellfire missiles over the country." This appears to explain how it then came to pass, in November 2002, that a Predator drone was in position to strike and kill a group of al Qaeda suspects—including one who was an American citizen—in a car moving through an isolated stretch of desert in Yemen.[47]

Even if we assume that some degree of consent to lethal strikes existed as of 2002, the extent to which it continues to exist today is subject to some uncertainty in light of the Yemeni government's understandable desire to minimize the public's appreciation for the extent of American-Yemeni security cooperation—that is, its desire for plausible deniability. As Priest recently summarized the situation, the "broad outlines of the U.S. involvement in Yemen" had become public knowledge at least by the end of 2009, but the full "extent and nature of the operations" did not become known until she reported in early 2010 that

> [i]n a newly built joint operations center, the American advisers are acting as intermediaries between the Yemeni forces and hundreds of U.S. military and intelligence officers working in Washington, Virginia and Tampa and at Fort Meade, Md., to collect, analyze and route intelligence. The combined efforts have resulted in more than two dozen ground raids and airstrikes.[48]

These revelations put the Saleh administration in a difficult position, as Priest acknowledged:

> The far-reaching U.S. role could prove politically challenging for Yemen's president, Ali Abdullah Saleh, who must balance his desire for American support against the possibility of a backlash by tribal, political and religious groups whose members resent what they see as U.S. interference in Yemen.[49]

Indeed, just a few weeks earlier, Yemen's Deputy Prime Minister for Security and Defense, Rashad al-Alimi, had insisted at a rare press conference "that there are limits to [Yemen's] military cooperation with the United States, warning that any direct U.S. action in this impoverished Middle East nation could bolster the popularity of Islamic militants."[50] He conspicuously did not exclude the possibility of further U.S.-directed airstrikes or drone strikes, however.

By late summer 2010, the actual state of affairs became somewhat clearer. The *New York Times* in August published an article describing a "shadow war" in Yemen in which at least four attacks ostensibly carried out by Yemeni government forces from late 2009 onward in fact had been conducted on a clandestine basis by U.S. military personnel and assets, including cruise missiles and Harrier fighter jets, with the consent in each instance of the Yemeni government.[51] And though these strikes in one instance apparently involved significant civilian casualties, and in another the death of a provincial deputy governor, American officials asserted to the *Times* that President Saleh was "not so angry as to call for a halt to the clandestine American operations."[52]

Notably, the same report suggested that an internal U.S. government debate was underway at that time with respect to the possibility of complimenting or even replacing the military's clandestine strikes—which are disclosed to and approved by Yemeni officials—with a CIA covert action program, at least in part in order to "allow the United States to carry out operations even without the approval of Yemen's government."[53] Then, after a failed attempt by AQAP to put bombs aboard cargo jets bound for the United States in October 2010, the media reported a new round of debate at the White House concerning the desirability of launching (or in this case, reviving)[54] a CIA-operated drone strike program in Yemen—including the possibility of seeking Yemeni government approval for such strikes.[55] Insofar as this reflected a decision not to use force without notification to the Yemeni government, it might well have stemmed from the fact that President Saleh, according to two anonymous government officials, "ha[d] shown a willingness to break off cooperation if the U.S. undertakes operations on Yemeni territory without his approval."[56] A separate, contemporaneous account added that drones actually had been redeployed to Yemen already, albeit under the military rather than the CIA's control, and that Yemen already had consented to their use should appropriate targets be located.[57]

Taken as a whole, these accounts provide substantial support for the following conclusions. First, the government of Yemen is eager for a host of understandable reasons to keep its cooperation with the United States out of the public's eye. Second, the government of Yemen nonetheless not only cooperates closely with U.S. personnel in mounting its own counterterrorism operations but also permits U.S. forces (the CIA in 2002, the military more recently) to use force directly, though subject to some form of notification-and-approval system. In those circumstances, it seems likely that the United States could make out a case for having consent to use force in the al-Awlaki scenario. A final consideration requires attention, however.

Let us assume that the United States government does receive *private* consent from the government of Yemen with respect to using force on its territory to kill al-Awlaki. Is a government's private consent, meant to be withheld from its own public, adequate to discharge the Article 2(4) concern?[58] Or does international law somehow require that consent be public?[59] Some scholars have argued that, as a matter of policy even if not legal obligation, consent ought to be given publicly and explicitly.[60] There are virtues to this position from a normative viewpoint, including that it would remove doubt as to whether consent had been given.[61] This is, after all, an evidentiary issue that proved problematic in the context of the International Court of Justice's (ICJ's) consideration of the collective self-defense argument in the *Nicaragua* decision.[62] Requiring public and explicit consent would also tend to make the government of the consenting state more accountable—both domestically and internationally—for such decisions. Such accountability certainly can be viewed in positive terms, yet it also can have a substantial and undesirable cost where, as in both Yemen and Pakistan, it is likely that the domestic response would render cooperation impossible or at least far more difficult. In any event, neither the U.N. Charter nor any other instrument addresses this issue, and the case has not been made that state practice supported by *opinio juris* establishes any such requirement; the argument sounds in policy, not legal obligation.[63]

B. Does the Right of Self-Defense Apply?

In the event that a consent argument is unavailing—or becomes so in the future[64]—the question becomes whether the United States nonetheless may act in Yemen pursuant to the right of self-defense preserved in Article 51.

We might begin by asking: self-defense against whom? Certainly not the government of Yemen, which is America's ally (however imperfect or constrained) in this endeavor. Rather, the argument is that the United States seeks to engage in self-defense against either al Qaeda or AQAP. But even if we assume that the United States has suffered an armed attack triggering Article 51 self-defense rights against al Qaeda or AQAP, does it follow that the United States can exercise those rights in Yemen's territory without Yemen's consent? I consider these issues in sequence below.

1. Self-Defense against al Qaeda

If the United States invokes the right to act in self-defense in the al-Awlaki scenario, is it best to understand this in terms of defense against AQAP in particular or, instead, against al Qaeda more generally? From the U.S. perspective, emphasizing core al Qaeda has advantages and disadvantages. The advantage is the relative ease of establishing that the United States has suffered an armed attack from that group. There seems to be widespread agreement that al Qaeda's 9/11 attacks constituted an "armed attack" against the United States,[65] notwithstanding much-criticized suggestions by the International Court of Justice in other contexts to the effect that Article 51 should be read textually to refer only to armed attacks committed by states.[66] Combined with overwhelming reason to believe that al Qaeda intends further attacks (and thus that a responsive use of force would not be a mere matter of revenge), these circumstances suffice to trigger the right of the United States to use force in self-defense vis-à-vis al Qaeda. Focusing on al Qaeda as such also presents a difficulty in the al-Awlaki scenario, however, in that it requires not just linking al-Awlaki to AQAP but also linking AQAP to al Qaeda.

In light of the facts recounted above, there is room for debate regarding AQAP's status as "part of" al Qaeda. Coming to grips with the organizational structure of a clandestine, nonstate actor network of this kind is famously difficult, and international law does not necessarily provide a substantive standard by which to resolve this inquiry.[67] Even if the standard to be applied were clear, moreover, the most pertinent evidence relevant to that task is not likely a matter of public record. In these conditions, it is simply not possible to say that the U.S. government is mistaken when it asserts that AQAP is indeed part and parcel of al Qaeda itself.

That said, focusing on AQAP as an extension of al Qaeda does not provide the strongest foundation for concluding that the United States may have Article 51 rights in this context. As I explain below, the argument is stronger when one focuses directly on the course of dealings between the United States and AQAP itself.

2. Self-Defense against AQAP

Set aside self-defense arguments based on core al Qaeda and the 9/11 attacks. Can a distinct self-defense argument be mounted based directly on AQAP's own activities? This approach avoids the difficulty of establishing an adequate link between al-Awlaki and core al Qaeda, but it introduces the need to point to AQAP-specific activities as sufficient triggers for Article 51.

Has AQAP engaged in an armed attack against the United States *already*, such that there is no need to broach the question of whether international law permits the *anticipatory* use of force in self-defense in this context? Here we are assuming a categorical distinction between AQAP and core al Qaeda, and hence must focus attention on a much narrower set of attacks. Yet questions of attribution still arise. For example, may we add to AQAP's account the bombing of the *USS Cole* in Yemen in 2000, or the mortar attacks on and car bombing of the U.S. embassy in Yemen conducted by AQAP prior to its adoption of this name? In light of the lineage discussed above in section I, it seems entirely appropriate to do the latter, and at least defensible to do the former. Combined with AQAP's attempts to destroy both passenger and cargo jets bound for the United States in 2009 and 2010,[68] the United States has a strong case for claiming that it has experienced multiple armed attacks at AQAP's hands. To be sure, none have reached the intensity of the 9/11 attacks in terms of the number of resulting deaths. That is hardly dispositive, however, unless one thinks that the 9/11 attacks somehow constitute a floor in terms of the necessary number of casualties involved in the armed attack calculus. The successful attacks emanating out of al Qaeda's Yemen operations have themselves been quite deadly, and those that were foiled would have been at least as destructive. This should suffice.

3. What of Yemen's Territorial Interests?

What has been said thus far supports no more than the claim that the United States has the right to act in self-defense against both al Qaeda and AQAP in *some* location. It does not automatically follow, however, that the United States can exercise this right in any location whatsoever without respect to the rights of the state in whose territory it proposes to act. Could Yemen properly object under Article 2(4) if it were to withdraw or refuse its consent but the United States nonetheless reached into its territory to kill al-Awlaki?

On current conditions, the answer is no. To be sure, some have argued that a defending state acting under Article 51 may not attack a nonstate actor in the territory of another state unless that other state is in a legal sense responsible for the predicate attack.[69] Mary Ellen O'Connell, for example, has written that "[e]stablishing the need for taking defensive action can only justify fighting on the territory of another state if that state is responsible for the on-going attacks," and that "[i]t may well be that . . . a group launching significant, on-going attacks has no link to a state and so no state can be the target of defensive counter-attack."[70] Let us call this the "strict" position.

Critics of the strict position, in contrast, argue either that: (1) there is no need to prove that the host state is legally responsible for the actions of a nonstate actor on its territory so long as the defending state confines its response to the personnel or assets of the nonstate actor rather than the host state, or (2) the responsibility of the host state is established in any event if it fails to take reasonable steps to suppress the threat posed by the nonstate actor.[71] The Chatham House Principles of International law on the use of force in self-defense provides an illustration:

> It may be that the state is not responsible for the acts of the terrorists, but it is responsible for any failure to take reasonable steps to prevent the use of its territory as a base for attacks on other states. Its inability to discharge the duty does not relieve it of the duty. . . . Thus, where a state is *unable or unwilling* to assert control over a terrorist organisation located in its territory, the state which is a victim of the terrorist attacks would, as a last resort, be permitted to act in self-defence against the terrorist organisation in the state in which it is located.[72]

This position, notably, is consonant in important respects with the law of neutrality. Under the laws of neutrality, "the region of war does not include the territories of neutral States, and no hostilities are permissible within neutral boundaries."[73] Among other things, this means that participants in hostilities must not use the territory of a neutral for troop transit, communications, or recruiting, while the neutral state itself has a corollary obligation to stop the parties from using its territory for operational purposes.[74]

In a similar spirit, Michael Schmitt emphasizes that a state's right of territorial inviolability must be construed in light of its corresponding duty to the international community to ensure that its territory is not used as a base from which to cause harm to others. On this view, the resulting capacity of the victim state to exercise its Article 51 rights on the territory of the host state is narrow, arguably requiring an ultimatum or *demarche* with a reasonable time period for response (though one can readily imagine circumstances where time does not permit this) and certainly requiring that force be limited if possible to the nonstate actor rather than the institutions of the host state itself (though the defending state would have to be able to respond with proportional force if the host state used its military to attempt to stop the Article 51 action).[75] At the risk of oversimplification, in any event, we might collect these views under the heading of the "broad" position.[76]

Significantly, the strict and broad approaches may not be as far apart as seems at first glance. To be sure, the strict state-responsibility position appears to preclude resort to force under Article 51 in another state's territory except in circumstances of state-sponsored terrorism or its practical equivalent, thus problematically failing to account for the need of states to respond militarily in some circumstances involving genuinely independent terrorist entities.[77] Yet at least some advocates of the strict state-responsibility position endorse a critical exception to that rule. As Mary Ellen O'Connell writes, a defending state may *also* use force against a nonstate entity on another state's territory where that state "cannot control the acts of groups on its territory," even if the state would not otherwise bear legal responsibility for the nonstate actors actions.[78] This exception—which she refers to as the "failed or impotent state"[79]—has direct application to Yemen's Shabwa province, where al-Awlaki was thought to be. The writ of the central government does not truly run there, thus providing ample grounds for the United States to argue that the "failed or impotent" state exception applies and that it can as a consequence use force in self-defense against AQAP even if one demands satisfaction of the strict state-responsibility standard in other contexts.

Even with the failed/impotent state exception, of course, important gaps remain between the strict and broad positions. First, the failed/impotent state exception does not necessarily encompass capable-but-unwilling host states (i.e., states that in theory could take effective action but choose not to do so out of sympathy, fear, or otherwise).[80] Even this difference may drop away, however, should one adopt the view that harboring or supporting a terrorist organization in any event suffices to satisfy any state responsibility requirement that might then attach; that is, the difference does not matter if one moves away from the control tests advanced in *Nicaragua* and *Tadić*.[81]

Second, and more problematical, evidentiary disagreements inevitably will arise as to whether the relevant conditions have been satisfied, whatever those conditions may be. As applied to the hypothetical scenario in which Yemen reduces or eliminates its cooperation with the United States, for example, the U.S. government no doubt would argue that Yemen's writ does not effectively run to the Shabwa province and that it is not truly interested in suppressing AQAP in any event, whereas Yemen surely would deny both claims and would point to various actions undertaken against AQAP in that province as evidence. All of which would drive home the point that such disputes ultimately may turn on who if anyone gets to decide them, what standard of proof that decision maker brings to bear, and what evidence is available. For better or worse, however, "[i]nternational law has no generally-accepted law of evidence" in this circumstance,[82] nor an authoritative forum for addressing such debates (except perhaps in the limited circumstances where the U.N. Security Council overcomes obstacles to its involvement or the International Court of Justice can properly assert jurisdiction).

4. Necessity and Proportionality as Elements of Article 51 Self-Defense

Assume for the sake of argument that the right of the United States to act in self-defense under Article 51 has been triggered (whether by core al Qaeda, AQAP, or both) and that any objection Yemen may have under Article 2(4) has been resolved. The next question is whether the manner in which the United States exercises that right is constrained by any considerations inherent in the self-defense concept itself (separate and apart from the international humanitarian law (IHL) and international human rights law (IHRL) considerations that would need to be discussed in a more holistic analysis).

Neither Article 51 nor any other aspect of the U.N. Charter specifies such restraints. There is substantial consensus, however, that the *customary* right to self-defense enshrined in Article 51 requires compliance with conditions of necessity and proportionality.[83] Just what these elements require, however, and how they relate to identically named requirements associated with IHL's *jus in bello* provisions and with IHRL, is less clear.

Consider first necessity. In the context of the customary right to self-defense, this element arguably entails two distinct inquiries. According to Murphy, "the International Court of Justice and scholars typically first consider whether there are peaceful alternatives to self-defense, such as pursuing available diplomatic avenues."[84] This aspect of the necessity inquiry is primarily a function of both the host state's *willingness* and its *actual capacity* to act effectively to suppress the threat. Both conditions must be satisfied. In most states they would be insofar as al Qaeda is concerned; France, for example, is both perfectly capable and willing to act against any al Qaeda threat that might turn out to be lurking within its borders, and hence the United States could not exercise Article 51 rights there. But not every state is both willing and capable of suppressing threats. Pre–9/11 Afghanistan provides an example of a government (albeit only a de facto regime) arguably able but certainly unwilling to act against al Qaeda. Current-day Somalia provides an example of a government (such as it is) that presumably is willing yet is entirely unable to act against al Qaeda. Pakistan arguably is a mixed case, with difficult questions regarding the extent of its willingness to act in the Federally Administered Tribal Areas and, in any event, substantial doubts surrounding its capacity to do so.[85]

Even if this first aspect of the necessity inquiry is satisfied, at least some observers contend that the analysis must continue with an inquiry into whether attacking a particular target would be useful as a means of preventing further attacks.[86] This inquiry at least partially overlaps with the separate requirement of proportionality, however, and is better dealt with under that heading. Proportionality in self-defense does not require a precise identity between the scale of the predicate attack and the scale of the force the defending state intends to use,[87] but it does require some reasonable degree of relationship between them.[88] Furthermore, some take the view that as time goes by the "proper *referent* of *ad bellum* proportionality changes with the nature and scope of the conflict."[89] On this view, the measure of proportionality at some point becomes not the original attack but, rather, "the object legitimately to be achieved."[90]

How might one analyze the killing of al-Awlaki under this framework? Consider first whether killing al-Awlaki could have been deemed "necessary" in both the senses described above—that is, necessary in the sense that Yemen is unable or unwilling to act effectively to suppress the threat he poses, and separately in the sense that targeting him would advance the goal of preventing further attacks. The case for necessity at the *individual* level is relatively strong, assuming that one credits the U.S. government's claims regarding al-Awlaki's "operational" role in AQAP's violent activities. The more difficult inquiry is the case for necessity in the broader sense in which we examine Yemen's willingness and capacity to suppress AQAP. It is unclear whether the Yemeni government actually is willing to arrest al-Awlaki and otherwise to act to suppress the threat of AQAP.[91] Even if we assume that it is willing, however, that it is not enough without a corresponding capacity to effectuate the arrest. At least for the time being Yemen's weak central government appears to lack the capacity to enforce its will reliably in Shabwa (the province where al-Awlaki was thought to be) and other relatively remote provinces where AQAP members enjoy the protection of local tribes.[92] Ironically, if the United States succeeds over time in its efforts to improve the capacities of Yemen's security forces, this answer could change, undermining the self-defense argument under Article 51 in the scenario in which Yemen continues to pursue AQAP but withdraws or refuses to give its consent to some particular uses of force by the United States.[93]

Next consider the proportionality question. As an initial matter, targeted killing of particular individuals is a relatively small-scale form of self-defense in comparison to, say, regime change and occupation, and certainly in proportion to the violence AQAP has directed and attempted to direct against the United States. But what of the objection that such strikes might be counterproductive in that they might generate sympathy for AQAP and hostility toward both the U.S. and Yemeni governments (because they generate collateral damage, for example)? This is a crucial consideration from a *policy* perspective, but it is difficult if not impossible to see how it could be operationalized with any degree of rigor as a *legal* constraint. Direct U.S. uses of force in Yemen—not to mention resulting civilian deaths—no doubt stoke local grievances and play into extremist propaganda narratives, but it is far from clear how one would translate the existence of this dynamic into a quantifiable output, let alone an output that could be compared with rigor to whatever benefits flow from such attacks.[94] None of this is to say that decision makers should ignore the possibility that short-term benefits may be outweighed by long-term costs, of course. That is an entirely appropriate consideration of policy judgment.

Establishing that the United States had the right to use force against al-Awlaki in Yemen by virtue of consent or Article 51 self-defense does not by any means resolve all pertinent legal questions, of course. It resolves no more—and no less—than the objections belonging to Yemen itself under the U.N. Charter.

In COIN, the side that learns faster and adapts more rapidly—the better learning organization—usually wins.[1]

[M]aritime law, in time of war, has long been the subject of deplorable disputes;...[T]he uncertainty of the law and of the duties in such a matter, gives rise to differences of opinion between neutrals and belligerents which may occasion serious difficulties, and even conflicts;...it is consequently advantageous to establish a uniform doctrine...[and] introduce into international relations fixed principles in this respect.[2]

10

THE GAZA FREEDOM FLOTILLA: POLITICIZING MARITIME LAW IN ASYMMETRIC CONTEXTS

Corri Zoli*

IN THE ANNALS of postmodern warfare, the spectacle of Israeli *Shayetet 13* naval commandos descending on the six-vessel "Gaza Freedom Flotilla" over seventy nautical miles from land in the early morning hours of May 31, 2010, has loomed large, especially for law, policy, and security specialists. Few recent incidents have so unilaterally united international opposition and condemnation.[3]

Yet, Israel's censure stemmed not only from the nine deaths of mainly Turkish passengers aboard the Comoros-flagged MV *Mavi Marmara*—the only vessel that did not practice passive resistance. Nor was the problem only the perception that Israel Defense Forces (IDF) had used excessive force against activists delivering aid to beleaguered populations in Gaza. Instead, the raid became indelibly linked to larger regional and geopolitical ills: the souring of the sixty-year-old Israel-Turkish relationship that offered ballast against chronic Mideast instabilities; shifts in the regional balance of power resulting in closer ties between Turkey, Iran, and Syria and new political inroads for Hamas and Hezbollah; the derailed Palestinian-Israeli peace process given the aftereffects of the 2008 Gaza incursion and the 2006 Lebanese War. The legality of the Gaza blockade itself became a kind of default *casus belli* upon which both Israeli allies and adversaries, international and nongovernmental organizations, and human rights groups issued worldwide policy pronouncements.[4]

Even Hamas and Hezbollah leveraged strategic gains from the incident. Stressing the international political advantages in "getting Israel to draw international condemnation through its own use of force, rather than by attacking the country," West Bank lawmaker Aziz Dweik claimed the "Gaza flotilla has done more for Gaza than 10,000 rockets."[5] Hamas officials began to publically extol the virtues of protest, civil disobedience, and lawsuits—once deemed "women's work"—over armed resistance. "When we use violence, we help Israel win international support," officials noted, whereas if Israel uses violence, international support favors Hamas.[6] Salah Bardawil added that Hamas had begun to appreciate the role of international support for its legitimacy, that it had not claimed responsibility for a suicide bombing in years, and even had begun to denounce rocket attacks as counterproductive. Hamas once thought international support was "just empty words," Bardawil explained, but today "it is very interested in international delegations... and in bringing Israeli officials to justice through legal proceedings."[7] This lawfare approach not only opened up a sea-based front for nonstate actors and activism, it presaged the bid by Palestinian authorities for full United Nations membership in the 2011 General Assembly session, just a year later.

The dizzying array of legal, policy, and strategic agendas attached to this single incident was neither fair nor realistic—though the intensity of the response signaled the incident itself had become a flashpoint in the complex asymmetric dynamics of postmodern warfare. The Gaza Freedom Flotilla raid, thus, serves as an appropriate closing case study in the diverse subjects of this book: the interdependent legal and policy challenges of postmodern warfare; countering actors adept at grafting force with cultural, social, religious, media, and economic factors; the fine-tuning of political pressure points to achieve goals far removed from the battlefield. Moreover, the ability of this incident to galvanize even incompatible perspectives underscores the distinctive features of contemporary insurgency: the redefined battlefield in political not military terms; the eclipse of kinetics in military operations; the masking and blending of identities with protected persons; the integration of all elements of civilian and military instruments of power to achieve political aims; and the core role of "hearts and minds" and global media to the new business of warfare.[8]

To give further shape to these challenges, this chapter makes three interdependent arguments. It first shows how inherent ambiguities in the law regulating armed conflict at sea have aided in the international politicization of this incident, with implications for counterinsurgency. Second, mindful of the diverse spectrum of asymmetric actors, it explores the flotilla incident as a species of insurgent savvy in which actors learn and adapt strategies from traditional insurgents to mobilize "support for its political interests (often religiously or ethnically based)" and "generate enough violence to achieve political consequences."[9] Insofar as politics and political support—not governance proper—were the main goals of the flotilla campaign, it has imbibed insurgent strategic aims: "to weaken the control and legitimacy of an established government, occupying power, or other political authority while increasing insurgent control" by the use of all available tools, including force.[10] Last, the chapter examines how counterinsurgency doctrine itself has been framed by the law of land warfare in ways that admit biases in rules governing hostilities at sea and, further, it projects, given the likely dominance of the air-sea warfare paradigm in the near future, the need for a more robust regime for armed conflict at sea.[11] Without the international community exerting more care in standardizing these rules and their application—the basic premise of the rule of law—incidents of this kind will continue apace, along with opportunistic uses of legal norms by those with the most to gain from politicizing them.

I. GAZA FREEDOM FLOTILLA: POLITICIZING INTERNATIONAL HUMANITARIAN LAW (IHL)

Few analysts—whatever their predilections about Israel, the protracted Israeli-Palestine conflict, the Gaza blockade, even the "freedom flotilla" campaign itself—would doubt the wholesale politicization of this early morning raid, or fail to notice how quickly it became the subject of polarized international discourse. What is less clear, however, are the underlying narratives that shaped its meaning or how these very different stories embedded broader expectations about IHL norms—with implications for future operations.

Below, I discuss the twin dueling narratives emanating from a broad swath of sources and commentators—many of whom viewed themselves as dispassionate (see Figure 10.1). While all stories require protagonists, in this case, both accounts share the same main characters: Israeli forces were figured as aggressors and pro-Palestinian activists were depicted as victims and, as such, a metonymy for the Palestinian people in general. There were, however, equally important players obscured in these narratives, though they occupied key roles in the actual incident: Egypt, an equal agent in the Gaza blockade; democratically elected Hamas, which stood to benefit most from international pressure on Israel; Turkey, the source of the only ship, the *MV Mavi Marmara*, that used armed resistance; a new controversial funding source and actor, the Foundation for Human Rights and Freedoms and Humanitarian Relief (IHH), that put millions of dollars into this blockade-busting effort, among others.[12]

Rule Breaker	Defensive State
1. Violation of international law	1. Compliance with international law
2. Use of force against noncombatant humanitarians	2. Legality of Gaza blockade
3. Excessive use of force	3. Politicization of humanitarian missions
4. Attack occurred in international waters	4. Regional and geopolitical strategic challenges
5. Illegal Gaza blockade	5. Missing policy tools to confront strategic challenges

FIGURE 10.1. Summary of Dueling Narratives.

The first prong of what I term "the rule-breaker" narrative, expressed immediately following the crisis from Turkish, European, and international representatives, treats the Israeli operation as a violation of international law, even a war crime.[13] Foreign Minister Ahmet Davutoglu of Turkey at the U.N. Security Council emergency meeting, where the event was officially "condemned" prior to its investigation, called the raid "tantamount to banditry and piracy" and "murder conducted by a state."[14] Many reactions by officials in the Arab and Muslim worlds repeated these often hyperbolic uses of international law: Syrian senior adviser to President Bashar al-Assad called the attack "a terrorist crime against innocent civilians and supporters of the Palestinian people who had no weapons, no armaments with them."[15] Representatives from the Kingdom of Bahrain "strongly" condemned the Israeli navy's "barbaric attack."[16] The Arab League called an emergency meeting on June 1, 2010, to address Israel's "terrorist act," and the Organization of Islamic Cooperation (OIC) "condemn[ed] Israel for this piracy which constitutes an act of state terrorism, a gross breach of international law, and serious violation of human rights."[17] Insofar as the action was viewed as a breach of law and Israel the inveterate rule breaker, Indonesia pledged to "work closely with the international community to ensure that Israel is held accountable for its action in accordance with international law."[18]

A second set of comments in the rule-breaker narrative focused on the use of force against "a civilian humanitarian convoy," as Palestinian Prime Minister Salam Fayyad put it, adding that such a crime indicates, "once again, Israeli disrespect for the lives of innocent civilians, and its disrespect for international law."[19] Saeb Erakat, chief Palestinian negotiator, painted the picture of the raid as an attack "on civilians who came from all over the world peacefully with wheelchairs and food supplies and medical supplies."[20] A Pakistani foreign minister "strongly" condemned the "use of brazen force by Israel against a humanitarian mission that was embarked on delivering humanitarian aid to the Palestinian people in Gaza," adding that the "killing of members of this humanitarian mission, which also included women, is brutal, inhuman, and constitutes a flagrant violation of international law and norms."[21] This view of a humanitarian mission derailed by Israeli armed violence held sway among most countries, despite the Free Gaza Movement's own description of its mission as a political one designed to break the Israeli blockade against Gaza—a goal, activists said, which must trump aid delivery when such goals conflict.[22] China and Russia reiterated these concerns about humanitarian targets: Yang Tao, a counselor of the Chinese permanent mission to the United Nations and present at the fifteen-nation council's emergency meeting, expressed shock "by Israel's attack on a multinational convoy carrying international aid to Gaza which resulted in heavy casualties," and, further, "condemn[ed] Israel's actions targeting humanitarian workers and civilians."[23]

A third prong of this narrative focused on how—not whether—force was used, in this case, disproportionately and, hence, illegally. A German government representative argued that "Israeli actions should observe the fundamental principle of proportionality," a claim that presumes IHL is the operative regime in this incident, and added that "a first glance suggests this basic principle was not adhered to."[24] French President Nicholas Sarkozy likewise condemned "the disproportionate use of force" against flotilla members, and the Secretary General of the Organization of American States (OAS) noted that "[n]othing can justify excessive use of military force to attack, with serious loss of human lives, a disarmed civil convoy."[25] Without full disclosure of facts or an extant investigation, Amnesty International, like many other prominent organizations and officials, argued that Israel used "excessive force" which was "out of proportion to any threat that could have been posed by the activists."[26]

A fourth concern took issue with the operation "in the international waters by Gaza," a point echoed across the spectrum of complainants—but with few specifying which law informed this opinion beyond broad reference to "freedom of the seas." Israeli lawyers, led by Avigdor Feldman, for instance, petitioned the Israeli High Court that Israel violated the U.N. Convention on the Law of the Sea (UNCLOS) by capturing and boarding vessels in international waters (though Israel is not a signatory).[27] George Bisharat, among other legal scholars, appealed to a generalized notion of the "law of the sea," noting the flotilla "enjoyed the right of free navigation in international waters, and Israel had no legal justification to interrupt its peaceful mission."[28] Richard Falk, U.N. human rights investigator for the Palestinian territories, argued for prosecuting Israeli leaders who ordered the raid, noting that Israel attacked unarmed civilians on ships "in the high seas where freedom of navigation exists."[29] In many respects, Russia's foreign ministry official, who also emphasized concerns about international waters, neatly summarized the first four prongs of the rule-breaker narrative while reiterating international feeling: "the assault by Israeli commandos [i]s a gross violation of international law" which requires a thorough and impartial investigation, and the "[u]se of weapons against civilians and detaining ships in the open sea without any legal reason constitute obvious and gross violations of generally accepted legal standards."[30]

Little mention was made of the fact that those who condemned Israel in terms of the rule-breaker narrative often posited Israel's role in Gaza as one of occupation—a point of serious contention given Israel's military disengagement from Gaza in 2005.[31] This issue had a bearing on the invocation of UNCLOS because occupation is treated, according to Common Article 2 of the revised Geneva Conventions of 1949, as a condition of international armed conflict.[32] In this case, if respondents wished to claim Israel occupied Gaza, then an international armed conflict must exist, and the operative legal regime governing the raid would not be law of the sea but the *lex specialis* of the law of armed conflict at sea. This issue of operative regimes is explored more fully in the following section, but suffice it to say here that such confusion shows blurred boundaries between legal regimes and their appropriate triggers and selective application for purposes of political impact, not analytical clarity or administrative order and accountability.

The last element of this narrative is most revealing in its purpose and function, namely, to advocate policy change: Israel's blockade against Gaza is illegal, critics argued, and constitutes a gross violation of international law and may even amount to collective punishment and retribution.[33] Catherine Ashton, European Union high representative for foreign affairs, called for "an immediate, sustained and unconditional opening" of the Gaza blockade.[34] Yang of the Chinese permanent mission to the United Nations raised "serious concerns" over the "grim" humanitarian situation in the Gaza Strip, one that requires that "[w]e urge Israel immediately and fully to implement Security Council resolution 1860, fully open border crossings, lift the blockade on the Gaza Strip so as to allow unimpeded flow of humanitarian aid, . . . and allow local people to achieve early recovery and restore a normal life."[35] Russia called for the "earliest possible lifting of the Israeli blockade of Gaza."[36] Even the U.S. ambassador, otherwise equanimous in response, found the blockade "unsustainable."[37] The international consensus—with few exceptions—focused on undoing the blockade, albeit within a distinctive narrative line: Israel, not Egypt, which had voluntarily relaxed its own blockade directly after the incident, must end "the inhumane embargo on Gaza."[38] As Jeffrey Goldberg indicated, "[m]issing so far in all the coverage of the flotilla disaster is a bit of strategic analysis," that is, understanding how "Egypt (and the Palestinian Authority) stand in opposition to Hamas's European and Turkish enablers (in their wish to end the blockade), and sincerely wish Israel would figure out smarter ways of neutralizing the existential threat (to moderate Arabs, not Israel) that Hamas poses."[39]

This first narrative, thus, illustrates several insights: first, the convenient, even opportunistic use of legal reasoning for purposes of policy change in ways that threaten the fundamental balance at the core of IHL between state security priorities and universal humanitarian values;[40] second, the tendency for even respected representatives of the international community to deviate from legal standards and procedures for handling disputes in ways that fuel the erosion of boundaries among law, politics, and policy in the context of armed conflict; third, a lack of concern or awareness about the adequacy of existing humanitarian law to govern emergent ambiguities in asymmetric settings; and, last, the increasingly expansive interpretation of IHL as a common vocabulary for international norms and less a strictly legal regime in ways more akin to the recent history of international human rights law.[41]

If the first narrative is a story of rule breakers, their disdain for purportedly clear and decisive international law with the goal of enacting policy constraints on the derelict state, the second narrative operates under completely different, even oppositional terms. Moreover, this "defensive state" narrative, applied to Israel here, captures the typical challenges for states facing irregular warfare and adversaries.[42] The first emphasis, in direct contrast to the first story, is on Israel's lawful right to interdict the Freedom Flotilla and, more generally, its status as a law-abiding nation. Israeli authorities appealed to Article 67(a) of the nonbinding *San Remo Manual on International Law Applicable to Armed Conflict at Sea*, which provides that "[m]erchant vessels flying the flag of neutral States may not be attacked unless they: are believed on reasonable grounds to be carrying contraband or breaching a blockade, and after prior warning they intentionally and clearly refuse to stop, or intentionally and clearly resist visit, search or capture."[43]

This application of the law is backed by equally prominent legal scholars. Ruth Wedgwood, for instance, argued that "under ordinary law of armed conflict, you have a right to prevent even neutrals from shipping arms to your opposing belligerents," adding that to "say that Israel itself has to be helpless in the face of the attempt to bring in unmonitored ocean traffic" is not right—"the right of visit and search under the law of the sea, or under the law of armed conflict, can be conducted on the high seas."[44]

The second focus is on the legality of Israel's Gaza blockade, including its naval portion. Again, the referent is to the *San Remo Manual*, Section II, "Methods of Warfare," Article 98, which states: "Merchant vessels believed on reasonable grounds to be breaching a blockade may be captured. Merchant vessels which, after prior warning, clearly resist capture may be attacked."[45] As noted in the Israeli press, "the government's decision to stop the vessels in the flotilla and avert a breach in the maritime blockade of Gaza was appropriate, correct and justified"—"[t]here are no half-sieges."[46] In fact, one requirement of the use of the blockade as an instrument of warfare is that it must be "effective."[47] Reputable legal scholars and policy makers, again, defended this entirely different view: Allen Weiner, former U.S. State Department lawyer and legal counselor at the American Embassy in The Hague, now a professor at Stanford Law School, noted that the "Israeli blockade against Gaza itself is not illegal, and it's okay for Israeli ships to operate in international waters to enforce it."[48] Beyond that, he added, "Israel has a legal obligation to allow humanitarian goods into Gaza and to exercise proportionality in the use of force."[49]

A third prong directs attention to the political motives of Hamas and blockade-busting advocates who—wittingly or not—helped politicize, even militarize, a humanitarian mission. Part of the move here was to group together diverse actors, including members of the Free Gaza Movement, itself an international coalition with very different agendas, but whose members agreed, as their name suggests, on undoing Israel's Gaza blockade.[50] In fact, Gaza Freedom supporters were explicit about their extra-humanitarian political goals, issued in public statements before the event: "A violent response from Israel will breathe new life into the Palestine solidarity movement, drawing attention to the blockade."[51] Much commentary has also pointed to the "qualitative change" in the Free Gaza Movement when the controversial Turkish charity İnsani Yardım Vakf (the Foundation for Human Rights and Freedoms and Humanitarian Relief, or IHH) became involved and as sources of support shifted from Australian, British, and American to Turkish and IHH donors, who provided more than $2 million and bought the *MV Mavi Marmara* outright. Former Israeli ambassador to the United States Itamar Rabinovich argued that the new Gaza mission was "a front for a radical Islamist organization, probably with links to the ruling party in Turkey" and called the aid mission "a provocation," drawing a contrast with earlier vessels that Israel's navy had let pass.[52] Henri Barkey of Carnegie Endowment described the IHH as "an Islamist organization, as it has been deeply involved with Hamas for some time," and noted that "[s]ome of [the IHH] members went on the boat saying that they had written their last will and testament."[53]

Many U.N. representatives also expressed dismay with Hamas's role in the tragedy insofar as Hamas officials denied Gazans the transported aid ultimately brought to the Israeli port—implying that Hamas support for the flotilla had little to do with actual aid delivery. Among critics and advocates of Israeli policy alike, then, a sobering sense emerged that "Hamas will benefit politically from its role as a supporter of the flotilla," will gain from "its public calls for Israel to not interfere with the effort to break the blockade," and will use the incident to increase its "fading 'resistance' credentials."[54] As senior political leader of Hamas Ismail Haniyeh predicted in a speech two days before the incident: "If ships reach Gaza—victory; if [the ship is] terrorized by Zionists—victory," implying the win-win prospects for Hamas, no matter how Israel responded.[55] Haniyeh also predicted the flotilla's message "will reach the entire world," that the "entire world opposes the siege on the Gaza Strip," and "if Israel behaves like pirates and sea-terrorists—we will win."[56]

A fourth emphasis in the defensive state narrative is on geopolitics, both the strategic obstacles facing Israel in the region and their broader repercussions in relations with neighboring states, allies, and beyond. In the first place, strategic threats for Israel are often framed in existential terms (with states and nonstates alike), such that even local incidents may be escalated to national security status. Second, those strategic threats are often shaped by ironic relations with neighbors who, at once, benefit from the U.S.-Israel security umbrella (i.e., Jordan, Egypt, Saudi Arabia, Iraq, Afghanistan, Pakistan), especially with respect to Islamist movements and terrorist activity, but who support anti-Israeli sentiments for reasons of short-term political expediency and domestic public opinion.[57] The United States in its refusal to censure Israel echoed this strategic conundrum: President Obama called for a "credible, impartial and transparent investigation of the facts surrounding this tragedy," and Secretary of State Clinton noted that "the situation from our perspective is very difficult and requires careful, thoughtful responses from all concerned."[58] Elsewhere Clinton raised concerns that the raid would distract from more pressing security priorities requiring intense coalition building—namely, Iran's nuclear ambitions. As Anthony Cordesman put it, where Israel is both a "strategic liability" and ally, such incidents hurt American security interests generally, including in Afghanistan and Iraq, costing "us significantly in terms of both blood and treasure," and imperiling the safety and effectiveness of American soldiers as they battle Islamic extremism.[59] Others emphasized the diplomatic cost for Israel with respect to Turkey, Arab nations, the United Nations, and world public opinion.

Fifth and last, the defensive state narrative anchors its security policies in a more general cultural and historical narrative about the limits of security policy tools available to states in Israel's situation. Analysts have noted the selective scrutiny of Israel as compared to others in the region, particularly with respect to alleged human rights violations, as well as Israel's exceptionalism in applying international law, specifically Geneva Convention IV, to its history of occupation in the Palestinian territories.[60] But this element of the defensive state posture also stems from the historical weight of broader survival experiences in the face of genocide, alienation, and diasporic struggle against overwhelming odds—resonant points of reference for post–World War II Jews who built the state of Israel and defined its modern terms within these experiences. This wellspring of historical struggle is threaded together with recent trials, resulting in a metanarrative that joins past and present stories and builds identity links across time and space. Both narratives, in fact, appeal to these deeply felt historical experiences of victimhood and struggle for transcendence, often against perceived intransigent opponents or inhospitable neighbors—with all the civilizational weight that "hospitality" carries for displaced communities.[61]

Two insights are, thus, worth emphasizing with respect to the defensive state narrative. First, despite their counterpoint structure, both stories share curious points of agreement in their emotional thrusts, in their actors, and in their historical function. But, ironically, the defensive state narrative often embraces less preferential terms—whether accurate or not—to maintain a veneer of power projection under conditions of insecurity. The second insight addresses the problem of synthesis. Given these polarized stories, backed by historical feeling and appeal to the same legal norms (albeit interpreted very differently), the most balanced approach would seem to be to find the truth somewhere in the middle, to see these stories as the first awkward steps in a dialectical process toward a later, greater synthesis. But the happy medium approach will not work here because it presumes common stakes of truth (and justice), not politics and law—that is, the attempt to trump law with politics—and, thus, misses how these narratives are set up not to achieve truth or conflict resolution but to ensure their own reproduction, with each new incident supplying fodder for their future dissemination.

Taken together, then, what makes these dueling narratives interesting is not only their ability to register emergent uses of humanitarian law for contested policies or the appeal by all "sides" to the same legal regimes to make completely antithetical cases, thus, making opaque what counts as clear lines of legal authority in international conflicts, but to show the degree to which politics has invaded the law, even by arbiter organizations. In all the focus on whether Israeli or flotilla actions were legal or effective what has not been given due attention is the ambiguities in the law and, consequently, the damage done to IHL when it is politically deployed—a strategy antithetical to the history and purpose of the *jus in bello*. At issue in the flotilla raid, then, is the larger implication of the law of war in conflict dynamics themselves—with consequences for managing future conflict and postconflict situations. I turn to these issues next.

II. OF INTERNATIONAL MARITIME LAW AND STRATEGY: THE LIMITS OF THE LAW OF NAVAL WARFARE

If the Gaza flotilla raid underscores ambiguities in IHL for regulating new conflict conditions, including how these new conflict conditions politicize the law, the best indication of such effects can be seen in the contradictory opinions reached by prominent international and national investigations. These included not one but two U.N.-affiliated reports that reached completely antithetical conclusions.

The Human Rights Council's *International Fact-Finding Mission to Investigate Violations of International Law, Resulting from the Israeli Attacks on the Flotilla of Ships Carrying Humanitarian Assistance* concluded, for instance, that Israel violated both human rights and humanitarian law; it found the blockade breached the laws of war, which itself made the flotilla interception unlawful, and it further documented excessive force used by the IDF and the subsequent mistreatment of detainees in custody.[62] By contrast, the *Report of the Secretary-General's Panel of Inquiry on the May 31, 2010 Flotilla Incident* concluded that the naval blockade was "imposed as a legitimate security measure" to prevent "weapons from entering Gaza by sea," found its implementation "complied with the requirements of international law," and determined that the flotilla not only "acted recklessly in attempting to breach the naval blockade" but raised "serious questions about the conduct, true nature and objectives of the flotilla organizers, particularly IHH."[63] Similarly, national reports by Israel and Turkey were opposed in their analytics and results:[64] in summarizing these, the *Palmer Report* noted in "setting out their own account of what they believe to be the state of public international law as it applies to the incident" both national investigations "differ as widely on the applicable law as they do on what actually happened."[65]

It would be easy to put such results down to political agendas plain and simple. But the fact of the matter is that even among the most faithful analysis, confusion prevailed over the law governing the incident, particularly in maritime settings. Significant areas of legal complexity included applicable law; threshold questions (if the law of armed conflict applied, whatever the theater, Israel had to be in a state of armed conflict with Hamas); the interplay between IHL and the law governing naval blockades and belligerent occupation; proportionality analysis for both the blockade and the IDF use of force; and the combatant status of participating actors.[66] Not only were many of these questions themselves legally unsettled, approaches to them tended to blur *jus in bello* and *jus ad bellum* analysis, not to mention IHL and international human rights law, so that proportionality assessments of the IDF's use of force, for instance, were often hinged upon the legality of the Gaza blockade itself.[67]

Registering the enormity of these challenges, even the measured *Palmer Report* abandoned the task of parsing existing norms, noting with evident frustration:

> The Panel will not add value for the United Nations by attempting to determine contested facts or by arguing endlessly about the applicable law. Too much legal analysis threatens to produce political paralysis. Whether what occurred here was legally defensible is important but in diplomatic terms it is not dispositive of what has become an important irritant not only in the relationship between two important nations but also in the Middle East generally.[68]

There is something extraordinary about the ability of this small raid—an incident far short of war, with minimal casualties, a simple, if botched aid delivery—to so easily prompt this capitulation from the United Nations and to separate the organization from its public international legal responsibilities, one source of its very legitimacy.[69]

It is for these reasons that the flotilla raid must be framed not only in legal but in strategic terms, using a framework that can account for the integration of political aims in the prosecution of conflicts. To put this point differently, it is worth considering how the flotilla tragedy might have gone differently if it were approached as a counterinsurgency (COIN) exercise, one mindful of emergent legal and security complexities addressed in this volume, and one engaged in a proactive—not dismissive—engagement with emerging IHL norms in postmodern warfare more generally. In this vein, there are two principle issues addressed below: first, as mentioned, the gaps in the law regulating conduct of hostilities at sea; and second, the law of land warfare bias evident in COIN strategy in ways that underestimate maritime or "joint" approaches to contemporary warfare. Since both subjects turn on the ample threshold dilemmas generated by the flotilla incident, I turn to these first.

A. Threshold Dilemmas in Postmodern Warfare

There are good reasons to question whether the flotilla raid took place against the backdrop of a clearly defined Geneva Convention Common Article 2 international conflict (the position of the Israeli *Turkel Report*, among others) and, further, an occupation situation (the position of the *UNHRC Flotilla Report*). Two immediate problems present themselves: first, Israel officially ended its occupation of the Palestinian Territories in 2005; and, second, the convergence trend between international and noninternational armed conflict, evident in the case law of the International Criminal Tribunal for the Former Yugoslavia (ICTY) and the International Committee of the Red Cross (ICRC) customary law study, increasingly makes differences in Common Article 2 and Common Article 3 rules academic.[70]

Nevertheless, the complex thicket of "threshold" questions in the case of the raid stems in large part from the traditionally separate law-triggering paradigms for armed conflict.[71] If IHL applies "when the Geneva Conventions and their Additional Protocols say it applies, namely, in the event of armed conflict," these instruments traditionally "cover and distinguish" between two conflict types: international and noninternational armed conflict.[72] International conflict, governed by Common Article 2, occurs when states use force against other states in a situation of de facto (not necessarily declared) war,[73] and Common Article 2 also applies in "all cases of partial or total occupation of the territory of a High Contracting Party, even if the said occupation meets with no armed resistance."[74] By contrast, Common Article 3 noninternational conflicts define hostilities between states and nonstate armed groups (or between such groups), do not carry the full law of war protections, and are triggered when a threshold of violence is reached, itself determined by a more complex calculus: the temporal and geographic scope of hostilities, the frequency and intensity of attacks, and the existence of identifiable parties to the conflict.[75] As Kenneth Watkin notes, identifying "the boundaries of non-international armed conflict has never been easy," and, increasingly, "attacks by nonstate actors challenge the view of a neat division of armed conflict into the two spheres of international and non-international conflict."[76]

There is, however, little dispute over the need in humanitarian law on a hard boundary between the presence and absence of armed conflict in recognition of the gravity of that which the law permits: namely, "killing and detention without all the customary protections of due process applicable in peacetime."[77] Yet, such neat boundaries—between the presence and absence of armed conflict, international and noninternational conflicts, and their implication for cardinal questions of the status of belligerents, targeting decisions, detention procedures—have eroded in new conflict contexts and in COIN operations more specifically.[78] This is particularly true for the flotilla raid for several reasons: the poor fit of the Geneva "conflict typing" model to asymmetric warfare with nonstate actors (i.e., Hamas); the unique sui generis nature of the protracted Israeli-Palestinian conflict which, like many new wars, shifts regularly between armed conflict types; and the limits of the default state-centric international armed conflict model when contemplating rules for nonconventional conflict scenarios, such as belligerent occupation, interventions into failed states, or responsibility to protect missions.

But by far the greatest obstacle to the consistent legal assessment of the flotilla raid and incidents like it is the compounded problem of the still unsettled noninternational armed conflict regime for asymmetric settings and the lack of modern rules for the conduct of hostilities at sea.

B. COIN and Noninternational Conflict

Until 9/11, noninternational armed conflict simply meant internal civil war, and a general consensus had formed as to its "predicate triggers" derived from Common Article 3 and ICRC Commentary, as mentioned.[79] Prior to the revised Conventions, international agreements regulating "conflicts of a non-international character" simply did not exist—a concession to states' sovereignty claims. They were governed by domestic law, no matter how violent, in a regulatory gap that lasted until World War II, when the Nazi regime demonstrated auspiciously that internal atrocities could pose just as "grave threats to humanitarian principles."[80] Today, determining when the law of noninternational conflict comes into force is challenged by the dominance and variety of "complex operations" that clearly exceed the scope of internal civil war, such as interventions into failed states, transnational armed conflicts, multicoalition forces in stabilization campaigns, and so forth.[81]

Many legal scholars have thus argued for applying the de facto resort to armed force standard from the *Tadić* decision to noninternational conflicts as the best indication that the threshold from internal disturbance to armed conflict has been crossed.[82] Others have argued, more broadly, that "the best reading" of Common Article 3 itself is that "it applies [minimal protections] to all armed conflicts," including those with transnational dimensions, thus closing the regulatory gap between interstate conflicts and civil wars.[83] If the Geneva Conventions of 1949 and their Additional Protocols saw two revolutionary developments in conflict mitigation—the laws of war began to govern de facto and de jure warfare, noninternational and international armed conflict—the post–9/11 decade is witness to an equally momentous paradigm shift: the convergence of Common Articles 2 and 3 norms and protections across all armed conflicts. In fact, this move is already de rigeur in U.S. defense policy in that all Defense Department personnel are required to "comply with the law of war during all armed conflicts, however such conflicts are characterized, and in all other military operations."[84]

Without the bifurcated interstate/intrastate paradigm or the narrowed scope of Common Article 3 conflicts, prescribed by Additional Protocol II, Article 1, the advantages of this expansive reading of Common Article 3 are obvious: the laws of war are brought to bear more consistently and thoroughly on today's broadened conflict spectrum.[85] But more important, this consistent and pragmatic approach to all new conflicts would go a long way in helping to regulate a new breed of humanitarian problems that emerge as both nonstate armed groups, often transnationally organized, and COIN responses, pressurize core IHL principles by operating within civilian populations, blending in with peaceable noncombatants, and, in the case of insurgents, by avoiding direct and proportional military confrontation in favor of asymmetric attacks.[86]

C. The *San Remo Manual*: IHL Applicable to Armed Conflicts at Sea

Even more challenging for assessing the flotilla raid and similar incidents is the fact that virtually no modern treaty law has been developed for the use of force at sea—least of all in asymmetric contexts.[87] This problem has not been overlooked. As early as 1986, Resolution 7 of the 25th International Conference of the Red Cross in Geneva noted that while IHL relating to land warfare had "been reaffirmed and developed"—namely, in the revised Geneva Conventions of 1949 and the two Additional Protocols of 1977—law relating to sea warfare was "in need of reaffirmation and clarification on the basis of existing fundamental principles of international humanitarian law."[88]

To complicate matters, the three main sources of this law—treaty, state practice, and nonbinding manuals—either reiterated its undeveloped nature or attempted to fill this gap. On the early treaty side, for instance, the only applicable agreement was the Convention for the Adaptation to Maritime Warfare of the Principles of the 1864 Geneva Convention, and of the seven treaties on naval operations adopted at the second Hague Peace Conference in 1907, only Convention (No. VIII) Relative to the Laying of Automatic Submarine Contact Mines, and portions of Nos. XI and XIII, relating to restrictions on the right of capture and the rights and duties of neutral powers, have continuing significance.[89] In fact, the Second Geneva Convention of 1949 for the Protection of the Wounded, Sick, and Shipwrecked, as supplemented by some Additional Protocol I provisions, constitutes the set of treaty-based rules governing conduct of war at sea today. In short, as J. T. Roach stresses, not only does the law of naval warfare from its formal beginnings to the present comprise largely customary international law, "[t]here is still no comprehensive treaty governing naval warfare."[90]

In the absence of treaties, nonbinding instruments have been relied upon heavily for guidance from the nineteenth century onward. The source for the effective blockade rule, for instance, derives from the 1856 Treaty of Peace ending the Crimean War, which included the *Paris Declaration Respecting Maritime Law* that abolished privateering, defined customary prize rules making neutral goods (except contraband) not subject to capture, and required a blockade to be effective in preventing access to the enemy coast in order to be binding.[91] The 1868 Geneva Conference extended the 1864 Geneva Convention for the Amelioration of the Condition of the Wounded in Armies in the Field to naval forces (acceded to by the United States in 1882 but never entered into force). The customary rules of prize law were later set out in the *1909 London Declaration concerning the Laws of Naval War*, but also never entered into force, and the Institute of International Law in 1913 crafted the most comprehensive nonbinding laws of naval warfare to date, *The Law of Naval War Governing the Relations between Belligerents* (known as the *Oxford Manual of Naval War*). Naval powers have, likewise, produced their own national manuals to fill the gap. The U.S. Navy's first manual on the law of naval warfare published in 1900, for instance, was updated for World War I and II, redeveloped in 1955, given the U.N. Charter and changes in postwar state practices and technologies, and, again, in light of UNCLOS in 1982.[92] Such updates reflected changes in state practice in light of technological advances (in aircraft, sea mines, long-range missiles, and, notably, unrestricted submarine warfare) that challenged existing law, as well as developments in other areas of international law, including the U.N. Charter, the law of the sea, and in the increased coordination of land, air, and sea force operations in air law and land warfare law.

Given the fact that no existing treaty rules "specifically make the general principles of the law of war on land applicable to war at sea,"[93] an international consensus recognized this law as long overdue for "reevaluation," as Louise Doswald-Beck noted,[94] and, hence, expert meetings were convened from 1987 to 1994, resulting in one of the "only comprehensive international instruments that has been drafted on the law of naval warfare since 1913."[95] Presuming general agreement that IHL principles, such as the requirements of distinction, military objectives, and precautions in attack should be applied, the *San Remo Manual* succeeded in "modernizing" the law of war at sea by summarizing existing treaty law, clarifying recent customary law, and attempting to frame controversial issues (i.e., exclusion zones) in light of recent state practice.[96] This last aim included some progressive development in the law, though the *San Remo Manual* was never intended to replace states' or intergovernmental organizations' responsibility for developing treaty law for armed conflict at sea.[97]

In any case, the *San Remo Manual* reaffirmed the updated IHL for hostilities at sea (taking into account the Geneva Conventions of 1949 and Additional Protocol I of 1977) and, importantly, introduced applied rules for maritime contexts from that established law. These included a clarified principle of distinction rule to limit attacks to warships, auxiliaries, and merchant vessels that directly helped the military action of the enemy, for instance, based on the concept of military objective in Article 52 of Additional Protocol I.[98] Likewise, the *San Remo Manual* incorporated provisions for protected persons from Geneva Convention IV and Additional Protocol I, including a section on the status and treatment of persons recovered at sea (an improvement on traditional law, which subjected them to the captor's discipline). The *San Remo Manual* also addressed complexities proper to naval operations, such as the involvement of neutral interests to a far greater degree than in land operations; changes in defining designated regions of operations (i.e., the high seas, internal waters, international sea lanes, exclusive economic zones);[99] and the introduction of the Additional Protocol I rule prohibiting starvation during blockades, which required blockading powers to allow relief shipments (another departure from traditional law).

Nevertheless, insofar as the *San Remo Manual* was nonbinding, it remains subject to fundamental questioning, particularly in the hard to categorize circumstances surrounding the flotilla raid. Notably, the *UNHCR Flotilla Report* questioned whether the law of armed conflict even applies to conflicts at sea as "[i]t has been suggested that the United Nations Convention on the Law of the Sea (UNCLOS), by reserving the use of the high seas for peaceful purposes, has effectively outlawed acts of naval warfare on the high seas."[100] Though the authors do finally agree that there was "no consensus" on this position during Convention negotiations, and "it was certainly not accepted by the major naval powers" whose military manuals "continue to include provisions on the law of naval warfare and blockade," and even "the majority of scholarly opinion" supports "the view that the law of naval warfare continues to be potentially applicable on the high seas," there is the sense that the *San Remo Manual* is only "[o]ne attempt at codifying this law," which is "not authoritative," and its most "significant impact" is on military manuals and that it "has been expressly relied upon by Israel."[101]

The threshold questions surrounding the flotilla raid, thus, illuminate this specific gap in IHL which has received too little attention in broader discussions of adapting this law to new asymmetric challenges—namely, naval warfare and the law of armed conflict at sea. This oversight will also implicate other hard-to-categorize operations of naval irregular warfare, ranging from the U.S. Naval Special Warfare Development Group's mission targeting Osama bin Laden on May 1–2, 2011, to responses to the Lashkar-e-Taiba Attack in Mumbai in 2008, launched by sea.[102] As Doswald-Beck notes, in light of developments in methods and means of warfare at sea and major changes in international law, the appeal to "governments to co-ordinate their efforts in appropriate fora in order to review the necessity and the possibility of updating the relevant texts of IHL relating to sea warfare" was never done.[103] Hence, in the case of the flotilla raid, Israeli government lawyers' reliance in their legal reasoning on the *San Remo Manual*—while sensible—is in effect a placeholder for the need to develop law regulating use of force at sea.

Conclusion: COIN and Naval Warfare

Tim Benbow argues in a prescient essay that the "role of maritime forces often tends to be overlooked and under-rated" in especially contemporary conflicts, and in "no area is this sea blindness more apparent than counter-insurgency."[104] While the "extent to which the maritime contribution is neglected in this literature is truly striking," even among maritime specialists, Benbow notes, the reason for its absence turns on the misunderstood role that sea power itself has played in COIN operations.[105] In fact, conventional wisdom still assumes that until irregular forces have navies, maritime forces will not be centrally involved in COIN. While this consensus is beginning to change, given the recent U.S. Air Force and Navy launch of the operational "air-sea" battle concept and efforts to develop expeditionary and irregular warfare offices at the Department of the Navy, still scholars and analysts have only begun to contemplate the strategic potential of maritime irregular warfare.

Benbow goes on to specify the strengths of maritime forces in counterinsurgency operations, directly and in support of other forces. To summarize, he emphasizes five elements: (1) maritime forces' low political and diplomatic footprint compared to land-based forces, both desirable for influencing local populations and for maintaining political will among governments anxious about investing large, vulnerable assets ashore; (2) these forces' inherent mobility, responsiveness, and flexibility within and across multiple theaters of operation, given naval training but also the nature of seaborne vessels; (3) the ability of these forces to be held ready—visibly or not—in international waters and to deploy rapidly from that mobile location; (4) the logistical ease with which warships from different nations partner, more so than land or air forces, thus allowing allies to share the eminent burden of COIN operations; and (5) the fact that afloat support is often less vulnerable to nonconventional tactics and typical insurgent threats (i.e., suicide bombers, improvised explosive devices (IEDs)).[106] While Benbow is careful to stress the "joint" military contribution of any successful COIN mission where "the three services [are] not simply cooperating but working together in a unified manner," such unity of effort has been "hindered by the lack of attention paid to the contribution of maritime forces."[107]

This analysis goes some way in laying the groundwork for the still underdeveloped air-sea battle concept which, despite criticisms, is at work in joint special operations campaigns today, and poised to replace the modern air-land theory of warfare. Yet, I invoke Benbow's analysis to raise a thornier problem: if purportedly innovative COIN thinking has been largely land-based, does the Gaza flotilla incident have something to teach us about this territorial bias with respect to new paradigms of warfare and their governing regimes? Are these legal gaps in IHL—what constitutes an armed conflict with nonstate belligerents, for instance—more intensified in the context of maritime operations? There is little doubt that part of the challenge of the Freedom Flotilla raid is that we have not yet sorted out the legal means by which to define and mitigate irregular warfare on land—let alone at sea, particularly with nonstate actors.

But this incident is also pedagogical with respect to certain naiveties about COIN itself. If warfare itself remains "a violent clash of interests between organized groups characterized by the use of force [and victory] still depends on a group's ability to mobilize support for its political interests," including violence to achieve those ends, COIN operations expose—far more than conventional warfare, with its centuries of accrued courtesies and conventions—the degree to which warfare is a political phenomenon based in notions of legitimacy and hegemony. It is not only that COIN operations strip down the artifice of warfare, leaving bare Clausewitz's political dictum: war is policy by other means. It is that political power—an insurgent group's political interests and their resonance for the ruled—are all that insurgents often have; they certainly do not have developed political institutions, systems of government, rule of law, and other features associated with functional governance. This sparse reality makes the fight turn on the population in ways that place great stress on civilian acceptance of an insurgent group's authority—by force, if need be. It is, thus, nearly impossible under these circumstances, as COIN theorists have long shown, for external forces to displace existing insurgents on these political terms and, more to the point, for outsiders to adapt nonkinetic tools (i.e., ideologies, including religious, ethnic, and cultural norms and beliefs, social networks) to make security, law, and order easily felt. This chapter began by showing the implication of the law of war in conflict dynamics; it ends by suggesting, equally importantly, that the flotilla incident remains a missed opportunity to capture next-generation COIN issues, to see their emergent shape and resulting challenges, and to identify legal and strategic terms that offer some guidance for future policy makers and practitioners. There is no question, that is, that the flotilla raid is a problem of persuasive warfare, where perceptions count as much if not more than reality, where law becomes a pawn in warfare, and where the best responses, as counterinsurgency doctrine shows, blend kinetic warfare with concerns over hearts and mind, as well as postconflict rebuilding initiatives, all in a wide range of legally ambiguous conflict settings where an armed conflict may not be taken for granted. In this way too it is also clear that the flotilla incident was, at once, an instance of asymmetric conflict and asymmetric activism where some quasi noncombatants engaged in hostilities just outside the reach of existing legal parameters. It is worth emphasizing that though counterinsurgency strategy has remained land-based, post-COIN operations are not likely to be relegated to land warfare in the future.[108]

Notes

INTRODUCTION

* Director, Institute for National Security and Counterterrorism; Board of Advisers Professor of Law; Professor of Public Administration and International Affairs, Syracuse University.

1. M. Tullius Cicero, *The Speech of M. T. Cicero in Defence of Titus Annius Milo, in* THE ORATIONS OF MARCUS TULLIUS CICERO (literally translated by C. D. Yonge & B. A. London, George Bell & Sons, York Street, Covent Garden 1891). For the original Latin, *see* M. Tullius Cicero. M. Tulli, Ciceronis Orationes: Recognovit breviqve adnotatione critica instrvxit Albertus Curtis Clark Collegii Reginae Socius (Albert Curtis Clark. Oxford. e Typographeo Clarendoniano 1918). Scriptorum Classicorum Bibliotheca Oxoniensis.

2. *See generally* JOHN KEEGAN, A HISTORY OF WARFARE (New York, Knopf 1993).

3. CARL VON CLAUSEWITZ, ON WAR 101 (A. Rapoport ed., 1982).

4. U.S. War Dep't, *The War of the Rebellion: A Compilation of the Official Records of the Union and Confederate Armies*, Series III, V3, 148–64 (Washington, D.C., Gov't Printing Office 1899).

5. Toni Pfanner, *Interview with General Sir Rupert Smith*, 864 INT'L REV. OF THE RED CROSS 720 (December 2006).

6. John Kiszely, *Post-Modern Challenges for Modern Warriors*, AUSTRALIAN ARMY JOURNAL 177 (Winter 2008).

7. John Kiszely, *Learning About Counterinsurgency*, 152 RUSI JOURNAL 19 (December 2006).

8. *See* J. S. PICTET, COMMENTARY OF THE FIRST GENEVA CONVENTION FOR THE AMELIORATION OF THE CONDITION OF THE WOUNDED AND SICK IN ARMED FORCES IN THE FIELD 49 (Geneva: Int'l Comm. of the Red Cross 1952). *See also* Marco Sassoli, *Transnational Armed Groups and International Humanitarian Law, Program on Humanitarian*

Policy and Conflict Research, HARVARD UNIV., *Occasional Paper Series*, no. 6, Winter 2006, at 8–9; But S. Boelaert-Suominen, *Humanitarian Law applicable to all Armed Conflicts*, 13 JOURNAL OF INT'L LAW 619 (2000) ("In light of the ICTY jurisprudence since 1995, it can be safely concluded that the threshold suggested by the ICRC Commentary has failed to crystallise into customary law.").

9. Hamdan v. Rumsfeld, 126 S. Ct. 2749 (2006). The United States had adopted the NIAC stance. *See* Michael N. Schmitt, *Drone Attacks under the Jus ad Bellum and Jus in Bello: Clearing the "Fog of War,"* 13 Y.B. INT'L HUM. LAW 311–326, n. 32 (December 2010).

10. Prosecutor v. Tadić, Case No. IT-94-1-I, Opinion and Judgment (Int'l Crim. Trib. for the Former Yugoslavia, May 7, 1997).

11. COMISAF's Counterinsurgency Guidance, International Security Assistance Force/ United States Forces-Afghanistan, Kabul, Afghanistan, APO AE 09356 (August 2010).

12. Michael N. Schmitt, *The Vanishing Law of War: Reflections on Law and War in the 21st Century*, HARVARD INT'L REV. 64, 67 (Spring 2009).

13. *See* Pfanner, *supra* note 5.

14. *See generally* PHILIP BOBBITT, TERROR AND CONSENT (2008).

15. Hugh Strachan, *Strategy or Alibi? Obama, McChrystal and the Operational Level of War*, 5 SURVIVAL 157, 177–78 (October–November 2010).

16. Kenneth Watkin, *Stability Operations: A Guiding Framework for "Small Wars" and Other Conflicts of the Twenty-First Century?*, 39 ISRAEL Y.B. ON HUM. RTS. 412, 420 (2009).

17. Gian P. Gentile, *A Strategy of Tactics: Population-centric COIN and the Army*, PARAMETERS 5, 6–7 (Autumn 2009).

18. *See* DAVID GALULA, COUNTERINSURGENCY WARFARE: THEORY AND PRACTICE (London, Pall Mall 1964).

19. *Id.* at 12–13.

20. *See* David H. Ucko, *Counterinsurgency after Afghanistan: A Concept in Crisis*, 3:1 PRISM 3–4 (2011).

21. Carl Fine, Crispin Burke, & Michael Few, *Evolving the COIN Field Manual: A Case for Reform*, SMALL WARS JOURNAL 1 (July 22, 2011).

22. Bradley Civins, *A Civilian's Comprehensive Critique of the U.S. Army/Marine Corps Counterinsurgency Field Manual... In 5–6 Pages*, SMALL WARS JOURNAL 1 (May 5, 2011).

23. Dep't of Defense, DOD Directive 3000.05, Military Support for Stability, Security, Transition, and Reconstruction (SSTR) Operations ¶¶ 4–1, 2 (2005); U.S. Army Stability Operations Field Manual, FM 3-07 (October 2008) [hereinafter FM 3-07].

24. Joint Chiefs of Staff, Join Pub. 1-02, Dep't of Defense Dictionary of Military and Associated Terms 84 (July 15, 2011).

25. *Id.* at 174.

26. *See* U.S. Dep't of the Army, *Counterinsurgency*, Field Manual (FM) 3-24/Marine Corps Warfighting Publication (MCWP) 3-33.5, Washington, D.C.: Department of the Army (Dec. 15, 2006) [hereinafter FM 3-24].

27. *Id.* ¶ 1–24.

28. *See* David Kilcullen, *Counter-insurgency Redux*, 48: 4 SURVIVAL 111 (Winter 2006–2007).

29. *Id.* at 112.

30. *Id.* at 113.

31. Audrey Kurth Cronin, *Cyber-Mobilization: The New Levée en Masse*, 36: 2 PARAMETERS 77–87 (Summer 2006).

32. FM 3-24, *supra* note 26; FM 3-07, *supra* note 23.

33. FM 3-24, *supra* note 26, ¶ 1–180.

34. *Id.* ¶¶ 1-150–153.

35. Ganesh Sitaraman, *Counterinsurgency, The War on Terror, and The Laws of War*, 95 VA. L. REV. 1745 (2009).

36. FM 3-24, *supra* note 26, ¶¶ 5-35–41.

37. Protocol Additional to the Geneva Conventions of August 12, 1949, and Relating to the Protection of Victims of International Armed Conflicts, June 8, 1977, 1125 U.N.T.S. 3.

38. FM 3-24, *supra* note 26, ¶ 7–33.

39. *Id.* ¶ 7–37.

40. Sarah Sewall, *Introduction to the Univ. of Chicago Press Edition, A Radical Field Manual*, *in* THE U.S. ARMY/MARINE CORPS COUNTERINSURGENCY FIELD MANUAL: U.S. ARMY FIELD MANUAL NO. 3–24, xxv (Univ. Chicago Press 2007).

41. *Id.* ¶ 1–142.

42. *See Counterinsurgency in the Post-COIN Era*, WORLD POLITICS REV., Jan. 24, 2012, *available at* http://www.worldpoliticsreview.com/features/78/counterinsurgency-in-the-post-coin-era.

CHAPTER 1

* Professor, Radzyner School of Law, Interdisciplinary Center, Herzliya, Israel.

1. *Sri Lanka: UN Rights Council Fails Victims*, HUMAN RIGHTS WATCH (May 27, 2009), *available at* http://www.hrw.org/news/2009/05/27/sri-lanka-un-rights-council-fails-victims.

2. *See, e.g., Secretary-General's Address to the Commission on Human Rights*, UNITED NATIONS (Apr. 7, 2005), *available at* http://www.un.org/sg/statements/?nid=1388.

3. ECOSOC Res. 5(I), at 5, U.N. Doc E/20 (Feb. 16, 1946).

4. U.N. Secretary-General, *In Larger Freedom: Towards Development, Security and Human Rights for All, Report of the Secretary-General*, ¶ 182, U.N. Doc A/59/2005 (Mar. 21, 2005).

5. U.N. Secretary-General, *In Larger Freedom: Towards Development, Security, and Human Rights for All, Report of the Secretary-General, Add., 1 Human Rights Council, Explanatory Note by the Secretary-General*, ¶ 2, U.N. Doc A/59/2005/Add.1 (May 23, 2005) [hereinafter *Larger Freedom Explanatory Note*].

6. *A More Secure World: Our Shared Responsibility: Report of the High-level Panel on Threats, Challenges and Change*, ¶ 291, U.N. Doc A/59/565 (Dec. 2, 2004).

7. *See Larger Freedom Explanatory Note*, *supra* note 5.

8. *Summary of the open-ended informal consultations held by the Commission on Human Rights pursuant to Economic and Social Council decision 2005/217*, ¶ 68, U.N. Doc. A/59/847-E/2005/73 (June 21, 2005) (document summarizing elements of the reform of the Commission on Human Rights following a day of informal consultations in Geneva with member states, observer states, U.N. specialized agencies, intergovernmental organizations, national human rights institutions, and NGOs).

9. *Id.* ¶ 68.

10. *See infra*, notes 85–88.

11. U.N. Human Rights Council, U.N. Doc. A/RES/60/251 (Apr. 3, 2006).

12. The resolution was adopted by 170 member states against 4 (Israel, Marshall Islands, Palau, and United States), with 3 abstentions (Belarus, Iran, and Venezuela). *See* G.A. Res. 60/251, U.N. Doc. A/60/L.48 (Mar. 15 2006).

13. *See* U.N. Doc. A/RES/60/251, *supra* note 11 (emphases added). In order to restore the body's credibility and prestige, the Council was established as a subsidiary of the General Assembly—not as a subsidiary of ECOSOC as its predecessor. In terms of membership, initially the Council was conceived as a more selective body, whose members would not include serious human rights violators. Despite concerted efforts, however, to set forth reasonable membership criteria, the Council ended up having almost as many members as the old Commission (47 for the Council instead of 53 for the Commission). On the difficulty in establishing membership criteria, *see* Lawrence Moss, *Will the Human Rights Council Have Better Membership than the Commission on Human Rights?*, 13 HUM. RTS. BRIEF 5 (2006); *see also* Philip Alston, *Richard Lillich Memorial Lecture: Promoting the Accountability of Members of the New UN Human Rights Council*, 15 J. TRANSNAT'L L. & POL'Y 49 (2005). Among the new features designed to make the Council more efficient than its predecessor in ensuring respect for human rights, a new mechanism known as "universal periodic review" requires all U.N. member states (i.e., not just state members of the Council) to submit reports every four years on actions that have been undertaken to improve respect for human rights. The universal periodic review exists in parallel to the special procedures, which already existed under the Commission, and allows the Council to address country-specific situations or thematic issues. For each situation or issue, an independent expert (known as a rapporteur) or a specialized working group is designated. "Special procedures" usually entail monitoring, advising, and publishing a report on human rights situations in specific countries or territories, known as country mandates, or on major phenomena of human rights violations, known as thematic mandates. *Human Rights Bodies*, OFFICE OF THE U.N. HIGH COMMISSIONER FOR HUMAN RIGHTS, http://www.ohchr.org/EN/HRBodies/Pages/HumanRightsBodies.aspx (last visited Feb. 28, 2012). Country mandates have included Cambodia and Sudan; thematic mandates have included arbitrary detention, education, drinking water and sanitation, and the right to food.

14. *Situation of human rights in the east of the Democratic Republic of the Congo*, U.N. Doc. A/HRC/S-8/1 (Dec. 1, 2008) (calling "upon all concerned parties to comply fully with their obligations under international law, including *international humanitarian law*, human rights law and refugee law, to ensure the protection of the civilian population and to facilitate the work of humanitarian agencies; Underlines that the Government has the primary responsibility to make every effort to strengthen the protection of the civilian population and to investigate and bring to justice perpetrators of violations of human rights and *of international humanitarian law*.") (emphasis added).

15. *Special session resolution Assistance to Sri Lanka in the promotion and protection of human rights*, U.N. Doc. A/HRC/S-11/1 (May 26, 2009) (the Council reaffirmed "the obligations of States to respect human rights law *and international humanitarian law* while countering terrorism" (emphasis added)).

16. *Situation of human rights in Myanmar*, U.N. Doc. A/HRC/S-8/14 (June 18, 2008) (the Council expresses concern "about reported widespread violations of human rights and *international humanitarian law* in the Kayin State and Bago Division" (emphasis added)); *Situation of Human Rights in Myanmar*, ¶ 11, U.N. Doc. A/HRC/16/L.11 (Mar. 18, 2011) ("Strongly calls upon the Government of Myanmar to take urgent measures to put an end to the continuing grave violations of international human rights and humanitarian law").

17. *See Human rights violations emanating from Israeli military incursions in the Occupied Palestinian Territory, including the recent one in northern Gaza and the assault on Beit-Hanoun*, Preamble, U.N. Doc. A/HRC/S-3/1 (Nov. 15, 2005); *Human rights violations emanating from Israeli military attacks and incursions in the Occupied Gaza Strip*, ¶ 9, U.N. Doc. A/HRC/S-6/1 (Jan. 24, 2008); *Human rights violations emanating from Israeli military attacks and incursions in the Occupied Palestinian territory, particularly the recent ones in the occupied Gaza Strip*, Preamble, U.N. Doc. A/HRC/7/1 (Mar. 6, 2008); *Israeli settlements in the Occupied Palestinian Territory, including East Jerusalem, and in the occupied Syrian Golan*, Preamble, U.N. Doc. A/HRC/7/18 (Mar. 27, 2008); *Follow-up to resolution S-3/1: human rights violations emanating from Israeli military incursions in the Occupied Palestinian Territory and the shelling of Beit Hanoun*, ¶ 5, U.N. Doc. A/HRC/9/18 (Sept. 8–24, 2008); *The grave violations of human rights in the Occupied Palestinian Territory, particularly due to the recent Israeli military attacks against the occupied Gaza Strip*, ¶¶ 6, 9, 14, U.N. Doc. A/HRC/S-9/1 (Jan. 9 and 12, 2009) (establishing and appointing the Fact-Finding Mission on the Gaza Conflict in order "to investigate all violations of international human rights law and international humanitarian law by the occupying Power, Israel, against the Palestinian people throughout the Occupied Palestinian Territory, particularly in the occupied Gaza Strip"); *Human Rights in Palestine and Other Occupied Arab Territories, Report of the Fact-Finding Mission on the Gaza Conflict*, U.N. Doc. A/HRC/12/48 (Sept. 15, 2009); *Report of the international fact-finding mission to investigate violations of international law, including international humanitarian and human rights law, resulting from the Israeli attacks on the flotilla of ships carrying humanitarian assistance*, U.N. Doc. A/HRC/15/21 (Sept. 27, 2010); U.N. Doc. A/HRC/14/1 (June 2, 2010) (report of the fact-finding mission established to "investigate violations of international law, including international humanitarian law and human rights law, resulting from the interception of the humanitarian aid flotilla bound for Gaza on May 31, 2010 during which nine people were killed and many others injured"—which places a strong emphasis on naval warfare and humanitarian law).

18. A special session can be requested by two-thirds of the Council's members. *See* A/RES/60/251, *supra* note 11, ¶ 10. The holding of a special session in July 2006 was requested by the Representative of Tunisia to the United Nations Office at Geneva, on behalf of the Group of Arab States. *See Human Rights Situation in the Occupied Palestinian Territory*, U.N. Doc. A/HRC/S-1/1 (July 6, 2006). The letter was accompanied by signatures in support of the above-mentioned request from the following twenty-one state members of the Council: Algeria; Azerbaijan; Bahrain; Bangladesh; Brazil; China; Cuba; Gabon; India; Indonesia; Jordan; Malaysia; Mali; Morocco; Pakistan; Russian Federation; Saudi Arabia; Senegal; South Africa; Sri Lanka; and Tunisia.

19. For more information on the first special session, *see Report of the First Special session of the Human Rights Council*, U.N. Doc. A/HRC/S1-3 (July 18, 2006).

20. U.N. Doc. A/HRC/S-1/1, *supra* note 18.

21. *The Grave Situation in Lebanon Caused by Israeli A/HRC/S1-3 Military Operations*, U.N. Doc. A/HRC/S-2/1 (Aug. 11, 2006).

22. U.N. Human Rights Council, *Commission of Inquiry on Lebanon, Report: Implementation of General Assembly Resolution 60/251 of March 15, 2006 entitled "Human Rights Council Report of the Commission of Inquiry on Lebanon pursuant to Human Rights Council Resolution S-2/1,"* ¶ 5, U.N. Doc. A/HRC/3/2 (Nov. 23, 2006).

23. *Id.* ¶¶ 8, 9.

24. *Id.* ¶ 9.

25. *Darfur*, U.N. Doc. A/HRC/DEC/2/115 (Nov. 28, 2006).

26. *Situation of human rights in Darfur*, U.N. Doc. A/HRC/DEC/S-4/10 (Dec. 13, 2006).

27. *Situation of human rights in the Sudan*, U.N. Doc. A/HRC/7/16 (Mar. 27, 2008) (the Council "[e]xpresses its deep concern at the seriousness of the ongoing violations of human rights and international humanitarian law in some parts of Darfur"); *Situation of human rights in the Sudan*, U.N. Doc. A/HRC/9/17 (Sept. 18, 2008) (the Council "[e]xpresses its deep concern at the serious violations of human rights law and international humanitarian law in Darfur").

28. *See Technical Assistance and Capacity-Building, Assistance to Somalia in the field of human rights*, ¶ 8, U.N. Doc. A/HRC/RES/12/26 (Oct. 12, 2009) ("Calls on the Transitional Federal Government to fulfill its obligations under international human rights and humanitarian law and for the implementation of the relevant provisions of the Transitional Federal Charter relating to human rights"); *Assistance to Somalia in the field of human rights*, ¶ 12, U.N. Doc. A/HRC/15/L.2 (Sept. 23, 2010) ("Urges all parties in Somalia to reject and cease all acts of violence, refrain from engaging in hostilities, prevent any act likely to increase tension and insecurity and respect fully their obligations under international human rights law and international humanitarian law"); *Assistance to Somalia in the field of human rights*, U.N. Doc. A/HRC/RES/15/28 (Oct. 7, 2010); *Assistance to Somalia in the field of human rights*, ¶ 3, U.N. Doc. A/HRC/RES/17/25 (July 14, 2011) ("Calls on Somalia to fulfill its obligations under international human rights law and international humanitarian law").

29. *See, e.g., Report of the independent expert on the situation of human rights in Somalia, Shamsul Bari*, Summary, U.N. Doc. A/HRC/18/48 (Aug. 29, 2011) ("Apart from the drought and famine, the armed conflicts between Islamist insurgents and the Transitional Federal Government, supported by the troops of the African Union Mission in Somalia (AMISOM), continues to cause deaths and injury to the civilian population. Indiscriminate shelling and firing in urban areas, and suicide and improvised explosive attacks by the insurgent group Al-Shabaab, are the main causes.").

30. *See, e.g., Technical Assistance and Capacity-Building, Report of the independent expert on the situation of human rights in Somalia, Shamsul Bari*, U.N. Doc. A/HRC/13/65 (Mar. 23, 2010) (making repeated references to humanitarian law at ¶¶ 5, 40, 55, 98, 99).

31. *Report of the independent expert on the situation of human rights in Somalia, Shamsul Bari*, ¶ 80, U.N. Doc. A/HRC/15/48 (Sept. 16, 2010) (such a commission has not yet been established).

32. *Id.* ¶ 17.

33. *Civil and Political Rights, Including the Questions of Disappearances and Summary Executions, Report of the Special Rapporteur on Extrajudicial, Summary or Arbitrary Executions*, ¶¶ 18, 19, U.N. Doc. A/HRC/4/20 (Jan. 29, 2007) (emphasis added). *See also Project on Extrajudicial Executions*, U.N. Special Rapporteur on Extrajudicial Executions Handbook 7 (N.Y.U. Univ. School of Law 2010) (noting that if the U.S. position was accepted, it "would represent a significant challenge not only to the work of this mandate but, more importantly, to a significant amount of the activities undertaken by the Human Rights Council. In brief, one of the consequences would be to disable the Council in relation to a large number of situations involving armed conflicts in which it has been actively involved over the past decade and more.").

34. *See* Daniel O'Donnell, *Trends in the Application of International Humanitarian Law by United Nations Human Rights Mechanisms*, 324 INT'L REV. RED CROSS 481, 497 (1998) ("U.N. human rights mechanisms also contribute to the development of international law by interpreting and consolidating the law through their practice. In so far as international humanitarian law is concerned, interesting developments can be observed regarding the evolution of certain standards from treaty law to customary law.").

35. Theodor Meron, *The Humanization of Humanitarian Law*, 94 AM. J. INT'L L. 239, 269–70 (2000).

36. 310 Eur. Ct. H.R. (ser. A) (1996).

37. *Id.* at 13.

38. *See also* Issa and Others v. Turkey, Eur. Ct. H.R. (Application No. 31821/96) (2004) (in the context of the Turkish operation which took place in Northern Iraq in 1995).

39. Marco Odello, *Fundamental Standards of Humanity: A Common Language of International Humanitarian Law and Human Rights Law*, *in* INTERNATIONAL HUMANITARIAN LAW AND HUMAN RIGHTS LAW: TOWARDS A MERGER IN INTERNATIONAL LAW 35 (Roberta Arnold and Noëlle Quénivet eds., 2008).

40. Banković and Others v. Belgium, Eur. Ct. H.R. (Application No. 522077/99) (2001).

41. Isayeva v. Russia, Eur. Ct. H.R. (Application No. 57950/00) (2005).

42. *See, e.g.*, Aisling Reidy, *The Approach of the European Commission and Court of Human Rights to International Humanitarian Law*, 324 INT'L REV. RED CROSS 513, 516, 519 (1998) (regretting that "neither Convention body has engaged in an extensive examination of the characterization of any public emergency in terms of humanitarian law (internal disturbances and tensions versus internal armed conflict)"); Françoise Hampson, *The Relationship Between International Humanitarian Law and Human Rights Law From the Perspective of a Human Rights Treaty Body*, 90 INT'L REV. RED CROSS 549, 570–71 (2008).

43. Case 11.137, Inter-Am. Ct. H.R. Report No. 55/97, OEA/Ser/L/V/II.97, doc. 38 (1997).

44. *Id.* ¶ 156.

45. *Id.* ¶ 157.

46. *Id.* ¶¶ 157–71. *See also* Cordula Droege, *The Interplay between International Humanitarian Law and International Human Rights Law in Situations of Armed Conflict*, 40 ISR. L. REV. 310, 322 (2007).

47. *See* Liesbeth Zegveld, *The Inter-American Commission on Human Rights and International Humanitarian Law: A Comment on the Tablada Case*, 324 INT'L REV. RED CROSS 505 (1998); *see also* Emiliano J. Buis, *The Implementation of International Humanitarian Law by Human Rights Courts: The Example of the Inter-American Human Rights System*, *in* INTERNATIONAL HUMANITARIAN LAW AND HUMAN RIGHTS LAW: TOWARDS A MERGER IN INTERNATIONAL LAW, *supra* note 39, at 269, 280.

48. Third Report on the Human Rights Situation in Colombia, Inter-Am. Comm'n H.R., OEA/Ser.L/V/II.102, Doc. 9 rev. 1, ch. IV, ¶ 1 (1999), *available at* http://www.cidh.oas.org/countryrep/colom99en/table%20of%20contents.htm (emphasis added).

49. *Id.* ¶ 10.

50. *Id.* ¶ 12.

51. Report on Terrorism and Human Rights, Inter-Am. Comm'n H.R., OEA/Ser.L/V/II.116, Doc. 5 rev. 1 corr. ¶ 29 (2002). The Inter-American Commission on Human Rights has

applied IHL in a number of other situations. *See* Meron, *supra* note 35, at 269 (referring to the Dominican Republic, El Salvador, and Haiti).

52. *IACHR Request to U.S., Detainees in Guantanamo Bay, Cuba*, Inter-Am. Comm'n H.R. (July 23, 2002) (emphasis added).

53. Las Palmeras Case, Judgment on Preliminary Objections, Inter-Am. Ct. H.R. (Ser. C) No. 67 (2000).

54. On the importance of cases preceding Las Palmeras in understanding the Court's position, *see* Buis, *supra* note 47, at 284.

55. Las Palmeras Case, *supra* note 53, ¶ 2.

56. *Id.* ¶ 16.

57. *Id.* ¶¶ 33, 34 (concluding that the American Convention on Human Rights "has only given the Court competence to determine whether the acts or the norms of the State are compatible with the Convention itself, and not with the 1949 Geneva Conventions" and that "[t]he fact that States members of the Organization of American States must observe the Geneva Conventions in good faith and adapt their domestic legislation to comply with those instruments does not give the Commission competence to infer State responsibility based on them.").

58. Inter-Am. Ct. H.R. (Ser. C) No. 70 (2000).

59. *Id.* ¶ 207.

60. *Id.* ¶ 209.

61. Christine Byron, *A Blurring of the Boundaries: The Application of International Humanitarian Law by Human Rights Bodies*, 47 Va. J. Int'l L. 839, 848 (2007).

62. *Id.* at 849.

63. *See infra*, note 81.

64. *See* Byron, *supra* note 61, at 849.

65. *See* Meron, *supra* note 35, at 270.

66. O'Donnell, *supra* note 34, at 484 (citing First Report of the ONUSAL Human Rights Division, ¶¶ 17–25, A/45/1055-S/23037 (1991)).

67. *Protection of cultural rights and property in situations of armed conflicts*, Preamble, U.N. Doc. A/HRC/RES/6/1 (Sept. 27, 2007).

68. *Protection of human rights of civilians in armed conflict*, Preamble, ¶¶ 1, 2, 4, U.N. Doc. A/HRC/9/9 (Sept. 18, 2008).

69. Office of the High Commissioner for Human Rights, *Question of the Violation of Human Rights and Fundamental Freedoms in any Part of the World, With Particular Reference to Colonial and Other Dependent Countries and Territories, Extra-Judicial, Summary or Arbitrary Executions, Report by the Special Rapporteur, Mr. Bacre Waly Ndiaye, submitted pursuant to Commission on Human Rights Resolution 1995/73*, ¶ 480, E/CN.4/1996/4 (Jan. 25, 1996).

70. *Civil and political rights, including the questions of: torture and detention, Letter dated April 2, 2003 from the Permanent Mission of the United States of America to the United Nations Office at Geneva addressed to the secretariat of the Commission on Human Rights*, U.N. Doc. E/CN.4/2003/G/73 (Apr. 7, 2003).

71. Office of the High Commissioner for Human Rights, *Civil and Political Rights, including the Questions of Disappearance and Summary Executions*, ¶ 43, U.N. Doc. E/CN.4/2005/7 (Dec. 22, 2004).

72. *Letter of May 4, 2006 from the United States to the Special Rapporteur on Extrajudicial, Summary, or Arbitrary Executions*, 344, U.N. Doc. A/HRC/4/20/Add.1 (Mar. 12, 2007).

73. *Id.* at 345.

74. *Id.* at 346.

75. *See also* in this sense, *Allegation letter sent on August 26, 2005*, ¶¶ 264–65, U.N. Doc. E/CN.4/2006/53/ADD.1 (Mar. 27, 2006); *Letter of April 22, 2003*, U.N. Doc. E/CN.4/2003/G/80 (Apr. 22, 2003).

76. *See U.S. Delegation Response to Oral Questions from the Members of the Committee*, U.S. Dep't of State (July 18, 2006) ("our view is simply that U.S. detention operations in Guantanamo, Afghanistan, and Iraq are part of ongoing armed conflicts and, accordingly, are governed by the law of armed conflict"); *see also* U.S. Periodic Report to the Human Rights Committee, ¶¶ 3, 130 (2000).

77. Coard et al. v. United States, Case 10.951, Inter-Am. Ct. H.R., Report No. 109/99, ¶ 38 (Sept. 29, 1999).

78. On Guantanamo Bay Precautionary Measures, Inter-Am. Comm'n H.R., Res. No. 2/06 (2006), *available at* http://www.cidh.oas.org/resolutions/resolution2.06.htm ("The United States has stated, among other things, that the IACHR lacks jurisdiction due to the fact that international humanitarian law and not international human rights law is the applicable regime in this matter."); *Statement by Andre Surena at a hearing of the Inter-American Commission on Human Rights on issues relating to prisoners detained at Guantanamo Bay, Cuba*, U.S. Dep't of State (Oct. 20, 2003), *available at* http://www.state.gov/s/l/2003/44385.htm; Precautionary Measures 2002, Inter-Am. Comm'n H.R., ¶ 80, *available at* http://www.cidh.org/medidas/2002.eng.htm. *See also* Brian T. Tittemore, *Guantanamo Bay and the Precautionary Measures of the Inter-American Commission on Human Rights: A Case for International Oversight in the Struggle against Terrorism*, 6 Hum. Rts. L. Rev. 378, 387 (2006).

79. *See, e.g., Summary Record of the 627th meeting: United Kingdom of Great Britain and Northern Ireland*, ¶ 22, U.N. Doc. CAT/C/SR.627 (Nov. 24, 2005); *Conclusions and Recommendations: United Kingdom of Great Britain and Northern Ireland*, ¶ 4(b), U.N. Doc. CAT/C/CR/33/3 (Dec. 10, 2004).

80. *See Implementation of the International Covenant on Economic, Social and Cultural Rights, Second periodic report submitted by States parties under articles 16 and 17 of the Covenant, Addendum, Israel*, ¶ 5, U.N. Doc. E/1990/6/Add.32 (Oct. 16, 2001) (the Committee added that "without prejudice to its basic position, Israel has been willing—and in fact, has done so in the context of its oral presentation of its initial report—to cooperate with the Committee and provide relevant information to the extent possible, with regard to the exercise of those powers and responsibilities, which according to the agreements reached with the Palestinians, continue to be exercised by Israel in the West Bank and the Gaza Strip.") *Id.* ¶ 8. *See also Implementation of the International Covenant on Economic, Social and Cultural Rights, Additional information submitted by States parties to the Covenant following the consideration of their reports by the Committee on Economic, Social and Cultural Rights, Addendum, Israel*, ¶ 2, U.N. Doc. E/1989/5/Add. 14 (May 14, 2001) ("Israel has consistently maintained that the Covenant does not apply to areas that are not subject to its sovereign territory and jurisdiction.... Accordingly, in Israel's view, the Committee's mandate cannot relate to events in the West Bank and the Gaza Strip, inasmuch as they are part and parcel of the context of armed conflict as distinct from a relationship of human rights."); *Committee on Economic, Social and Cultural Rights, Report on the Twenty-Fifth, Twenty-Sixth and Twenty-Seventh Session (April 23–May 11, 2001, August 13–31, 2001, November 12–30, 2001)*, Annex VI, U.N. Doc. E/2002/22-E/C.12/2001/17

(Nov. 20, 2001); *Statement by the Permanent Representative of Israel to the United Nations Office at Geneva addressed to the Committee on Economic, Social and Cultural Rights* (Aug. 17, 2001, Twenty-Sixth session) ("In addition to the serious procedural reservations I have emphasized, I would like to conclude by reiterating Israel's reservation with regard to the Committee's handling of events in the West Bank and the Gaza Strip. It has consistently been Israel's position that consideration of questions of human rights in these territories is not within the mandate of the Committee, as such territories fall within the context of armed conflict and international humanitarian law."); *Consideration of reports submitted by states parties under Articles 16 and 17 of the Covenant, Concluding observations of the Committee on Economic, Social and Cultural Rights, Israel,* ¶ 12, U.N. Doc. E/C.12/1/Add. 69 (Aug. 31, 2001) ("The Committee rejects the State party's assertion regarding the distinction between human rights and humanitarian law under international law to support its argument that the Committee's mandate 'cannot relate to events in the Gaza Strip and West Bank.' The Committee reminds the State party that even during armed conflict, fundamental human rights must be respected and that basic economic, social and cultural rights as part of the minimum standards of human rights are guaranteed under customary international law and are also prescribed by international humanitarian law."); *Consideration of reports submitted by States parties under articles 16 and 17 of the Covenant, Concluding observations of the Committee on Economic, Social and Cultural Rights, Israel,* ¶ 15, U.N. Doc. E/C.12/1/Add.90 (May 23, 2003) ("[t]he Committee is deeply concerned at the insistence of the State party that, given the circumstances in the occupied territories, the law of armed conflict and humanitarian law are considered as the only mode whereby protection may be ensured for all involved, and that this matter is considered to fall outside the sphere of the Committee's responsibility.").

81. *Consideration of reports submitted by States parties under article 40 of the Covenant, Concluding Observations of the Human Rights Committee, Israel,* ¶ 10, U.N. Doc. CCPR/C/79/Add.93 (Aug. 18, 1998); *Consideration of reports submitted by States parties under article 40 of the Covenant, Second Periodic Report, Addendum, Israel,* U.N. Doc. CCPR/C/ISR/2001/2 (Dec. 4, 2001) (reiterating the same position); *Consideration of Reports Submitted by States Parties Under article 40 of the Covenant, Concluding Observations of the Human Rights Committee, Israel,* ¶ 11, U.N. Doc. CCPR/CO/78/ISR (Aug. 21, 2003) ("The Committee has noted the State party's position that the Covenant does not apply beyond its own territory, notably in the West Bank and in Gaza, especially as long as there is a situation of armed conflict in these areas.").

82. *Documents submitted in compliance with a special decision of the Committee, Israel,* U.N. Doc. CERD/C/282 (May 3, 1995).

83. *Committee on the Rights of the Child, Summary record of the 829th meeting: Israel,* ¶¶ 39–42, U.N. Doc. CRC/C/SR.829 (Oct. 10, 2002) ("The international law on armed conflicts was the body of law best suited to regulating the situation in the West Bank and the Gaza Strip.").

84. Philip Alston, Jason Morgan-Foster, and William Abresch, *The Competence of the UN Human Rights Council and its Special Procedures in Relation to Armed Conflicts: Extrajudicial Executions in the "War on Terror,"* 19 Eur. J. Int'l L. 183, 199 (2008).

85. Situation of human rights in Kuwait under Iraqi occupation, Comm'n H.R., Res. 1992/60 (Mar. 3, 1992).

86. *See, e.g.,* The situation of human rights in the territory of the former Yugoslavia, Comm'n H.R., Res. 1992/S-1/1 (Aug. 14, 2002); Situation in the territory of the former Yugoslavia, Comm'n H.R., Res. 1993/7 (Feb. 23, 1994); The situation of human rights in the territory of

the former Yugoslavia: Violation of human rights in Bosnia and Herzegovina, Croatia and the Federal Republic of Yugoslavia (Serbia and Montenegro), Comm'n H.R., Res. 1994/72 (Mar. 9, 1994).

87. *See, e.g.*, Situation of human rights in Rwanda, Comm'n H.R., A/RES/49/206 (1994); *Id.* Res. 1995/91 (1995).

88. Human Rights Situation in Southern Lebanon and West Bekaa, Comm'n H.R., Res. 1996/68 (Apr. 23, 1996).

89. *See* Meron, *supra* note 35, at 269 and n. 180 (arguing that special rapporteurs have turned to international humanitarian law—in particular to Common Article 3 of the Geneva Conventions—when alleging violations against nonstates, and giving the example of the Special Rapporteur's consideration of the Sudan People's Liberation Army in Sudan).

90. Alston, *supra* note 84.

91. *See supra* note 65.

92. Int'l Comm. of the Red Cross, *28th International Conference of the Red Cross and Red Crescent, International Humanitarian Law and the Challenges of Contemporary Armed Conflicts* (Dec. 2–6, 2003), at 22, *available at* http://www.icrc.org/eng/assets/files/other/ihlcontemp_armedconflicts_final_ang.pdf [hereinafter ICRC Report].

93. Byron, *supra* note 61, at 847; *see also* Arnold and Quénivet, *supra* note 39, at 9.

94. *See* Reidy, *supra* note 42, at 528.

95. *See* ICRC Report, *supra* note 92, at 58–59 (also noted as an issue was the "lack of adequate knowledge of international humanitarian law by members of the human rights bodies").

96. Office of the United Nations High Commissioner for Human Rights, *Letter dated November 30, 2006*, 355, *reproduced in Communications to and from Governments*, A/HRC/4/20. Add.1 (emphasis added).

97. Noam Lubell, *Challenges in Applying Human Rights Law to Armed Conflict*, 87 Int'l Rev. Red Cross 737, 742 (2005).

98. *See* Lubell, *supra* note 97, at 743.

99. Human Rights Council, *Code of Conduct for Special Procedures Mandate-holders of the Human Rights Council*, art. 7, U.N. Doc. A/HRC/5/2 (June 18, 2007) (emphasis added) (unsurprisingly, this article was criticized by Philip Alston in Philip Alston, *Hobbling the Monitors: Should U.N. Human Rights Monitors be Accountable?*, 52 Harv. J. Int'l L., 561, 601 (2011)).

100. *See, e.g.*, Lubell, *supra* note 97.

101. *See HRC President of the 6th Cycle*, Office of the U.N. High Commissioner for Human Rights, http://www.ohchr.org/EN/HRBodies/HRC/Pages/Presidency.aspx (last visited Feb. 28, 2012).

102. *Id.* (President of the Council (Laura Dupuy Lasserre) and one of the members of the Bureau (H.E. Mr. Christian Strohal from Austria)).

103. *Id.* The bios of the Council's officers are only available for 2011–2012 and 2010–2011. As noted above, for the year 2011–2012, one member of the Council's bureau had experience in IHL. The year before, none of the Bureau's members had any experience in IHL.

104. U.N. Doc. A/59/565, *supra* note 6, ¶ 286 ("In the first half of its history, the Commission was composed of heads of delegation who were key players in the human rights arena and who had the professional qualifications and experience necessary for human rights work. Since then this practice has lapsed. We believe it should be restored, and we propose members of the

Commission on Human Rights designate prominent and experienced human rights figures as the heads of their delegations.").

105. Byron, *supra* note 61, at 882.

106. *See* Byron, *supra* note 61, at 882–83 ("if human rights bodies are to continue to deal with situations involving armed conflict, and furthermore to apply humanitarian law…then it is vital that an effort is made to employ more members with humanitarian law expertise."); Hampson, *supra* note 42, at 571.

107. Presentation by Daniel Reisner, *What You See Depends on Where You Sit* (June 2009) (on file with author).

108. In this sense, *see also* W. Michael Reisman, *The View from the New Haven School of International Law*, Faculty Scholarship Series, Paper 867 (1992) (emphasizing the importance of the decision-maker's "observational standpoint").

109. *See, e.g.*, O'Donnell, *supra* note 34, at 493 ("There appears to be a risk that UN human rights mechanisms may arrive at different conclusions regarding the question of whether an armed conflict exists in a given country."). In this sense, *see also* Byron, *supra* note 61, at 893 (highlighting the risk that "this lack of expertise may lead a human rights body to hold that a non-international armed conflict exists when humanitarian law experts would deny that the situation was more than internal disturbances."); Lubell, *supra* note 97, at 743.

110. *See supra*, section II.

111. Yuval Shany, *Human Rights and Humanitarian Law as Competing Legal Paradigms for Fighting Terror*, *in* HUMAN RIGHTS AND HUMANITARIAN LAW, COLLECTED COURSES OF THE ACADEMY OF EUROPEAN LAW 13 (Orna Ben-Naftali ed., Oxford Univ. Press 2011).

112. *Id.* at 29.

113. *Id.* (emphasis in original).

114. *Id.* at 30.

115. *Id.* at 31.

116. Laurie R. Blank, *The Application of IHL in the Goldstone Report: A Critical Commentary*, 12 Y.B. OF INT'L HUMAN. LAW 347 (2009).

117. *Id.* at 394.

118. *Id.* at 397.

119. *See supra* note 106; Hampson, *supra* note 42, at 571 (suggesting to provide training in IHL in nonspecialized bodies).

120. Hampson, *supra* note 42, at 572. In this sense, *see also* O'Donnell, *supra* note 34, at 485 ("What does not yet exist, and would be useful, is a forum where UN human rights experts and specialists in international humanitarian law could periodically engage in an informal exchange of views on issues of common concern.").

121. *See* Arnold and Quénivet, *supra* note 39, at 1.

122. *See* Blank, *supra* note 116, at 377. *See also* Robert Sloane, *The Cost of Conflation: Preserving the Dualism of Jus ad Bellum and Jus in Bello in the Contemporary Law of War*, 47 YALE J. INT'L L. 34 (2009); Eyal Benvenisti, *The Law on Asymmetric Warfare*, *in* LOOKING TO THE FUTURE: ESSAYS ON INTERNATIONAL LAW, *in* HONOR OF W. MICHAEL REISMAN (Mahnoush H. Arsanjani, et al. eds., 2010).

123. *See* Yuval Shany, *Human Rights and Humanitarian Law as Competing Legal Paradigms for Fighting Terror*, *in* XIX/1 COLLECTED COURSES OF THE ACADEMY OF EUROPEAN LAW, HUMAN RIGHTS AND INTERNATIONAL HUMANITARIAN LAW 13 (Orna Ben-Naftali ed.,

Oxford Univ. Press 2011); *see also* Evan Criddle, *Proportionality in Counterinsurgency: A Relational Theory*, 87 NOTRE DAME L. REV. (forthcoming 2012).

124. O'Donnell, *supra* note 34, at 501 ("The increasing application of humanitarian law by U.N. human rights mechanisms is, perhaps, the inevitable consequence of years of promoting the idea that human rights law and humanitarian law are complementary and dedicated to the same ultimate objective." In addition, he notes, it is in situations of armed conflicts that the most serious and widespread types of human rights violations occur.).

125. Cordula Droege refers to these common goals as a "common humanist ideal." Droege, *supra* note 46, at 312 (though she also highlights the different theoretical foundations and motivations at the roots of the two systems). *See also* Buis, *supra* note 47, at 269 ("[b]oth branches of law share a basic hardcore of well-defined rules, which aim at the protection of the human person and constitute a homogenous legal *corpus*.").

126. *See, e.g.*, Adam Roberts, *Transformative Military Occupation: Applying the Laws of War and Human Rights*, 100 AM. J. INT'L L. 580, 589 (2006).

127. *See* Orna Ben-Naftali and Yuval Shany, *Living in Denial: The Application of Human Rights in the Occupied Territories*, 37 ISR. L. REV. 17, 47 (2003) (quoting the President of the Geneva Conference during the signing ceremony "our texts are based on certain fundamental rights . . . the Universal Declaration of Human Rights and the Geneva Conventions are both derived from one and the same ideal.").

128. *See Human Rights in Armed Conflicts, Resolution XXIII adopted by the International Conference on Human Rights, Tehran*, GA Res. 23 (May 12, 1968) ("peace is the underlying condition for the full observance of human rights and war is their negation;" "even in periods of armed conflict, humanitarian principles must prevail.").

129. Legality of the Threat or Use of Nuclear Weapons, Advisory Opinion, 1996 I.C.J. ¶ 25 (July 8).

130. *See generally* Criddle, *supra* note 123.

131. *Id.*

132. Exceptions include Shany, *supra* note 111; Blank, *supra* note 116.

133. *See generally* Thomas Buergenthal, *The Evolving International Human Rights System*, 100 AM. J. INT'L L. 783 (2006).

134. *See States accepting the competence of the International Humanitarian Fact Finding Commission based on Article 90 of the Protocol Additional to the Geneva Conventions of August 12, 1949, and relating to the Protection of the Victims of International Armed Conflicts of June 8, 1977 (Protocol I), available at* http://www.ihffc.org/Files/en/country_list_29.11.2011.pdf (including Canada, Germany, Australia, and the United Kingdom) (in chronological order).

135. *See The International Humanitarian Fact-Finding Commission (the Article 90 Commission), available at* http://www.ihffc.org/Files/en/pdf/article%2090_oct2005_keith_engl.pdf.

136. On this point, *see Report of the Work of the IHFFC on the Occasion of its 20th Anniversary* (Oct. 21, 2011), *available at* http://www.ihffc.org/Files/en/news/ihffc_report2011_en.pdf, at 17 [hereinafter *Report*].

137. *See About us—In a few words*, INT'L HUMAN. FACT-FINDING COMM'N, *available at* http://www.ihffc.org/index.asp?Language=EN&page=aboutus_general. This is true even though the Commission has offered its services to states on a number of occasions. *See Report*, *supra* note 136, at 14.

138. *See supra*, section III, discussion of the case law of the European Court of Human Rights.

139. *See* Criddle, *supra* note 123; Katarina Mansson, *Implementing the Concept of Protection of Civilians in the Light of International Humanitarian Law and Human Rights Law: the Case of MONUC*, *in* INTERNATIONAL HUMANITARIAN LAW AND HUMAN RIGHTS LAW: TOWARDS A NEW MERGER IN INTERNATIONAL LAW, *supra* note 39, at 563.

140. Buis, *supra* note 47, at 292.

141. Geoffrey Corn, *Mixing Apples and Hand Grenades: The Logical Limit of Applying Human Rights Norms to Armed Conflict*, 1 J. INT'L HUMAN. LEGAL STUDIES 52, 56 (2010); *see* William Banks, Introduction to this volume.

142. Though the creation of an IHL-specific enforcement mechanism has been raised, it unfortunately does not offer an immediate and urgently needed solution to the issues brought up in this chapter. *See* Hampson, *supra* note 42, at 572.

143. Buis, *supra* note 47, at 293.

CHAPTER 2

* Associate Professor of Law, Syracuse University College of Law. This chapter is a condensed version of *Proportionality in Counterinsurgency: A Relational Theory*, 87 NOTRE DAME L. REV. 1073 (2012).

1. For a general introduction to these debates, *see* INTERNATIONAL HUMANITARIAN LAW AND INTERNATIONAL HUMAN RIGHTS LAW (Orna Ben-Naftali ed., 2011).

2. Additional Protocol I to the Geneva Conventions of August 12, 1949, Relating to the Protection of Victims of International Armed Conflicts art. 51(5)(b), June 8, 1977, 1125 U.N.T.S. 3 [hereinafter Additional Protocol I].

3. *See* U.N Human Rights Council, Rep. of the Special Rapporteur on Extrajudicial, Summary or Arbitrary Executions, at 11, U.N. Doc. A/HRC/14/24/Add.6 (May 28, 2010) (prepared by Philip Alston) [hereinafter U.N. Report on Targeted Killing] ("Lethal force under human rights law is legal if it is strictly and directly necessary to save life.").

4. Whether IHL's current formulation of the principles of distinction, necessity, and proportionality satisfy the relational principles of noninstrumentalization and domination is a question I defer for future consideration.

5. *See* Michael N. Schmitt, *Fault Lines in the Law of Attack*, *in* TESTING THE BOUNDARIES OF INTERNATIONAL HUMANITARIAN LAW 277 (Susan C. Breau & Agniesszka Jachec-Neale eds., 2006).

6. Additional Protocol I, *supra* note 2, art. 51(2).

7. *Id.* art. 57(4).

8. *Id.* art. 51(5)(b).

9. *See* Geneva Convention (III) Relative to the Treatment of Prisoners of War art. 3, Aug. 12, 1949, 6 U.S.T. 3316, 75 U.N.T.S. 135 [hereinafter Geneva Convention III]; Additional Protocol I, *supra* note 2, arts. 40–41.

10. *Id.* art. 1(4).

11. Additional Protocol II to the Geneva Conventions of August 12, 1949, Relating to the Protection of Victims of Non-International Armed Conflicts art. 1(1), June 8, 1977, 1125 U.N.T.S. 609 [hereinafter Additional Protocol II].

12. *See* Leslie C. Green, The Contemporary Law of Armed Conflict 83 (3d ed. 2008). The United States is one of several states that have not ratified Additional Protocol II.

13. Additional Protocol II, *supra* note 11, at 1449–50, ¶ 4772.

14. *Id.* at 1490, ¶ 4772.

15. *See, e.g.*, Michael N. Schmitt et al., The Manual on the Law of Non-International Armed Conflict with Commentary § 2.1.1.4 (2006), *reprinted in* 36 Isr. Y.B. Hum. Rts. (Special Supplement) (2006).

16. *See* Prosecutor v. Tadić, Case No. IT-94-1-l, Decision on the Defence Motion for Interlocutory Appeal on Jurisdiction, ¶¶ 119, 127 (Int'l Crim. Trib. for the Former Yugoslavia Oct. 2, 1995).

17. Int'l Comm. of the Red Cross, Customary International Humanitarian Law Rules, at xxxv (Jean-Marie Henckaerts & Louise Doswald-Beck eds., 2005) [hereinafter ICRC Report], *available at* http://www.icrc.org/customary-ihl/eng/docs/v1 (last visited Nov. 23, 2011).

18. *See, e.g.*, William Abresch, *A Human Rights Law of Internal Armed Conflict: The European Court of Human Rights in Chechnya*, 16 Eur. J. Int'l L. 741, 748–49 (2005); John B. Bellinger III & William J. Haynes II, *A U.S. Government Response to the International Committee of the Red Cross Study Customary International Humanitarian Law*, http://www.icrc.org/eng/assets/files/other/irrc_866_bellinger.pdf (raising objections to the ICRC's methodology).

19. *See* Council of Europe, Convention for the Protection of Human Rights and Fundamental Freedoms art. 2(2), Nov. 4, 1950, E.T.S. No. 5, 213 U.N.T.S. 221 [hereinafter European Convention on Human Rights]; Human Rights Committee, General Comment No. 6, ¶ 3 HRI/GEN/1/Rev.6 (1982) ¶ 129.

20. *See, e.g.*, Khatsiyeva and Others v. Russia, App. No. 5108/02 Judgment, Jan. 17, 2008.

21. Additional Protocol I, *supra* note 2, art. 51(5)(b) (emphasis added).

22. *See* David S. Koller, *The Moral Imperative: Toward a Human Rights-Based Law of War*, 46 Harv. Int'l L.J. 231, 260–61 (2005); Kenneth Watkin, *Controlling the Use of Force: A Role for Human Rights Norms in Contemporary Armed Conflict*, 98 Am. J. Int'l L. 1, 32 (2004).

23. Isayeva v. Russia, 41 Eur. Ct. H.R. 847, ¶ 175 (2005).

24. *Id.*

25. *See* Yoram Dinstein, The Conduct of Hostilities Under the Law of International Armed Conflict 29 (2d ed. 2010); Int'l Comm. of the Red Cross, *Interpretive Guidance on the Notion of Direct Participation in Hostilities Under International Humanitarian Law*, 90 Int'l Rev. Red Cross 991, 1007 (2008).

26. *See* McCann and Others v. United Kingdom, 21 Eur. Ct. H.R. 97, ¶¶ 149–50 (1995).

27. Legality of the Threat or Use of Nuclear Weapons, Advisory Opinion, 1996 I.C.J. 226 (July 8).

28. *Id.* at 240, ¶ 25.

29. Abella v. Argentina, Case 11.137, Inter-Am. Comm'n H.R., Report No. 55/97, OEA/Ser.L./V/II.9, doc. 6 rev. ¶ 271 (1998).

30. *Id.* ¶¶ 1–3.

31. *Id.* ¶ 161.

32. *Id.* ¶ 154.

33. *Id.* ¶ 179.

34. Legal Consequences of the Construction of a Wall in the Occupied Palestinian Territory, Advisory Opinion, 2004 I.C.J. 136, 178 (July 9). [hereinafter Palestinian Wall]

35. *See* Jean-Marie Henckaerts, *Concurrent Application of International Humanitarian Law and Human Rights Law: A Victim Perspective*, *in* INTERNATIONAL HUMANITARIAN LAW AND HUMAN RIGHTS LAW 237, 249 (Roberta Arnold & Noelle Quenivet eds., 2008); William A. Schabas, *Lex Specialis? Belt and Suspenders? The Parallel Operation of Human Rights Law and the Law of Armed Conflict, and the Conundrum of* Jus ad Bellum, 40 ISR. L. REV. 592, 596–97, 611 (2007).

36. *See, e.g.*, Isayeva, Yusupova, and Bazayeva v. Russia, 41 Eur. H.R. Rep. 39 (2005) [hereinafter *Isayeva I*]; Isayeva v. Russia, 41 Eur. H.R. Rep. 38 (2005) [hereinafter *Isayeva II*]; Bámaca-Velásquez v. Guatemala, Merits and Judgment, Inter-Am Ct. H.R. (ser. C) No. 70, ¶ 209 (Nov. 25, 2000); Las Palmeras v. Colombia, Merits and Judgment, Inter-Am. Ct. H.R. (ser. C) No. 67 ¶ 33 (Feb. 4, 2000); Kaya v. Turkey (no. 22535/93), 2000-III, Eur. Ct. H.R. (2000).

37. *Isayeva I*, 41 Eur. H.R. Rep. 39, ¶ 13.

38. *Id.* ¶ 29.

39. *Id.* ¶¶ 167, 169, 171 (citing, inter alia, McCann and Others v. United Kingdom (ser. A) No. 324, 21 Eur. H.R. Rep. 39, 45–46, 57 (1995)).

40. *Id.* ¶¶ 178–79.

41. *See, e.g.*, Human Rights Committee, Aug. 4–5, 2003, *Concluding Observations*, ¶ 11, U.N. DOC. CCPR/CO/78/ISR (Aug. 21, 2003), *available at* http://www.unhchr.ch/tbs/doc.nsf/(Symbol)/CCPR.CO.78.ISR.En?OpenDocument (last visited Nov. 23, 2011); Christian Tomuschat, *Human Rights and International Humanitarian Law*, 21 EUR. J. INT'L L. 15, 20 (2010).

42. *See, e.g.*, Geoffrey Corn, *Mixing Apples and Hand Grenades: The Logical Limit of Applying Human Rights Norms to Armed Conflict*, 1 INT'L HUMANITARIAN LEGAL STUD. 52 (2010).

43. *See, e.g.*, Evan J. Criddle & Evan Fox-Decent, *A Fiduciary Theory of Jus Cogens*, 34 YALE J. INT'L L. 331 (2009); Evan Fox-Decent & Evan J. Criddle, *The Fiduciary Constitution of Human Rights*, 15 LEGAL THEORY 301 (2009); Helen Stacy, *Relational Sovereignty*, 55 STAN. L. REV. 2029 (2003).

44. *See* Paul B. Miller, *A Theory of Fiduciary Liability*, 56 MCGILL L.J. 235 (2011).

45. *See, e.g.*, Criddle & Fox-Decent, *supra* note 43; Fox-Decent & Criddle, *supra* note 43.

46. *See* IMMANUEL KANT, GROUNDING FOR THE METAPHYSICS OF MORALS 39–42 (James W. Ellington trans., 1981).

47. IMMANUEL KANT, THE METAPHYSICS OF MORALS 64 (Mary Gregor trans., 1997).

48. *Id.*

49. *See* ARTHUR RIPSTEIN, FORCE AND FREEDOM (2009) (providing a general introduction to Kant's theory of legal order).

50. *See, e.g.*, G.A. Res. 60/1, ¶ 138–40, U.N. Doc. A/RES/60/1, at 30 (Sept. 16, 2005); INT'L COMM'N ON INTERVENTION AND STATE SOVEREIGNTY, THE RESPONSIBILITY TO PROTECT (December 2001), http://www.responsibilitytoprotect.org/ICISS%20Report.pdf [hereinafter ICISS].

51. ICISS, *supra* note 50, at viii.

52. *See* Criddle & Fox-Decent, *supra* note 43; Fox-Decent & Criddle, *supra* note 43.

53. *See* Evan J. Criddle & Evan Fox-Decent, *Human Rights, Emergencies, and the Rule of Law*, 34 HUM. RTS. Q. 39 (2012), *available at* http://papers.ssrn.com/sol3/papers.cfm?abstract_id=1591970.

54. *See* Criddle & Fox-Decent, *supra* note 43.

55. *See* American Convention on Human Rights: "Pact of San Jose, Costa Rica," Nov. 22, 1969, 1144 U.N.T.S. 123; International Covenant on Civil and Political Rights, arts. 4, 6, Dec. 19, 1966, 999 U.N.T.S. 171; Convention for the Protection of Human Rights and Fundamental Freedoms, art. 2, Nov. 4, 1950, 213 U.N.T.S. 221.

56. Geneva Convention III, *supra* note 9, art. 3(1).

57. *Id.*

58. *Id.* art. 3(1)(a), (c)–(d).

59. *Id.* art. 3(2).

60. *See* U.N. Charter arts. 1, ¶¶ 1–2, 4.

61. Additional Protocol I, *supra* note 2, art. 51(4).

62. Geneva Convention (IV) Relative to the Protection of Civilian Persons in Time of War, art. 3, Aug. 12, 1949, 75 U.N.T.S. 287.

63. Geneva Convention III, *supra* note 9.

64. Geneva Convention (I) for the Amelioration of the Condition of the Wounded and Sick in Armed Forces in the Field, Aug. 12, 1949, 75 U.N.T.S. 31 [hereinafter Geneva Convention I]; Geneva Convention (II) for the Amelioration of the Condition of Wounded, Sick and Shipwrecked Members of Armed Forces at Sea, Aug. 12, 1949, 75 U.N.T.S. 85 [hereinafter Geneva Convention II].

65. Geneva Convention III, *supra* note 9, art. 3.

66. *See* Criddle & Fox-Decent, *supra* note 43; Fox-Decent & Criddle, *supra* note 43.

67. *See, e.g.*, Geneva Convention IV, *supra* note 62, art. 40 (protecting persons by limiting the extent of work they are compelled to perform).

68. *See* Corn, *supra* note 42, at 57–94.

69. *Id.* at 83.

70. *See id.* at 83, 89–90.

71. E-mail from Geoffrey Corn, Professor of Law, South Texas Coll. of Law, to Evan J. Criddle, Assoc. Professor of Law, Syracuse University Coll. of Law (May 26, 2011) (on file with author).

72. Corn, *supra* note 42, at 89.

73. *See* YORAM DINSTEIN, THE INTERNATIONAL LAW OF BELLIGERENT OCCUPATION 35 (2009).

74. *Id.* at 286.

75. HCJ 769/02 The Pub. Comm. Against Torture in Israel v. Gov't of Israel [2005] (Isr.) [hereinafter *Targeted Killings*].

76. *Id.* ¶ 2.

77. *Id.* ¶ 18.

78. *Id.* ¶ 45. *But see* Hamdan v. Rumsfeld, 548 U.S. 557, 629–31 (2006) (holding that hostilities between state and nonstate actors did not constitute *international* armed conflict).

79. *Targeted Killings*, HCJ 769/02, ¶ 60.

80. *Id.*

81. *Id.*

82. *Id.*; *see also Palestinian Wall*, 2004 I.C.J. at 180 (holding that the International Covenant on Civil and Political Rights "is applicable in respect of acts done by a State in the exercise its of jurisdiction outside its own territory").

83. *Targeted Killings*, HCJ 769/02, ¶ 4 (separate opinion of Vice President Rivlin) (citing Eyal Benvenisti, *Human Dignity in Combat: The Duty to Spare Enemy Civilians*, 39 ISR. L. REV. 81, 88 (2006)).

84. HCJ 7015/02 Ajuri v. IDF Commander [2002] (Isr.); *cf. Palestinian Wall*, 2004 I.C.J. at 178 (recognizing Israel as an occupying power in the West Bank).

85. *See* HCJ 7957/04 Mara'abe v. Prime Minister of Israel [2005] IsrSC 38(2) 393, ¶¶ 24, 29; Martti Koskenniemi, *Occupied Zone—A Zone of Reasonableness*, 41 ISR. L. REV. 13, 17–18 (2008) (suggesting that the Court on various occasions has characterized the military as a "trustee of the Arab population").

86. *Palestinian Wall*, 2004 I.C.J. at 181.

87. Isayeva v. Russia, 41 Eur. Ct. H.R. 847, ¶ 175 (2005).

88. *See* David Kilcullen, *Twenty-Eight Articles: Fundamentals of Company-level Counterinsurgency*, 1 SMALL WARS J. 1, 5 (2006); Peter R. Mansoor & Mark S. Ulrich, *Linking Doctrine to Action: A New COIN Center-of-Gravity Analysis*, MILITARY REV., September–October 2007, at 46.

89. CHAIRMAN OF THE JOINT-CHIEFS OF STAFF, COUNTERINSURGENCY OPERATIONS, JOINT PUBLICATION 3–24, xvi (2009).

90. *Id.* at III-18.

91. *See* Michael N. Schmitt, *Targeting and International Humanitarian Law in Afghanistan*, 39 ISR. Y.B. HUM. RTS. 307, 315–20 (2009) (discussing the "likely and identifiable threat" standard).

92. Press Release, Int'l Security Assistance Force H.Q. Pub. Affairs, Gen. Petraeus Issues Updated Tactical Directive, Emphasizes "Disciplined Use of Force" (Aug. 1, 2010), *available at* http://www.dvidshub.net/news/53931/gen-petraeus-issues-updated-tactical-directive-emphasizes-disciplined-use-force (providing excerpts from the classified tactical directive).

93. *See id.*

94. *See* HEADQUARTERS, DEPARTMENT OF THE ARMY & HEADQUARTERS, MARINE CORPS COMBAT DEVELOPMENT COMMAND, FM3-24/MCWP 3-33.5, THE U.S. ARMY/MARINE CORPS COUNTERINSURGENCY FIELD MANUAL 48–49 (Univ. of Chicago Press, 2007).

95. *See* David W. Barno, *Fighting "The Other War": Counterinsurgency Strategy in Afghanistan, 2003–2005*, MILITARY REV., September–October 2007, at 90.

CHAPTER 3

* Associate Professor, Brigham Young University Law School. Previously Chief, International Law Branch, Office of The Judge Advocate General, U.S. Army; Military Legal Advisor to U.S. forces in Baghdad, Iraq and in Tuzla, Bosnia; Legal Advisor to the U.S. contingent of U.N. Forces in Macedonia. This chapter is based on an article that was originally published in the *Chicago Journal of International Law* (2012), and on a presentation by the author at the World Summit on Counter-Terrorism, International Institute for Counter-Terrorism, Herzliya, Israel in September, 2010.

1. Al-Bihani v. Obama, 590 F.3d 866, 882 (D.C. Cir. 2010) (Brown, J., concurring).

2. Convention for the Amelioration of the Condition of the Wounded and Sick in Armed Forces in the Field (1949), 75 U.N. Treaty Ser. 31 (1950) (First Geneva Convention); Convention for the Amelioration of the Condition of the Wounded, Sick, and Shipwrecked Members of Armed Forces at Sea (1949), 75 U.N. Treaty Ser. 85 (1950) (Second Geneva Convention); Geneva Convention Relative to the Treatment of Prisoners of War (1949), 75 U.N. Treaty Ser. 135 (1950) (Third Geneva Convention); Convention Relative to the Protection of Civilian Persons in Time of War (1949), 75 U.N. Treaty Ser. 287 (1950) (Fourth Geneva Convention) [collectively hereinafter Geneva Conventions].

3. Protocol Additional to the Geneva Conventions of August 12, 1949, and Relating to the Protection of Victims of International Armed Conflicts (1977), 1125 U.N. Treaty Ser. 3 (1979) (AP I) [hereinafter Additional Protocol I]; Protocol Additional to the Geneva Conventions of August 12, 1949, and Relating to the Protection of Victims of Non-international Armed Conflicts (1977), 1125 U.N. Treaty Ser. 609 (1979) (AP II) [hereinafter Additional Protocol II] [collectively hereinafter Protocols].

4. *See* M. Cherif Bassiouni, *The New Wars and the Crisis of Compliance with the Law of Armed Conflict by Non-State Actors*, 98 J CRIM. L. & CRIMINAL 711, 807–08 (2008) (arguing that states' ability to manipulate conflict classification encourages noncompliance by nonstate actors).

5. Memorandum from Jay Bybee, Office of Legal Counsel, U.S. Dep't of Justice, Application of Treaties and Laws to al-Qaeda and Taliban Detainees 5 (Jan. 22, 2002).

6. *See* HUMAN RIGHTS WATCH, *Colombia: Investigate Spate of Killings by Armed Groups* (July 8, 2011), *available at* http://www.hrw.org/news/2011/07/08/colombia-investigate-spate-killings-armed-groups (cataloguing recent attacks on civilians by armed groups and calling on the Colombian government to investigate and intervene).

7. *See* Paola Gaeta, *The Armed Conflict in Chechnya Before the Russian Constitutional Court*, 7 EUR. J. INT'L L. 563 (1996) (discussing the decision by the Russian Constitutional Court to declare Russia's conflict with Chechnya as subject to AP II).

8. *See* Carina Bergal, *The Mexican Drug War: The Case for Non-International Armed Conflict Classification*, 34 FORDHAM INT'L L. J. 1042, 1088 (2011) (arguing that Mexico has not officially declared its situation against the drug cartels as a NIAC, but that it should do so).

9. Ken Watkin, *Controlling the Use of Force: A Role for Human Rights Norms in Contemporary Armed Conflict*, 98 AJIL 1, 34 (2004); Geoffrey Corn, *Mixing Apples and Hand Grenades: The Logical Limit of Applying Human Rights Norms to Armed Conflict*, 1 INT'L HUMANITARIAN LEGAL STUD. 52, 52–94 (2010).

10. *See* W. Michael Reisman, *Application of Humanitarian Law in Noninternational Armed Conflicts: Remarks by W. Michael Reisman*, 85 ASIL PROCEEDINGS 83, 85 (1991).

11. Emily Crawford, *Unequal Before the Law: The Case for the Elimination of the Distinction Between International and Non-international Armed Conflicts*, 20 LEIDEN J. INT'L L. 441, 443–45 (2007).

12. James G. Stewart, *Towards a Single Definition of Armed Conflict in International Humanitarian Law: A Critique of Internationalized Armed Conflict*, INT'L REV. OF THE RED CROSS 313, 316–17 (2003) (citations omitted), *available at* http://www.icrc.org/Web/eng/siteengo.nsf/htmlall/5PYAXX/$File/irrc_850_Stewart.pdf.

13. Conference of Government Experts on the Reaffirmation and Development of International Humanitarian Law Applicable in Armed Conflicts, *Volume V: Protection of Victims of Non-International Armed Conflicts*, INT'L COMMITTEE OF THE RED CROSS 1

(1971), *available at* http://www.loc.gov/rr/frd/Military_Law/pdf/RC-conference_Vol-5.pdf [hereinafter Conference of Government Experts]. *See also* Antonio Cassese, *The Status of Rebels Under the 1977 Geneva Protocol on Non-International Armed Conflict*, 30 ILCQ 415 (1981).

14. Conference of Government Experts, *supra* note 13, at 2.

15. Article 2 states:

> In addition to the provisions which shall be implemented in peacetime, the present Convention shall apply to all cases of declared war or of any other armed conflict which may arise between two or more of the High Contracting Parties, even if the state of war is not recognized by one of them. The Convention shall also apply to all cases of partial or total occupation of the territory of a High Contracting Party, even if the said occupation meets with no armed resistance.
>
> Although one of the Powers in conflict may not be a party to the present Convention, the Powers who are parties thereto shall remain bound by it in their mutual relations. They shall furthermore be bound by the Convention in relation to the said Power, if the latter accepts and applies the provisions thereof.

Geneva Conventions, *supra* note 2, art. 2.

16. *Id.* art. 3.

17. Stewart, *supra* note 12, at 317.

18. Conference of Government Experts, *supra* note 13, at 7.

19. The Report of the Government Experts proposed:

> When, in the case of non-international armed conflict, the Party opposing the authorities in power presents the component elements of a State—in particular if it exercises public power over a part of the territory, disposes of a provisional government and an organized civil administration, as well as of regular armed forces—the Parties to the conflict shall apply the whole of the international humanitarian law applicable in international armed conflicts.

Id. at 15. The same report also concluded that when a third state becomes involved in the conflict, the full LOAC should apply. *Id.* at 21.

20. Commentary on the Additional Protocols of June 8, 1977, to the Geneva Conventions of 1949, ¶ 4387 (Y. Sandoz et al. eds., 1987).

21. Additional Protocol I, *supra* note 3. AP I, Article 1, paragraphs 3 and 4 state:

> 3. This Protocol, which supplements the Geneva Conventions of 12 August 1949 for the protection of war victims, shall apply in the situations referred to in Article 2 common to those Conventions.
>
> 4. The situations referred to in the preceding paragraph include armed conflicts in which peoples are fighting against colonial domination and alien occupation and against racist régimes in the exercise of their right of self-determination, as enshrined in the Charter of the United Nations and the Declaration on Principles of International Law concerning Friendly Relations and Co-operation among States in accordance with the Charter of the United Nations.

22. Additional Protocol II, *supra* note 3. Article 1 and 2 state:

> 1. This Protocol, which develops and supplements Article 3 common to the Geneva Conventions of 12 August 1949 without modifying its existing conditions of application, shall apply to all armed conflicts which are not covered by Article 1 of the Protocol Additional to the Geneva Conventions of 12 August 1949, and relating to the Protection of Victims of International Armed Conflicts (Protocol I) and which take place in the territory of a High Contracting Party between its armed forces and dissident armed forces or other organized armed groups which, under responsible command, exercise such control over a part of its territory as to enable them to carry out sustained and concerted military operations and to implement this Protocol.
> 2. This Protocol shall not apply to situations of internal disturbances and tensions, such as riots, isolated and sporadic acts of violence and other acts of a similar nature, as not being armed conflicts.

23. Crawford, *supra* note 11, at 446.

24. When President Reagan sent the Protocols to the Senate, his letter of transmittal made exactly that point. He characterized Protocol I as "fundamentally and irreconcilably flawed" and stated that "we cannot allow other nations of the world, however numerous, to impose upon us and our allies and friends an unacceptable and thoroughly distasteful price for joining a convention drawn to advance the laws of war." Ronald Reagan, Letter of Transmittal to the U.S. Senate (Jan. 29, 1987), *reprinted in* 81 AM. J. INT'L L. 910, 911 (1987).

President Reagan also transmitted Additional Protocol II to the Senate. Michael Schmitt describes the view of the president and the State Department:

> Despite the altered balance symbolized by Additional Protocol II, President Reagan submitted the instrument to the Senate in 1987 for advice and consent. In his letter of transmittal, the President opined that the agreement was, with certain exceptions, a positive step toward the goal of "giving the greatest possible protection to the victims of [non-international] conflicts, consistent with legitimate military requirements." The Legal Adviser to the State Department characterized the instrument's terms as "no more than a restatement of the rules of conduct with which United States military forces would almost certainly comply as a matter of national policy, constitutional and legal protections, and common decency."

Michael N. Schmitt, *Military Necessity and Humanity in International Humanitarian Law: Preserving the Delicate Balance*, 50 VA. J. INT'L L. 795, 811 (2010) (citations omitted).

25. ANTHONY CULLEN, THE CONCEPT OF NON-INTERNATIONAL ARMED CONFLICT IN INTERNATIONAL HUMANITARIAN LAW 110 (2010).

26. Paola Gaeta, *The Armed Conflict in Chechnya Before the Russian Constitutional Court*, 7 EJIL 563, 568 (1996).

27. HCJ 769/02 Public Committee against Torture in Israel v. Government of Israel [2006] IsrSC, *available at* elyon1.court.gov.il/Files_ENG/02/690/007/a34/02007690.a34.pdf; Curtis A. Bradley, *The United States, Israel & Unlawful Combatants*, 12 Green Bag 2d 397, 401 (2009).

28. *See* section III (B), *infra*.

29. As James Stewart recently wrote: "Under these circumstances, and in the absence of an impartial body charged with authoritatively determining the status of armed conflicts, it is fair to assume that parties will characterize conflicts in terms that best suit their own interests." Stewart, *supra* note 12, at 344.

30. *Id.* at 342.

31. *Id.* at 313.

32. Prosecutor v. Tadic, Case No. IT-94-1-1, Decision on the Defence Motion for Interlocutory Appeal on Jurisdiction, Appeals Chamber of the International Tribunal for the Prosecution of Persons Responsible for Serious Violations of International Humanitarian Law Committed in the Territory of Former Yugoslavia since 1991, ¶ 119 (Oct. 2, 1995), *available at* http://www.icty.org/x/cases/tadic/acdec/en/51002.htm.

33. *See, e.g.*, William Bradford, *Barbarians at the Gates: A Post-September 11th Proposal to Rationalize the Laws of War*, 73 MISS. L. J. 639, 697–710 n.12 (2004); SUN TZU, THE ART OF WAR 76 (Samuel Griffith trans., Oxford Univ. 1963); Thomas C. Wingfield, *Chivalry in the Use of Force*, 32 U. TOL. L. REV. 111, 114 (2001); Gregory P. Noone, *The History and Evolution of the Law of War Prior to World War II*, 47 NAVAL L. REV. 176, 182–85 (2000).

34. THE HANDBOOK OF INTERNATIONAL HUMANITARIAN LAW 15–22 (Dieter Fleck ed., 2nd ed., 2008) [hereinafter Handbook of International Humanitarian Law].

35. *Id.* at 19 (quoting Grotius 28 (Prolegomena)).

36. *But see* Jordan Paust, *Nonstate Actor Participation in International Law and the Pretense of Exclusion*, 51 VA. J. INT'L L. 977 (2011) (where the author argues that though states play a primary role, there is clearly a strong role for nonstate actors).

37. PHILIP BOBBITT, THE SHIELD OF ACHILLES 81–90, 96–118 (2002); Frederic Gilles Sourgens, *Positivism, Humanism, and Hegemony: Sovereignty and Security for Our Time*, 25 PENN ST. INT'L L. REV. 433 (2006) (where the author quotes sixteenth-century writer Bodin as defining sovereignty as the "absolute and perpetual power of commonwealth resting in the hands of the state").

38. Rousseau stated:

> War then is a relation, not between man and man, but between State and State, and individuals are enemies only accidentally, not as men, nor even as citizens, but as soldiers; not as members of their country, but as its defenders.... The object of the war being the destruction of the hostile State, the other side has a right to kill its defenders while they are bearing arms; but as soon as they lay them down and surrender they become once more merely men, whose life no one has any right to take.

HANDBOOK OF INTERNATIONAL HUMANITARIAN LAW, *supra* note 34, at 19–20 (quoting Jean-Jacques Rousseau).

39. BOBBITT, *supra* note 37, at 509–519; Remarks by Ambassador Richard S. Williamson, Policy Salon Sponsored by The Stanley Foundation, Washington, D.C., May 18, 2009.

40. *A Treatise on the Juridical Basis of the Distinction Between Lawful Combatant and Unprivileged Belligerent*, THE JUDGE ADVOCATE GENERAL'S SCHOOL, U.S. ARMY 14 (1959) (on file with author); Allison Marston Danner, *Beyond the Geneva Conventions: Lessons from the Tokyo Tribunal in Prosecuting War and Terrorism*, 46 VA. J. INT'L L. 83, 101 (2005).

41. *Instructions for the Government of Armies of the United States in the Field (Lieber Code)*, General Orders No. 100, art. 57, U.S. WAR DEP'T (Apr. 24, 1863), *available at* http://www.icrc.org/ihl.nsf/FULL/110?OpenDocument.

42. *Id.* art. 82.

43. Memorandum from Alberto R. Gonzales, White House Counsel, to U.S. President George W. Bush, Decision re Application of the Geneva Convention on Prisoners of War to the Conflict with al Qaeda and the Taliban 2 (Jan. 25, 2002).

44. *See, e.g.*, Benjamin Wittes, *Law and the Long War: The Future of Justice in the Age of Terror* (Penguin Books 2008). *Assessing Damage, Urging Action: Report of the Eminent Jurists Panel on Terrorism, Counter-terrorism, and Human Rights*, INT'L COMM. OF JURISTS (2009), *available at* http://ejp.icj.org/IMG/EJP-Report.pdf.

45. For an excellent analysis of this issue, *see* Bergal, *supra* note 8.

46. *Attorney General Leading War on Mexico Drug Cartels Resigns,* FOX NEWS, Sept. 8, 2009, http://www.foxnews.com/printer_friendly_story/0,3566,547568,00.html.

47. Sara A. Carter, *100,000 foot soldiers in Mexican Cartels*, WASH. TIMES, March 3, 2009, at A1, *available at* http://www.washingtontimes.com/news/2009/mar/03/100000-foot-soldiers-in-cartels/; *U.S. to boost Mexico border defence*, BBC NEWS, Mar. 25, 2009, http://news.bbc.co.uk/go/pr/fr/-/2/hi/americas/7961670.stm.

48. Mexico has not signed Protocol II. *See Ratification of International Human Rights Treaties—Mexico*, UNIV. MINNESOTA, http://www1.umn.edu/humanrts/research/ratification-mexico1.html.

49. For various decisions and statements concerning the characterization of the conflict in Israel, *see Rule of Law in Armed Conflicts Project*, GENEVA UNIVERSITY, http://www.adh-geneva.ch/RULAC/applicable_international_law.php?id_state=113.

50. HCJ 769/02 Public Committee against Torture in Israel.

51. Memorandum from Gordon England, Deputy Secretary of Defense on the Application of Common Article 3 of the Geneva Conventions to the Treatment of Detainees in the Department of Defense to Department of Defense officials (July 7, 2006), *available at* http://www.fas.org/sgp/othergov/dod/geneva070606.pdf.

52. *See* Hamdan v. Rumsfeld, 548 U.S. 557 (2006).

53. Crawford, *supra* note 11, at 443–45.

54. Office of the Staff Judge Advocate, 101st Airborne Division (Air Assault), OPERATION IRAQI FREEDOM (OIF) AFTER ACTION REPORT 64 (on file with author) (reflecting lack of information from national level authorities on Rules of Engagement); Office of the Staff Judge Advocate, 1st Cavalry Division, OPERATION IRAQI FREEDOM (OIF) AFTER ACTION REPORT 19–20 (on file with author) (reflecting lack of information from national level authorities on Detention Operations).

55. John P. Cerone, *Status of Detainees in Non-International Armed Conflict, and their Protection in the Course of Criminal Proceedings: The Case of Hamdan v. Rumsfeld*, AMERICAN SOCIETY OF INTERNATIONAL LAW, July 14, 2006, *available at* http://www.asil.org/insights060714.cfm#_edn9; Harold Hongju Koh, Speech at the Annual Meeting of the American Society of International Law, Mar. 25, 2010, *available at* http://www.state.gov/s/l/releases/remarks/139119.htm.

56. U.K. Ministry of Defence, MANUAL OF THE LAW OF ARMED CONFLICT ¶ 1.33.6 (2004).

57. Marc L. Warren, *The First Annual Solf-Warren Lecture in International and Operational Law*, 196 MIL. L. REV. 129, 138 (2008).

58. Tadic, Case No. IT-94-1-1, ¶ 73.

59. *See* section II, *supra*.

60. Convention on the Prohibition of the Development, Production, Stockpiling and Use of Chemical Weapons and on Their Destruction, G.A. Res. 47/39, U.N. GAOR, 47th Sess., 74th plen. mtg., U.N. Doc. A/RES/47/39 (Nov. 30, 1992).

61. *See* Bergal, *supra* note 8.

62. Rules of Engagement (ROE) are orders by which commanders at all levels control the use of force by their subordinates. For the United States, the primary ROE document is the Chairman of the Joint Chiefs of Staff's Standing Rules of Engagement, commonly referred to as the SROE. The SROE is classified SECRET, but the basic instruction and Enclosure A titled "Standing Rules of Engagement for U.S. Forces" are UNCLASSIFIED. Chairman, Joint Chiefs of Staff Instruction 3121.01B, STANDING RULES OF ENGAGEMENT/STANDING RULES FOR THE USE OF FORCE FOR U.S. FORCES, Encl. A (Jun. 13, 2005). The SROE details basic concepts of ROE that apply generally and then sets out a methodology for establishing mission specific ROE. The document is designed to "establish fundamental policies and procedures governing the actions to be taken by U.S. commanders and their forces during all military operations and contingencies and routine Military Department functions occurring outside U.S. territory." *Id.* at 1. There are additional rules for the application of force within the United States, which are contained in later enclosures.

63. The tactic of using the legal compliance of a nation as a tactical advantage has been termed "lawfare" and is a characteristic of recent conflicts. Charles J. Dunlap, Jr., *Law and Military Interventions: Preserving Humanitarian Values in 21st Century Conflicts* (Carr Center for Human Rights, John F. Kennedy Sch. of Gov't, Harvard U., Working Paper, 2001), *available at* http://www.ksg.harvard.edu/cchrp/Web%20Working%20Papers/Use%20of%20Force/Dunlap2001.pdf; Charles J. Dunlap, Jr., *Lawfare Today: A Perspective*, 3 YALE J. INT'L. AFF. 146, 146 (2008); Charles J. Dunlap, Jr., *Lawfare: A Decisive Element of 21st-Century Conflicts?*, 54 JOINT FORCE QUARTERLY 34 (2009), *available at* http://www.dtic.mil/cgi-bin/GetTRDoc?AD=ADA515192&Location=U2&doc=GetTRDoc.pdf; Charles J. Dunlap, Jr., *Is "LAWFARE" a Useful Term?: Does Lawfare Need an Apologia?*, 43 CASE W. RES. J. INT'L L. 121 (2010). The author has argued elsewhere that current international law and policy are already encouraging the use of lawfare by disadvantaged parties to a conflict.

64. Anthea Roberts & Sandesh Sivakumaran, *Hybrid Sources of Law: Armed Groups and the Creation of International Law*, 37 YALE J. INT'L L. 107 (2012).

65. HANNIS TAYLOR, A TREATISE ON INTERNATIONAL PUBLIC LAW 454 (1901) (quoted in LINDSAY MOIR, THE LAW OF INTERNAL ARMED CONFLICT 3 (2002)).

66. DEPARTMENT OF PEACE AND CONFLICT RESEARCH, UPPSALA UNIVERSITY, UPPSALA CONFLICT DATA PROGRAM (2009), *available at* http://www.pcr.uu.se/research/ucdp/publications/data_publications/.

67. *See, e.g.*, JAN RÖMER, KILLING IN A GRAY AREA BETWEEN HUMANITARIAN LAW AND HUMAN RIGHTS 2 (noting that in 2007, Colombian military and police "officially killed 2,703 members of different 'guerrilla groups,' 'self-defense groups,' and 'criminal bands'"). In 2008, the military and police killed 1,564. *Id.*

68. *See, e.g.*, Erica Werner & Jacques Billeaud, *Obama Set to Send 1,200 Troops to Border*, HUFFINGTON POST, May 25, 2010, *available at* http://www.huffingtonpost.com/2010/05/25/obama-set-to-send-1200-tr_n_589208.html; William Booth, *Mexico's drug gangs go on the offensive against authorities; Criminals increasingly target police officials and government figures, changing the dynamics of violence in a deadly new phase of the narcotics fight*, WASH. POST, May 2, 2010, at A10, *available at* http://www.washingtonpost.com/wp-dyn/content/article/2010/05/01/AR2010050102869.html.

69. *See, e.g.*, Cornelius Friesendorf, *The Military and the Fight against Serious Crime: Lessons from the Balkans*, 9 CONNECTIONS 45, 52–53 (2010) (showing that, while ineffective, the military still was asked to take on this mission in Bosnia); *Our Mission*, UNITED STATES PACIFIC COMMAND, http://www.pacom.mil/web/site_pages/staff%20directory/jiatfwest/jiatfwest.shtml (last visited Nov. 2, 2011) ("Joint Interagency Task Force West combats drug-related transnational organized crime to reduce threats in the Asia-Pacific region in order to protect national security interests and promote regional stability."). *See generally Strategy to Combat Transnational Organized Crime*, NATIONAL SECURITY COUNCIL, *available at* http://www.whitehouse.gov/administration/eop/nsc/transnational-crime (last visited Nov. 2, 2011) (talking about using all the elements of national power, including the military, to combat transnational crime).

70. *See* S.C. Res. 1291, ¶¶ 1, 4, 7–8, U.N. Doc S/RES/1291 (2000) (establishing the U.N. Organization Mission in the Democratic Republic of the Congo (MONUC) in order to facilitate the parties' fulfillment of their Ceasefire Agreement obligations as well as authorizing MONUC to take "the necessary action, in the areas of deployment of its infantry battalions and as it deems it within its capabilities, to protect United Nations and co-located [Joint Military Commission] personnel, facilities, installations and equipment,... and protect civilians under imminent threat of physical violence").

71. *See generally* Chairman of the Joint Chiefs of Staff, Joint Publication 3-0, *Joint Operations* (Mar. 22, 2010); Anne E. Story & Aryea Gottlieb, *Beyond the Range of Military Options*, JOINT FORCE QUARTERLY (1995), *available at* http://www.dtic.mil/doctrine/jel/jfq_pubs/2309.pdf (both discussing the current range of military operations).

72. *See generally* Chairman of the Joint Chiefs of Staff, Joint Publication 3-7, *Joint Doctrine for Military Operations Other Than War* (June 16, 1995).

73. RICHARD F. GRIMMETT, CONG. RESEARCH SERV., R41677, INSTANCES OF USE OF UNITED STATES ARMED FORCES ABROAD 1798–2010 (2011), *available at www.fas.org/sgp/crs/natsec/R41677.pdf.*

74. U.S. Military Assistance Command, Vietnam, Directive No. 381-46, Military Intelligence: Combined Screening of Detainees (Dec. 27, 1967), *reprinted in* Howard Levie, *Documents on Prisoners of War*, 60 INT'L L. STUD. 748 (1979). The directive identified the following groups' members as qualifying for POW status: Viet Cong Main Forces; Viet Cong Local Forces; North Vietnamese Army Units; and Organized Forces of Irregular Guerillas, and Self-Defense Forces who had not engaged in terrorism, sabotage, or spying. *Id.*, annex A.

75. Dep't of Defense Directive 5100.77, Dep't of Defense Program for the Implementation of the Law of War (Short Title: DoD Law of War Program), Nov. 5, 1974.

76. Dep't of Defense Directive 5100.77, Dep't of Defense Law of War Program, Dec. 9, 1998.

77. Dep't of Defense Directive 2311.01E, Dep't of Defense Law of War Program, ¶ 4.1, May 9, 2006.

78. *See* Customary International Humanitarian Law, Vol. I and II (Jean-Marie Henckaerts & Louise Doswald-Beck eds., 2005) (where the authors describe rules governing the law of armed conflict in volume 1, which are supported by annotated state practice in volume II).

79. For a list of the Rules that includes a designation as to which rules apply to IAC, NIAC, or both, *see Annex List of Customary Rules of International Humanitarian Law*, 87 Int'l Review of the Red Cross (2005), *available at* http://www.icrc.org/eng/assets/files/other/customary-law-rules.pdf [hereinafter Customary Rules of International Law].

80. However, the same court also held "this extension [of IAC rules] has not taken place in the form of a full and mechanical transplant of those rules to internal conflicts; rather, the general essence of those rules, and not the detailed regulation they may contain, has become applicable to internal conflicts." Tadic, Case No. IT-94-1-1.

81. Stewart, *supra* note 12, at 322. *But see id.* at 323 (where the author quotes the Tadic Jurisdiction Appeal to say "this extension has not taken place in the form of a full and mechanical transplant of those rules into internal conflicts; rather, the general essence of those rules, and not the detailed regulation they may contain, has become applicable to internal conflicts.").

82. Michael N. Schmitt, Yoram Dinstein & Charles H. B. Garraway, The Manual on the Law of Non-International Armed Conflict: With Commentary (International Institute of Humanitarian Law 2006).

83. *Id.*, at preface.

84. Customary International Humanitarian Law, Vol. I and II (Jean-Marie Henckaerts & Louise Doswald-Beck eds., 2005).

85. The provisions include the prohibition of attacks on civilians (Rule 1); the obligation to respect and protect medical personnel, units, and transports, and religious personnel (Rules 25–26, 28–30); the obligation to protect medical personnel (Rules 26, 30); the prohibition of starvation as a method of warfare (Rule 53); the prohibition of attacks on objects indispensable to the survival of the civilian population (Rule 54); the obligation to respect the fundamental guarantees of civilians and persons hors de combat (Rules 87–105); the obligation to search for and respect and protect the wounded, sick, and shipwrecked (Rules 109–111); the obligation to search for and protect the dead (Rules 112–113); the obligation to protect persons deprived of their liberty (Rules 118–119, 121, and 125); the prohibition of forced movement of civilians (Rule 129); and protections afforded to women and children (Rules 134–137). Customary Rules of International Law, *supra* note 79, at 198–210.

86. *See* Nuclear Tests Case (Australia & New Zealand v. France), 1974 I.C.J. 253 (Dec. 20) (where the ICJ held that unilateral acts can have full legal effect between states).

CHAPTER 4

* Professor of law, South Texas College of Law.

1. The term "belligerent" is used throughout this chapter to designate a member of an armed group who performs the type of function historically performed by lawful combatants who are members of the regular armed forces of a state. *See generally* Richard R. Baxter, *So-Called "Unprivileged Belligerency": Spies, Guerillas, and Saboteurs*, 28 Brit. Y.B. Int'l L. 323 (1951) (describing the legal treatment of those who fall outside of the traditional two classes of persons entitled to protection under international law during times of war). One of the challenges associated with selecting a term to designate such individuals is that the law of armed conflict

(LOAC) definition of "combatant" is limited to the context of international armed conflicts. *See* Protocol I Additional to the Geneva Conventions of August 12, 1949, and Relating to the Protection of Victims of International Armed Conflicts art. 43, June 8, 1977, 1125 UNTS 3 [hereinafter Protocol I] ("[m]embers of the armed forces of a Party to a conflict other than medical personnel and chaplains covered by Article 33 of the Third Convention [are combatants], that is to say, they have the right to participate directly in hostilities."). As a result, various terms have been offered for this designation, including unlawful combatant, unprivileged belligerent, fighter, nonstate actor, and nonstate opponent. All of these terms reflect a common underlying meaning: designation of an individual who, as the result of his relationship with enemy belligerent leadership and function as an enemy belligerent operative, should be treated for purposes of attack authority no differently than a combatant within the meaning of Protocol I.

2. *See* Gary D. Solis, The Law of Armed Conflict: International Humanitarian Law in War 188 (2010) ("Combatants may be attacked at any time until they surrender or are otherwise hors de combat, and not only when actually threatening the enemy."); *see* Yoram Dinstein, The Conduct of Hostilities under the Law of International Armed Conflict 27 (2nd ed. 2010) ("[c]ombatants in an international armed conflict fall into two alternative categories: (i) Members of the armed forces of a belligerent Party.... (ii) Any other persons who take an active part in the hostilities."); *see also* Nils Melzer, *Interpretive Guidance on the Notion of Direct Participation, in* Hostilities Under International Humanitarian Law, 11–12 (2009), *available at* http://www.icrc.org/eng/assets/files/other/icrc-002-0990.pdf (discussing civilian participation in hostilities).

3. *See, e.g.*, Charles J. Dunlap, Jr., *Lawfare Today: A Perspective*, 3 Yale J. Int'l Aff. 146, 148 (2008) (highlighting the use of modern media by insurgents to exploit civilian casualties in order to use U.S. forces' adherence to law as a weapon against them); *see also, e.g.*, Eric Talbot Jensen, *Combatant Status: It Is Time for Intermediate Levels of Recognition for Partial Compliance*, 46 Va. J. Int'l L. 209 (2005) (reviewing the history of combatant status and arguing for a modification of the law of combatants to allow for some intermediate level of recognition in order to incentivize some type of self-identification); *see also Attack on the 507th Maintenance Company, March 23, 2003, An Nasiriyah, Iraq, Executive Summary*, Riciok.com, http://www.riciok.com/attack_summary.htm (outlining the attack on the 507th Maintenance Company by an enemy using asymmetric tactics); Michael L. Gross, Moral Dilemmas of Modern War: Torture, Assassination, and Blackmail in an Age of Asymmetric Conflict (2009) (evaluating modern changes to combatant relationships and identifying the moral issues inherent in asymmetric warfare).

4. *See* Joint Chiefs of Staff, Joint Pub 5-03.1: Joint Operation Planning and Execution System Volume I (Planning Policies and Procedures) ch. 2, ¶ 8 (1993), *available at* http://edocs.nps.edu/dodpubs/topic/jointpubs/JP5/JP5-03.1_930804.pdf (discussing the use of threat identification and assessment in the larger context of the joint planning process); *see also* Dep't. of the Army, FMI 3–07.22, Counterinsurgency Operations ¶ 2–50 (2004) (calling for the coordination of ARSOF operations in order to avoid potential unwanted casualties).

5. *See* Counterinsurgency Operations, *supra* note 4, ¶ 2–10 (proposing that protection of the civilian population is essential to a successful counterinsurgency operation).

6. *Compare* Melzer, *supra* note 2, at 41–42 ("In practice, civilian direct participation in hostilities is likely to entail significant confusion and uncertainty in the implementation of the

principle of distinction. In order to avoid the erroneous or arbitrary targeting of civilians entitled to protection against direct attack, it is therefore of particular importance that all feasible precautions be taken in determining whether a person is a civilian and, if so, whether he or she is directly participating in hostilities. In case of doubt, the person in question must be presumed to be protected against direct attack."), *with* Protocol I, *supra* note 1, art. 51.2–3 ("The civilian population as such, as well as individual civilians, shall not be the object of attack ... unless and for such time as they take a direct part in hostilities.").

7. Protocol I, *supra* note 1, art. 48 ("In order to ensure respect for and protection of the civilian population and civilian objects, the Parties to the conflict shall at all times distinguish between the civilian population and combatants.... ").

8. *See generally* Melzer, *supra* note 2.

9. *See id.* at 27–36 (discussing the distinction of civilians, armed forces, and organized armed groups under the LOAC, and the application of the CCF concept).

10. Jacobellis v. Ohio, 378 U.S. 184, 197 (1964) (Stewart, J., concurring) ("I shall not today attempt further to define the kinds of material I understand to be embraced within that shorthand description [of hard-core pornography]; and perhaps I could never succeed in intelligibly doing so. But I know it when I see it, and the motion picture involved in this case is not that.").

11. *See* Protocol I, *supra* note 1, art. 43–2 (defining combatants as members of the armed forces of any party to a conflict).

12. *See id.* art. 50 (declaring a civilian to be any person other than those defined in Article 43 of the Protocol and Article 4(A) of the Convention).

13. Commentary on the Additional Protocols of June 8, 1977 to the Geneva Conventions of August 12, 1949 610–12 (Claude Pilloud et al. eds., 1987) [hereinafter Commentary on 1977 Additional Protocols].

14. *See* Gary D. Solis, The Law of Armed Conflict: International Humanitarian Law in War 41–42 (2010) (discussing the history of the combatant's privilege); *see also* Geneva Convention Relative to the Treatment of Prisoners of War art. 4, Aug. 12, 1949, 75 U.N.T.S. 135 (defining the term "prisoner of war" and outlining categories into which prisoners of war fall, under this provision).

15. *See* The Hague Convention (II) with Respect to the Law and Customs of War on Land and its Annex: Regulations Concerning the Laws and Customs of War on Land art. 1, July 29, 1899, 31 Stat. 1803, 1 Bevans 247, *available at* http://www.icrc.org/ihl.nsf/FULL/150?OpenDocument [hereinafter Hague (II) Regulations for War on Land] (providing for the issuing of instructions to armed land forces).

16. *Id.* sec. 1, ch. 1, art. 1 (outlining the rights and duties of armies, militia, and volunteer corps).

17. The Hague Convention (IV) with Respect to the Law and Customs of War on Land, Annex to the Convention: Regulations Respecting the Laws and Customs of War on Land art. 1, Oct. 18, 1907, 36 Stat. 2277, 1 Bevans 270 [hereinafter Hague (IV) Regulations for War on Land] (establishing conventions regulating the debt collection, the rules of war, and the rights and obligations of neutral countries).

18. *See id.*

19. Geneva Convention (III) Relative to the Treatment of Prisoners of War, *supra* note 14, art. 4.

20. *See* Hague Convention II and Annexed Regulations, *supra* note 15, sec. 1, ch. 2 (detailing correlation between treatment of prisoners of war and that of belligerents); Hague IV and

Annexed Regulations, *supra* note 17, sec. 1, ch. 2 (outlining the context within which a belligerent may be considered a prisoner of war).

21. *See* Geneva Convention Relative to the Treatment of Prisoners of War, *supra* note 14, art. 1. (identifying the characteristics of a lawfully qualified belligerent).

22. *Id.* art. 4.

23. *See id.*

24. *See* THE GENEVA CONVENTIONS (III) OF AUGUST 12, 1949: COMMENTARY, VOL. 3 45–48 (Jean Pictet ed., 1952) (describing how the 1899 and 1907 Conferences' lengthiest and most important discussions focused on the definition of "belligerent status"); *see also* Geoffrey S. Corn, *Unarmed but How Dangerous? Civilian Augmentees, the Law of Armed Conflict, and the Search for a More Effective Test for Permissible Civilian Battlefield Functions*, 2 J. NAT'L SEC. L. & POL'Y 257 (2008) (detailing the history of and qualifying characteristics that compose prisoner of war status, as outlined in part by the Geneva Conventions).

25. "In armed conflict with an international character, a person of enemy nationality who is not entitled to prisoner-of-war status is, in principle, a civilian protected by the fourth Convention, so that there are no gaps in protection. However, things are not always so straightforward in the context of the armed conflicts of Article 1 (General principles and scope of application), paragraph 4, as the adversaries can have the same nationality. Moreover, the concept of alien occupation often becomes rather fluid in guerrilla operations as no fixed legal border delineates the areas held by either Party, and this may result in insurmountable technical difficulties with regard to the application of some of the provisions of the fourth Convention." COMMENTARY ON 1977 ADDITIONAL PROTOCOLS, *supra* note 13, at 558.

26. Additional Protocols I and II of 1977 were developed to supplement and bring up to date the four Geneva Conventions of 1949. Although the United States played a significant role in the drafting of these treaties, Additional Protocol I was withdrawn from Senate consideration by President Reagan. *See* Letter of Transmittal from President Ronald Reagan, Protocol II Additional to the 1949 Geneva Conventions, and Relating to the Protections of Victims of Non International Armed Conflicts, S. Treaty Doc. No. 2, 100th Cong., 1st Sess., at III (1987) (stating the U.S. government's reasoning as to why Protocol I would not be submitted for senatorial advice and consent). This was in large measure because of the conclusion that several provisions of Additional Protocol I extended LOAC protections to terrorists; and that the scope provision of Additional Protocol II was too restrictive. *Id.*

Nonetheless, prior to U.S. military response to the attacks of September 11, 2001, characterizing the bulk of Additional Protocol I as a reflection of customary international law binding on the United States would have been relatively uncontroversial among legal experts responsible for advising U.S. military planners and commanders. The post–September 11, 2001, legal determinations made by President George W. Bush regarding the applicability of law-of-war provisions to the conflict with al Qaeda, however, radically altered this practice. A much more textual approach prevailed when interpreting law-of-war treaty obligations. *See* Memorandum from Jay S. Bybee, Assistant Att'y Gen., Office of Legal Counsel, Dep't of Justice, to Alberto R. Gonzales, Counsel to the President, and William J. Haynes II, Gen. Counsel of the Dep't. of Def., Application of Treaties and Laws to al Qaeda and Taliban Detainees (Jan. 22, 2002), *available at* http://washingtonpost.com/wp-srv/nation/documents/012202bybee.pdf (responding to the issue of whether certain treaties included in the laws of armed conflict apply to trial procedures and detention conditions for members of al Qaeda and the Taliban).

This revised approach to interpreting the status of provisions of Additional Protocol I is reflected by comparing treatment of this treaty in the law-of-war chapter of the *Operational Law Handbook*, which is perhaps the most widely relied upon reference for military legal practitioners supporting ongoing operations. The current version of the *Handbook* provides the following:

> 1977 Geneva Protocols (ref. (7)). Although the U.S. has not ratified [Additional Protocol I] and II, 155 nations have ratified [Additional Protocol I]. U.S. Commanders must be aware that many allied forces are under a legal obligation to comply with the Protocols. This difference in obligation has not proved to be a hindrance to U.S./allied or coalition operations since promulgation of AP I in 1977.

INT'L & OPERATIONAL LAW DEP'T, THE JUDGE ADVOCATE GENERAL'S LEGAL CENTER & SCHOOL, JA 422, OPERATIONAL LAW HANDBOOK, August 2006.

Although this excerpt does not explicitly indicate a rejection of prior interpretations of obligation vis-à-vis Additional Protocol I, it clearly does not explicitly assert such an obligation. The full significance of this excerpt is only apparent when compared to the description of Additional Protocol I in prior editions of the *Operational Law Handbook*. For example, the 2003 edition states the following:

> 1977 Geneva Protocols (ref. (7)). Although the U.S. has not ratified [Geneva Protocol] I and II, judge advocates must be aware that approximately 150 nations have ratified the Protocols (thus most of the 185 member states of the [United Nations]). The Protocols will come into play in most international operations. U.S. Commanders must be aware that many allied forces are under a legal obligation to comply with the Protocols. *Furthermore, the U.S. considers many of the provisions of the Protocols to be applicable as customary international law.*

Comparison of these two versions of the *Operational Law Handbook* indicates a general "rollback" by the executive branch of the treatment of Additional Protocol I provisions. Numerous experts and government legal advisers have argued for years that many of these provisions reflect binding norms of customary international law. *See, e.g.*, Michael J. Matheson, *The United States Position on the Relation of Customary International Law to the 1977 Protocols Additional to the 1949 Geneva Conventions*, 2 AM. U. J. INT'L & POL'Y 419 (1987) (demonstrating that the 1977 Protocols have a solid basis in customary international law); *see also, e.g.*, Memorandum from W. Hays Parks, LCDR Michael F. Lohr, Dennis Yoder, and William Anderson, to John H. McNeill, Assistant Gen. Counsel (Int'l), Office of the Sec'y of Def., 1977 Protocols Additional to the Geneva Conventions: Customary International Law Implications (May 8, 1986). Unfortunately, opponents of this proposition have relied on the repudiation of Additional Protocol I by President Reagan. These opponents assert this repudiation is particularly relevant vis-à-vis the armed conflict with al Qaeda because it was motivated in large part by the U.S. concern that Additional Protocol I unjustifiably extended law-of-war protections to terrorist operatives.

27. COMMENTARY ON 1977 ADDITIONAL PROTOCOLS, *supra* note 13, at 515.

28. Protocol I, *supra* note 1, art. 43(2).

29. *Id.* art. 43(1).

30. COMMENTARY ON 1977 ADDITIONAL PROTOCOLS, *supra* note 13, at 515 (emphasis added).

31. *See id.* at 514, 599–600.

32. *See* Protocol I, *supra* note 1, art. 43(2) ("Members of the armed forces of a Party to a conflict (other than medical personnel and chaplains covered by Article 33 of the Third Convention) are combatants, that is to say, they have the right to participate directly in hostilities.").

33. *E.g.*, COMMENTARY ON 1977 ADDITIONAL PROTOCOLS, *supra* note 13, at 515 ("All members of the armed forces are combatants, and *only* members of the armed forces are combatants.") (emphasis added).

34. "[AP I] contains significant advances in LOAC. The core LOAC concepts of distinction, unnecessary suffering, and proportionality, formerly found only in customary law, are codified and described in [AP I], if only in broad terms." However, some of the provisions of AP I pushed the boundaries of LOAC too far and rendered it objectionable to the United States (as well as many other nations) leading the United States not to ratify it. For example, Article 1(3) makes the treaty applicable to international armed conflicts. Then, Article 1(4) "goes on to expand the definition of what constitutes an international armed conflict" making the treaty applicable to CARs—conflicts purporting to resist colonial domination, alien occupation, or a racist regime. The United States was unable to accept this definition, arguing that it "blurs national and international conflicts, making the applicability of [LOAC], turn on the asserted motive of a rebel force." The United States has not ratified AP I; the AP I version on an international armed conflict is not accepted as customary international law. *See* Protocol I, *supra* note 1, art. 1(3)–(4) (noting that the protections of international customary law are available to persons whose cases are not covered by the Protocol); SOLIS, *supra* note 2, at 123–24 (2010).

35. *See generally* Protocol Additional to the Geneva Convention of August 12, 1949, and relating to the Protection of Victims of Non-International Armed Conflicts (Protocol II), Dec. 7, 1978, 1125 U.N.T.S. 610 [hereinafter Protocol II] (containing no definition of the term "combatant").

36. *See* Protocol I, *supra* note 1, art. 43(1) ("The armed forces of a Party to a conflict consist of all organized armed forces, groups and units which are under a command responsible to that Party for the conduct of its subordinates, even if that Party is represented by a government or an authority not recognized by an adverse Party. Such armed forces shall be subject to an internal disciplinary system which, *inter alia*, shall enforce compliance with the rules of international law applicable in armed conflict.").

37. Although AP II does not contain an analogous definition of combatant as that in AP I, it does require that civilians be "protected against the dangers arising from military operations." *See* Protocol II, *supra* note 35, art. 13.

38. *See* COMMENTARY, CONVENTION (I) FOR THE AMELIORATION OF THE CONDITION OF THE WOUNDED AND SICK IN ARMED FORCES IN THE FIELD. GENEVA, AUG. 12, 1949 43–44 (Jean Pictet ed., 1949) [hereinafter WOUNDED AND SICK COMMENTARY] ("From the very outset, in the course of the first discussions of a general character, divergences of view became apparent. A considerable number of delegations were opposed, if not to any and every provision in regard to civil war, at any rate to the unqualified application of the Convention to such conflicts. The principal criticisms of the Stockholm draft may be summed up as follows. It was said that it would cover in advance all forms of insurrection, rebellion, anarchy, and the break-up of States, and even plain brigandage. Attempts to protect individuals—might well prove to be at the expense of the equally legitimate protection of the State. To compel

the Government of a State in the throes of internal convulsions to apply to these internal disturbances the whole body of provisions of a Convention expressly concluded to cover the case of war would mean giving its enemies, who might be no more than a handful of rebels or common brigands, the status of belligerents, and possibly even a certain degree of legal recognition. There was also a risk of common or ordinary criminals being encouraged to give themselves a semblance of organization as a pretext for claiming the benefit of the Conventions, representing their crimes as 'acts of war' in order to escape punishment for them. A party of rebels, however small, would be entitled under the Conventions to ask for the assistance and intervention of a Protecting Power.").

39. *See, e.g.*, *id.* (explaining how the differences in opinion led to criticism of the Stockholm draft).

40. *See* Grand Jury Indictment, at 11–12, United States v. Walker Lindh, No. 02-37a (E.D. Va. Feb. 5, 2002), *available at* http://news.findlaw.com/hdocs/docs/lindh/uswlindh020502cmp.html (indicting a nonstate actor for conspiracy to kill Americans based on his involvement with al Qaeda); Memorandum from President Bush on the Humane Treatment of Taliban and al Qaeda to the Vice President et al. (Feb. 7, 2002) *available at* http://www.gwu.edu/~nsarchiv/NSAEBB/NSAEBB127/02.02.07.pdf (expressing his view that the Geneva Convention does not apply to the United States' conflict with al Qaeda because al Qaeda is not a state actor); MINISTRY OF DEFENSE, THE MANUAL OF THE LAW OF ARMED CONFLICT c. 15 (2004) (U.K.) ("Unlike combatants in an international armed conflict, members of dissident armed forces remain liable to prosecution for offences under domestic law."); Chris Jenks & Eric Talbot Jensen, *Indefinite Detention Under the Laws of War*, 22 STAN. L. & POL'Y REV. 41, 54 (2011) ("Increasingly prevalent in today's conflicts involving non-state actors and asymmetric warfare, are law of war provisions which govern detention of those who directly participate in hostilities, perform a continuous combat function for an organized armed group, or otherwise pose a security risk or threat.").

41. *See* Protocol I, *supra* note 1, art. 43 (defining and describing "armed forces" and "combatants").

42. However, hostilities are not always necessary; the law applicable to IAC also applies to uncontested occupations. An occupation is simply "taking firm possession of enemy territory for the purpose of holding it." *See* DEPT. OF THE ARMY, FM27-10, THE LAW OF LAND WARFARE, ¶ 352(a) (1956) *available at* http://www.aschq.army.mil/gc/files/fm27-10.pdf (defining the term "occupation" and outlining the effectiveness, maintenance, and termination of an occupation). Occupations, whether contested or not, are governed by the Geneva Convention for the protection of civilians. *E.g.*, Geneva Convention Relative to the Protection of Civilian Persons in Time of War art. 2, Aug. 12, 1949, 6 U.S.T. 3516, 3518 ("The Convention shall also apply to all cases of partial or total occupation of the territory of a High Contracting Party, even if the said occupation meets with no armed resistance.").

43. *But see* Geoffrey S. Corn, *Hamdan, Lebanon, & the Regulation of Hostilities: The Need to Recognize a Hybrid Category of Armed Conflict*, 40 VAND. J. TRANSNAT'L L. 295, 309 (2007) (arguing that the either/or approach to determining whether a conflict is international or noninternational fails to consider the possibility of a hybrid category of armed conflict: extraterritorial noninternational armed conflict).

44. *See* WOUNDED AND SICK COMMENTARY, *supra* note 38, at 43–44 ("Insurgents . . . are not all brigands.").

45. *E.g.*, *id.* (stating that the reason states did not want to extend the scope of Common Article 3 was so as not to extend legitimacy to internal dissident groups).

46. *See* Protocol II, *supra* note 35, art. 13(1)–(2) (detailing protections afforded to civilian populations).

47. *See id.* ("Civilians shall enjoy the protection afforded by this Part, unless and for such time as they take a direct part in hostilities.").

48. *See* COMMENTARY ON 1977 ADDITIONAL PROTOCOLS, *supra* note 13, at 1449–53.

49. *See id.* at 1449 (emphasis added).

50. *See id.* at 1452 (emphasis added).

51. *See* Melzer, *supra* note 2, at 36 (explaining that civilians cease to be identified as such when they engage in continuous combat as members of organized armed groups).

52. *Id.* at 27–28 (emphasis added).

53. The authors use the term "belligerent" in lieu of "combatant" for the purpose of distinguishing nonstate belligerent operatives (who lack any legal privilege to participate in hostilities) with state belligerent operatives, who are designated "combatants" in accordance with Article 43 of AP I. However, this is not intended to suggest any fundamental difference in the authority to subject individuals of either designation to deliberate attack.

54. *See, e.g.*, CTR. FOR ARMY LEADERSHIP, COMMAND AND GEN. STAFF COLL., ARMY LEADERSHIP: COMPETENT, CONFIDENT, AND AGILE FM 6-22 (2006), *available at* http://usacac.army.mil/cac2/Repository/Materials/fm6-22.pdf (explaining the fundamental leadership competencies required of all army officers and army civilians); *see also* Yamashita v. Styer, 66 S. Ct. 340, 353–58 (1946) (Murphy, J., dissenting) (arguing that the majority's reasoning for charging the Japanese general with war atrocities committed by troops under his command but which he did not order was without precedent in international law and a violation of due process); Corn, *supra* note 24, at 259–61 (proposing the functional discretion test as an alternate approach for determining the legal limits of using civilian personnel in battlefield support functions).

55. *See* COMMENTARY, *supra* note 24, at 45–48 (discussing the historic dialogue on the meaning of a belligerent and final rule that "[b]elligerents are persons belonging to organized military forces . . . provided that such [military forces] fulfill the following conditions: (a) that of being commanded by a person responsible for his subordinates.").

56. The Westphalian system refers to the balance-of-power system that arose out of the Peace of Westphalia in 1648, in which the power of a state was determined by the number of other states with "substantially equal strength." *See* HENRY KISSINGER, DIPLOMACY 21 (1994); *see also* WOUNDED AND SICK COMMENTARY, *supra* note 38, at 43–44 (providing commentary to Common Article 3 in which states refused to recognize grant legal authority to rebel groups engaged in an armed conflict against a state); GIULIANA ZICCARDI CAPALDO, THE PILLARS OF GLOBAL LAW, 310 tbl. 1 (2008), *available at* http://www.asil.org/teachingmaterial/material/public/other01.pdf (describing the Westphalian system).

57. *See* Geneva (III) Treatment of Prisoners, *supra* note 19, art. 4 (describing prisoners of war as inter alia either members of the armed forces, members of the militia, or persons who accompany members of the armed forces or the militia, and thus suggesting that a prisoner of war is someone who likely exercised a belligerent function).

58. *See* Melzer, *supra* note 2, at 31–32 (describing the way in which to determine membership in an organized armed group).

59. *See* Geneva (III) Treatment of Prisoners, *supra* note 19, art. 3 ("In the case of armed conflict not of an international character occurring in the territory of one of the High Contracting Parties, each Party to the conflict shall be bound to apply, as a minimum, the following provisions. . . .").

60. *See* Melzer, *supra* note 2, at 33.

61. *See* Geoffrey S. Corn, *Thinking the Unthinkable: Has the Time Come to Offer Combatant Immunity to Non-State Actors?*, 22 STAN. L. & POL'Y REV. 253, 262–63 (2011) (comparing the treatment of the Taliban and al Qaeda to illustrate that "POW qualification has been and continues to be reserved for individuals fighting on behalf of a state, and therefore this qualification is inapplicable in the context of non-international armed conflicts").

62. *Compare* Protocol I, *supra* note 1, art. 43 (defining combatant status through group association in the context of an IAC) *with* Protocol II, *supra* note 35 (providing no definition of combatant but mandating that civilians be protected from attack unless they take a direct part in hostilities, thereby requiring a conduct-biased analysis in the context of a NIAC).

63. *See Appendix B, A Soldier's Task: Use Force Appropriately, ROE TRAINING, NEWSLETTER No. 96-6*, GLOBALSECURITY.ORG, http://www.globalsecurity.org/military/library/report/call/call_96-6_roeappb1.htm (last visited Apr. 1, 2011) (describing the procedures army personnel should follow in anticipating an attack and determining the amount of force needed under the circumstances).

64. *See, e.g., Obama Declares America "At War" with Al Qaeda, Offers New Security Initiatives*, CNSNEWS (Jan. 7, 2010), http://www.cnsnews.com/node/59446 (recognizing that America is "at war against al Qaeda").

65. A genuine DPH situation refers to a situation in which a civilian is engaged in "acts, which by their nature and purpose are intended to cause actual harm to the personnel and equipment of the armed forces. Thus [becoming a] legitimate target, [but] only for as long as he takes part in hostilities." *See* COMMENTARY ON 1977 ADDITIONAL PROTOCOLS, *supra* note 13, at 618.

66. "Consequently, under IHL, the decisive criterion for individual membership in an organized armed group is whether a person assumes a continuous function for the group involving his or her direct participation in hostilities (hereafter: 'continuous combat function'). Continuous combat function does not imply *de jure* entitlement to combatant privilege. Rather, it distinguishes members of the organized fighting forces of a non-state party from civilians who directly participate in hostilities on a merely spontaneous, sporadic, or unorganized basis, or who assume exclusively political, administrative or other non-combat functions." *See* Melzer, *supra* note 2, at 33–34.

67. *See id.* at 36 (providing the standard used for determining the targetability of belligerents in a NIAC: "In non-international armed conflict, organized armed groups constitute the armed forces of a non-state party to the conflict and consist only of individuals whose continuous function it is to take a direct part in hostilities ('continuous combat function').").

68. *See id.* at 45 (explaining that DPH "refers to specific hostile acts carried out by individuals," but not mentioning CCF).

69. *See id.* at 34 (emphasis added).

70. *See id.* ("Individuals who continuously accompany or support an organized armed group, but whose function does not involve direct participation in hostilities, are not members of that group within the meaning of IHL. Instead, they remain civilians assuming support functions, similar to private contractors and civilian employees accompanying state armed forces.").

71. *See id.* at 35 ("In practice, the principle of distinction must be applied based on information which is practically available and can reasonably be regarded as reliable in the prevailing circumstances. A continuous combat function may be openly expressed through the carrying of uniforms, distinctive signs, or certain weapons. Yet it may also be identified on the basis of conclusive behavior, for example where a person has repeatedly directly participated in hostilities in support of an organized armed group in circumstances indicating that such conduct constitutes a continuous function rather than a spontaneous, sporadic, or temporary role assumed for the duration of a particular operation.").

72. *See generally* Bill Boothby, *"And for Such Time As": The Time Dimension to Direct Participation in Hostilities*, 42 N.Y.U. J. INT'L L. & POL. 741 (2010) (finding civilians lose protection under the Guidance when engaging in acts of preparation for direct participation); Ryan Goodman & Derek Jinks, *The ICRC Interpretive Guidance on the Notion of Direct Participation in Hostilities Under International Humanitarian Law: An Introduction to the Forum*, 42 N.Y.U. J. INT'L L. & POL. 637 (2010) (critiquing the four critical perspectives on the interpretation of DPH); Nils Melzer, *Keeping the Balance between Military Necessity & Humanity: A Response to Four Critiques of the ICRC's Interpretive Guidance on the Notion of Direct Participation in Hostilities*, 42 N.Y.U. J. INT'L L. & POL. 831 (2010) (arguing that the ICRC's position on the definition of membership in an organized armed force can be accurately understood when Guidance sections are conjunctively read); W. Hays Parks, *Part IX of the ICRC "Direct Participation in Hostilities" Study: No Mandate, No Expertise, and Legally Incorrect*, 42 N.Y.U. J. INT'L L. & POL. 769 (2010) (providing a critical historical analysis of the Guidance in regard to protecting civilians during war); Michael N. Schmitt, *The Interpretive Guidance on the Notion of Direct Participation in Hostilities: A Critical Analysis*, 1 HARV. NAT'L SEC. L.J. 5 (2010) (noting that the Guidance formula for international armed conflict defines civilians negatively); Michael N. Schmitt, *Deconstructing Direct Participation in Hostilities: The Constitutive Elements*, 42 N.Y.U. J. INT'L L. & POL. 697 (2010) (explaining that civilians who "directly participate" in hostilities are not protected from attack); Kenneth Watkin, *Opportunity Lost: Organized Armed Groups & the ICRC "Direct Participation in Hostilities" Interpretive Guidance*, 42 N.Y.U. J. INT'L L. & POL. 641 (2010) (arguing that "organized armed groups" are classified with unique criteria and are not considered civilian).

73. *See id.*

74. *See* Melzer, *supra* note 2, at 33–34.

75. *See* Protocol II, *supra* note 35, art. 13(3) (allowing for the targeting of civilians "for such time as they take a direct part in hostilities"); *see also* Melzer, *supra* note 2, at 36 (allowing for the targeting of individuals who serve a "continuous combat function").

76. *See* Protocol I, *supra* note 1, art. 43(1)–(2) ("(1) The armed forces of a Party to a conflict consist of all organized armed forces, groups and units which are under a command responsible to that Party for the conduct of its subordinates, even if that Party is represented by a government or an authority not recognized by an adverse Party. Such armed forces shall be subject to an internal disciplinary system which, inter alia, shall enforce compliance with the rules of international law applicable in armed conflict. (2) Members of the armed forces of a Party to a conflict (other than medical personnel and chaplains covered by Article 33 of the Third Convention) are combatants, that is to say, they have the right to participate directly in hostilities.").

77. *See* Protocol II, *supra* note 35, art. 13(3) (allowing for the targeting of civilians "for such time as they take a direct part in hostilities.").

78. *See* COMMENTARY ON 1977 ADDITIONAL PROTOCOLS, *supra* note 13, at 618–19 ("The immunity afforded individual civilians is subject to an overriding condition, namely, on their abstaining from all hostile acts. Hostile acts should be understood to be acts which by their nature and purpose are intended to cause actual harm to the personnel and equipment of the armed forces. Thus a civilian who takes part in armed combat, either individually or as part of a group, thereby becomes a legitimate target, though only for as long as he takes part in hostilities.").

79. *See id.* at 618–19 (distinguishing between civilians directly participating in hostilities and those who are merely participating in the war effort).

80. *See, e.g.*, Prosecutor v. Boškoski, Case No. IT-04-82-A, Judgment on Appeal, ¶¶ 79–86 (Int'l Crim. Trib. for the Former Yugoslavia May 19, 2010) (noting that the burden of proof is not shifted from the armed force making the attack just because there is insufficient evidence in regard to the appearance of the victim).

81. *See* DEPT. OF THE ARMY, *supra* note 42, ¶ 3(a) (emphasis added).

82. The qualifier "not otherwise prohibited by international law" indicates that the authority to inflict harm terminates once an opponent is rendered hors de combat (out of combat by reason of wounds, sickness, or capture). *See* Protocol I, *supra* note 1, art. 41.

83. *See* Protocol I, *supra* note 1, art. 41(2)(a)–(c) (describing the three types of persons hors de combat as being those who are captured by an adverse party, show an intention to surrender, or have been incapacitated from wounds or illness).

84. *See* Melzer, *supra* note 2, at 27–36 (commenting on the "mutual exclusiveness of the concepts of civilian, armed forces, and organized armed groups").

85. *See id.* at 36 ("For the purposes of the principle of distinction in non-international armed conflict, all persons who are not members of state armed forces or organized armed groups of a party to the conflict are civilians and, therefore, entitled to protection against direct attack unless and for such time as they take a direct part in hostilities in non-international armed conflict, organized armed groups constitute the armed forces of a non-state party to the conflict and consist only of individuals whose continuous function it is to take a direct part in hostilities ('continuous combat function').").

86. Firing a weapon at a military force would be an act "which by [its] nature and purpose are intended to cause actual harm to the personnel and equipment of the armed forces" and therefore would result in loss of protection for the civilian. *See* COMMENTARY ON 1977 ADDITIONAL PROTOCOLS, *supra* note 13, at 618.

87. *See* Melzer, *supra* note 2, 27–28 (remarking that unclear boundaries between civilians and nonstate armed forces will result in potential harms for the civilian population).

88. *See id.* at 28 ("[T]his approach would seriously undermine the conceptual integrity of the categories of persons underlying the principle of distinction, most notably because it would create parties to non-international armed conflicts whose entire armed forces remain part of the civilian population.").

89. *See id.* at 77–82 (suggesting an imperative necessity to maintain restraint and proportionality even when dealing with an organized, albeit nonstate, armed force).

90. Proportionality requires that armed forces refrain from any "attack which may be expected to cause incidental loss of civilian life, injury to civilians, damage to civilian objects, or a combination thereof, which would be excessive in relation to the concrete and direct military advantage anticipated." *See* Protocol I, *supra* note 1, art. 51(4)(b).

91. *See id.* art. 52(2) ("Attacks shall be limited strictly to military objectives. In so far as objects are concerned, military objectives are limited to those objects which by their nature, location, purpose or use make an effective contribution to military action and whose total or partial destruction, capture or neutralization, in the circumstances ruling at the time, offers a definite military advantage.").

92. *See id.* art. 35(2) ("It is prohibited to employ weapons, projectiles and material and methods of warfare of a nature to cause superfluous injury or unnecessary suffering.").

93. *See id.* art. 35(1) (giving Parties the right to "*choose* methods and means of warfare") (emphasis added).

94. "'Warfare... justifies subjecting an enemy to massive and decisive force, and the suffering that it brings. Military necessity only justifies the infliction of suffering upon an enemy combatant.... [H]owever... military necessity only justifies the infliction of as much suffering as is necessary to bring about the submission of an enemy.' Military necessity is the balance between destruction of the enemy and humanity." SOLIS, *supra* note 2, at 270 (2010) (citing Geoffrey Corn, *International & Operational Law Note: Principal 4: Preventing Unnecessary Suffering*, THE ARMY LAWYER 1, 50–51 (1998)).

95. *See Disarmament: The Convention on Certain Conventional Weapons*, UNOG, http://www.unog.ch/80256ee600585943/(httppages)/4f0def093b4860b4c1257180004b1b30?opendocument (last visited Apr. 2, 2011) (outlining the prohibition on certain types of weapons which cause unnecessary suffering).

96. Protocol I, *supra* note 1, art. 51(5).

97. *Id.* art. 51(5)(b) ("An attack which may be expected to cause incidental loss of civilian life, injury to civilians, damage to civilian objects, or a combination thereof, which would be *excessive in relation* to the concrete and direct military advantage anticipated.") (emphasis added).

98. Article 51 of AP I is entitled "Protection of the Civilian Population." *Id.* art. 51.

99. *Id.* art. 51(1) (stating that Article 51 was written to increase the protection of civilians under LOAC).

100. *See generally* Protocol I, *supra* note 1.

101. *See, e.g., id.* art. 48, 51, 57 (providing restrictions on otherwise lawful attacks for the protection of the civilian population but providing no protection or formula for restraint for purely military objectives).

102. *See, e.g.*, Sue Chan, *Iraq Faces Massive U.S. Missile Barrage*, CBS NEWS (Jan. 24, 2003), http://www.cbsnews.com/stories/2003/01/24/eveningnews/main537928.shtml (describing the war plan prior to the invasion of Iraq that would have called for an attack of eight hundred cruise missiles over two days against Baghdad in order to bring about the "psychological destruction of the enemy's will to fight").

103. Protocol I, *supra* note 1, art. 41 ("A person... *hors de combat* shall not be made the object of attack.").

104. COMMENTARY ON 1977 ADDITIONAL PROTOCOLS, *supra* note 13, at 482 ("It is a fundamental principle of the law of war that those who do not participate in the hostilities shall not be attacked. In this respect harmless civilians and soldiers *hors de combat* are *a priori* on the same footing.").

105. Melzer, *supra* note 2, at 77 ("Loss of protection from direct attack, whether due to direct participation in hostilities (civilians) or continuous combat function (members of organized armed groups), does not mean that the persons concerned fall outside the law.").

106. Protocol I, *supra* note 1, art. 51(5)(b) (providing protection only for civilians).

107. Melzer, *supra* note 2, at 80–82.

108. *Id.* at 80 (conceding that "it is impossible to determine... the precise amount of force to be used").

109. *Id.*

110. *Id.* at 80–81 (emphasis added).

111. *Id.* at 81.

112. Melzer, *supra* note 2, at 81.

113. *Id.* at 80–82 (implying that proportionality might require that identifiable members of belligerent armed groups be nonlethally arrested, whether by the state's armed or police forces).

114. *Id.* at 81 (proffering example of civilians directly participating in hostilities).

115. *Id.* at 79 (forbidding causing injury or destruction when not necessary in order to accomplish a legitimate military objective).

116. Commentary on 1977 Additional Protocols, *supra* note 13, at 482 (providing the standard for lawful belligerents. "It is a fundamental principle of the law of war that those who do not participate in the hostilities shall not be attacked. In this respect harmless civilians and soldiers *hors de combat* are *a priori* on the same footing.").

117. In a traditional IAC, combatants are required to distinguish themselves from the civilian population, but in a modern NIAC, nonstate forces have no desire or incentive to do so. *See* Protocol I, *supra* note 1, art. 44(3), (obliging combatants to distinguish themselves in combat in order to protect civilian populations); *see also* Geoffrey S. Corn, *Thinking the Unthinkable: Has the Time Come to Offer Combatant Immunity to Non-State Actors?*, 22 Stan. L. & Pol'y Rev. 253 (2011) (arguing for the extension of lawful combatant status as a means of ensuring compliance with humanitarian law).

118. *See* Melzer, *supra* note 2, at 82.

119. *Id.* at 82 n. 221.

120. *See supra* note 72 and accompanying text.

121. *See* Andrew Curry, *Don't Shoot Until You See the Whites of Their Eyes!*, Military Channel, http://military.discovery.com/history/revolutionary-war/bunker-hill/bunker-hill-2.html (noting the colonial armies forced delay in opening fire in order to maximize British casualties during the early stages of the American Revolution); *see also* Department of the Army, FM3–24 Counterinsurgency, 5–12 (2006) (discussing the moral and strategic need for restraint in counterinsurgency operations, such as the need to avoid turning the populace against the counterinsurgency effort due to collateral or intentional injuries to innocents).

122. In other words, conflating what the law is with what the law should be. *See "Beck's Law Dictionary": A Compendium of International Law Terms and Phrases*, http://people.virginia.edu/~rjb3v/latin.html#lex_lata.

123. *See* Chairman of the Joint Chiefs of Staff, Instruction 3121.01A, Standing Rules of Engagement for U.S. Forces (2000) (establishing guidelines for the use of force by U.S. forces operating in different environments). The standing rules, or SROE, provide not only an example of how rules of engagement reflect LOAC but also operational and policy constraints.

124. For example, successive commanders of NATO's International Security Assistance Force in Afghanistan have issued a "tactical directive," which limited where and under what circumstances coalition militaries may use force. The version General Stanley McChrystal issued

is notable for its limitations on the use of airstrikes. Memorandum from the Headquarters of the NATO International Security Assistance Force (July 6, 2009) *available at* http://www.nato.int/isaf/docu/official_texts/Tactical_Directive_090706.pdf. *See also* Andrew Curry, *supra* note 121 (noting the colonial armies forced delay in opening fire in order to maximize British casualties during the early stages of the American Revolution); DEPT. OF THE ARMY, *supra* note 121, at 5–12 (discussing the need and reasons for restraint in counterinsurgency operations).

125. *See* Melzer, *supra* note 2, at 27–28.

126. COMMENTARY ON 1977 ADDITIONAL PROTOCOLS, *supra* note 13, at 482 (remarking that persons hors de combat may not be deliberately attacked).

127. *See Life in Afghanistan: Taliban Soldiers*, GUARDIAN UNLIMITED, http://www.guardian.co.uk/pictures/image/0,8543,-10304255641,00.html (last visited Apr. 2, 2011) (showing an image of Taliban soldiers in civilian clothing).

128. *See* Melzer, *supra* note 2, at 33–36 (positing that membership to an armed group "must depend on whether the continuous function assumed by the individual corresponds to that collectively exercised by the group as a whole").

129. *See* Geneva (III) Treatment of Prisoners, *supra* note 14, art. 4.

130. CCF does not require that an individual satisfy the four requirements in the Third Geneva Convention but merely requires that they be involved in a belligerent group on more than just a "spontaneous, sporadic or unorganized basis." *Id. See also* Melzer, *supra* note 2, at 33–34 (noting that the loss of protection is coextensive with the individuals, direct participation in hostilities).

131. Melzer, *supra* note 2, at 39.

132. *Id.*

133. *See* Protocol I, *supra* note 1, art. 51(3) ("Civilians shall enjoy the protection afforded by this Section, unless and for such time as they take a direct part in hostilities."); COMMENTARY ON 1977 ADDITIONAL PROTOCOLS, *supra* note 13, at 618 ("The immunity afforded individual civilians is subject to . . . their abstaining from all hostile acts.").

134. *See* COMMENTARY ON 1977 ADDITIONAL PROTOCOLS, *supra* note 13, at 618; Melzer, *supra* note 2, at 33.

135. This is evidenced through the limited scope of authority granted in Article 51. *See* COMMENTARY ON 1977 ADDITIONAL PROTOCOLS, *supra* note 13, at 615–28.

136. *See* Melzer, *supra* note 2, at 82.

137. *Id.* at 43 (noting the necessity of point to "specific acts" when establishing direct participation).

138. *See id.* (focusing solely on individual participation in hostilities).

CHAPTER 5

* Professor of Law, Roger Williams University School of Law. I thank Ken Anderson, Sam Estreicher, Monica Hakimi, Dick Jackson, Bill Kuebler, Mike Lewis, David Luban, and Matt Waxman, and participants at the 2012 National Security Law Conference at the South Texas College of Law for comments on earlier drafts.

1. *See* Protocol Additional to the Geneva Conventions of Aug. 12, 1949, and relating to the Protection of Victims of International Armed Conflicts, Annex I, Art. 48, June 8, 1977, 1125 U.N.T.S. 3 [hereinafter Additional Protocol I]; *see also* STEPHEN L. CARTER, THE VIOLENCE

OF PEACE: AMERICA'S WARS IN THE AGE OF OBAMA 58 (2011) (referring to principle of "discrimination" that bars targeting of noncombatants); YORAM DINSTEIN, THE CONDUCT OF HOSTILITIES UNDER THE LAW OF INTERNATIONAL ARMED CONFLICT 89 (2d ed., 2010); Int'l Inst. Humanitarian L., Manual on the Law of Non-International Armed Conflict art. 1.2.2 (Michael N. Schmitt, Charles H. B. Garraway, & Yoram Dinstein, Drafting Committee 2006) (describing principle of distinction as "foundation" of LOAC) (citation omitted).

2. *See* Additional Protocol I, arts. 52 and 57.

3. *See* MICHAEL WALZER, JUST AND UNJUST WARS: A MORAL ARGUMENT WITH HISTORICAL ILLUSTRATIONS 154–55 (1977) (IHL proportionality is insufficient; instead, attacker must "minimize [harm to civilians]... accepting costs to himself"); Avishai Margalit & Michael Walzer, *Israel: Civilians and Combatants*, N.Y. REV. BKS., May 14, 2009, *available at* http://www.nybooks.com/articles/archives/2009/may/14/israel-civilians-combatants/(same); David Luban, *Risk Taking and Force Protection*, *in* READING WALZER (Itzak Benhaji & Naomi Sussman eds., Routledge, forthcoming 2013), *available at* http://ssrn.com/abstract=1855263 (same); Paul W. Kahn, *The Paradox of Riskless Warfare*, 22(3) PHIL. & PUB. POL'Y Q. 2, 4 (Summer 2002) (claiming that "[w]ithout the imposition of mutual risk, warfare is not war at all"). Since much of this chapter critiques Luban's account, I should note my great respect for Luban's work in the areas of legal ethics and international criminal law. *See, e.g.*, David Luban, *The Torture Lawyers of Washington*, *in* LEGAL ETHICS AND HUMAN DIGNITY 162, 176–80, 200–02 (2007) (analyzing flaws in Justice Department legal memoranda authorizing "enhanced interrogation techniques" such as waterboarding); DAVID LUBAN, JULIE R. O'SULLIVAN, & DAVID P. STEWART, INTERNATIONAL AND TRANSNATIONAL CRIMINAL LAW 14–15 (2010) (discussing legal principles that predicate punishment on fair notice that conduct violates international norms).

4. *See* Christine Jolls, Cass R. Sunstein, & Richard H. Thaler, *A Behavioral Approach to Law and Economics*, *in* BEHAVIORAL LAW AND ECONOMICS 13, 45–46 (Cass R. Sunstein ed., 2000) (noting tendency to inappropriately discount future costs); Daniel Read, *Intertemporal Choice*, *in* BLACKWELL HANDBOOK OF JUDGMENT AND DECISION MAKING 424, 428–29 (Derek J. Koehler & Nigel Harvey eds., 2004) (noting tendency of individuals to prefer "smaller-sooner reward"); *cf.* DANIEL KAHNEMAN, THINKING, FAST AND SLOW 245–48 (2011) (noting that even expert planners often base time estimates for completion of a project on their current enthusiasm and fail to consider less vivid factors that hinder a project over time, such as personal crises, bureaucratic interference, or conflicts with other goals).

5. *See* Jeffrey J. Rachlinski, *A Positive Psychological Theory of Judging in Hindsight*, *in* BEHAVIORAL LAW AND ECONOMICS, *supra* note 4, at 95, 95 (maxims such as "hindsight... is '20/20'" indicate that "[l]earning how the story ends... [distorts] our perception of what could have been predicted"); Neal J. Roese, *Twisted Pair: Counterfactual Thinking and the Hindsight Bias*, *in* HANDBOOK OF JUDGMENT AND DECISION MAKING, *supra* note 4, at 258, 260–61 (hindsight bias is "the tendency to believe that an event was predictable before it occurred, even though for the perceiver it was not").

6. *See* Leslie Green, *What Is—Why Is There—the Law of War?*, *in* THE LAW OF ARMED CONFLICT: INTO THE NEXT MILLENNIUM 141, 148 (Michael N. Schmitt & Leslie C. Green eds., 1998), also cited as 71 INT'L L. STUDIES (U.S. Naval War Coll.) (quoting Edward Gibbon in *Decline and Fall of the Roman Empire* as citing Roman leaders' appreciation of the "permanent benefits [of]... a moderate use of conquest; and a just apprehension lest the desolation which we reflect

on the enemy's country... may be retaliated on our own"); *see also* Peter Margulies, *The Fog of War Reform: Change and Structure in the Law of Armed Conflict After September 11*, 95 MARQUETTE L. REV. 1417 (2012) (discussing roots of law of war in temporal judgment and signaling).

7. *See* Laurie R. Blank, *Finding Facts But Missing the Law: The Goldstone Report, Gaza and Lawfare*, 43 CASE W. RES. J. INT'L L. 279 (2010) (criticizing factual and legal errors in Goldstone Report, which contains disputed and partially repudiated criticisms of Israeli Defense Force's 2008–2009 incursion into the Gaza Strip to stop persistent rocket attacks); Chris Jenks & Geoffrey Corn, *Siren Song: The Implications of the Goldstone Report on International Criminal Law*, at 3, *available at* http://ssrn.com/abstract=1788542 (noting that Goldstone Report engaged in "*ex post facto* evaluation of targeting decisions... in contravention of a long established rule of war crimes liability" which requires "evaluation of targeting decisions... through the perspective of the military commander at the time the judgment at issue was made").

8. *See* ERIC A. POSNER, LAW AND SOCIAL NORMS 18–20 (2000) (discussing signaling in public and private law); Michael Spence, *Signaling in Retrospect and the Informational Structure of Markets*, 92 AM. ECON. REV. 434, 436–38 (2002) (signaling in markets); Robert Jervis, *Deterrence and Perception*, 7 INT'L SECURITY 3, 6–7 (1982–1983) (noting that in international affairs, misunderstood signals about intentions can lead to war).

9. *See* WALZER, JUST AND UNJUST WARS, *supra* note 3, at 156.

10. *Cf.* Michael Schmitt, *Bellum Americanum: The U.S. View of Twenty-First-Century War and Its Possible Implications for the Law of Armed Conflict*, *in* THE LAW OF ARMED CONFLICT: INTO THE NEXT MILLENNIUM, *supra* note 6, at394 (providing history of recent efforts to spare civilians through development of IHL); *see also* GARY SOLIS, THE LAW OF ARMED CONFLICT: INTERNATIONAL HUMANITARIAN LAW IN WAR 51 (2010) (noting enhanced destructive power of armaments, including machine guns and air power); Michael N. Schmitt, *Military Necessity and Humanity in International Humanitarian Law: Preserving the Delicate Balance*, 50 VA. J. INT'L L. 795, 796–800 (2010) (describing efforts to harmonize IHL with destructive technology).

11. *See* Additional Protocol I, arts. 52 and 57. Article 57(2)(a)(iii) provides that a party to an armed conflict should "refrain from... any attack which may be expected to cause incidental [harm to civilians]... excessive in relation to the concrete and direct military advantage thereof." The United States accepts Article 57 as binding customary international law, although it has not ratified AP I.

12. *See* WALZER, JUST AND UNJUST WARS, *supra* note 3, at 152–54.

13. *See* Margalit & Walzer, *supra* note 3 (emphasis in original); *cf.* Michael Walzer, *Kosovo*, *in* ARGUING ABOUT WAR 99, 101 (2004) (arguing that conducting hostilities through air power with concomitant reluctance to send in ground troops is immoral); Luban, *supra* note 3, at 24–25 (discussing Walzer and Margalit's account).

14. Luban concedes that his approach would not require "risk [that] is too great," Luban, *supra* note 3, at 24. However, Luban does not define what he means by "too great," beyond implying that the soldier would have no duty to warn in the cellar scenario if the soldier *knew* that German soldiers were lurking. *Id.* Elsewhere, Luban argues that soldiers must accept tactical changes yielding increases in risk that are substantially greater than these changes' reductions in risk to civilians. *Id.* at 28.

15. *Id.* at 25; *cf.* Charles A. Cantu, *Distinguishing the Concept of Strict Liability for Ultra-Hazardous Activities from Strict Products Liability Under Section 402A of the Restatement*

(Second) of Torts: Two Parallel Lines of Reasoning That Should Never Meet, 35 AKRON L. REV. 31, 34–43 (2001) (discussing rationales for strict liability); James A. Henderson, Jr. & Aaron D. Twerski, *Closing the American Products Liability Frontier: The Rejection of Liability without Defect*, 66 N.Y.U. L. REV. 1263, 1276 n. 54 (1991) (discussing limited scope of strict liability in U.S. courts).

16. WALZER, JUST AND UNJUST WARS, *supra* note 3, at 156.

17. *Id.* and 156 n. *; *cf.* Adil Ahmad Haque, *Protecting and Respecting Civilians: Correcting the Substantive and Structural Aspects of the Rome Statute*, 14 NEW CRIM. L. REV. 519 (2011) (arguing that the Rome Statute, which describes crimes against humanity triable by the International Criminal Court, fails to require adequate care by states in preventing harm to civilians during armed conflict).

18. *See* Samuel W. Buell, *What Is Securities Fraud?*, 61 DUKE L.J. 511, 577 (2011) (discussing gradations of liability). One could argue that theorists espousing the duty to risk merely want a *heightened* standard of reasonableness beyond what proportionality requires, not strict liability *tout court*. However, even this more cabined standard introduces finer gradations into the commander's calculus. That change increases the risk of hindsight bias that impairs command decisions.

19. *See* Michael A. Newton, *Drone Warfare and the Changing Face of Humanitarian Law: Keynote Address to the 2010 Sutton Colloquium*, 39 DENVER J. INT'L L. & POL'Y 601, 606 (2011) ("successful military operations require that commanders...are educated and empowered to make important and accurate decisions").

20. *See* MARK OSIEL, THE END OF RECIPROCITY: TERROR, TORTURE, AND THE LAW OF WAR 312–14 (2009) (discussing IHL's importance to military identity, culture, and discipline).

21. *See* Jolls, *supra* note 4.

22. *See* Baruch Fischhoff, *For Those Condemned to Study the Past: Heuristics and Biases in Hindsight*, *in* JUDGMENT UNDER UNCERTAINTY: HEURISTICS AND BIASES 335, 342 (Daniel Kahneman, Paul Slovic, & Amos Tversky eds., 1982) ("Consider decision makers who have been caught unprepared by some turn of events and who try to see where they went wrong...If, in retrospect, the event appears to have been relatively likely, they can do little more than blame themselves for not taking the action that their knowledge seems to have dictated"); *cf.* Rachlinski, *Judging in Hindsight*, *supra* note 5, at 95; Roese, *supra* note 5, at 260–61.

23. *See* Jeffrey J. Rachlinski, *Heuristics, Biases, and Governance* (hereinafter *Heuristics and Governance*), *in* BLACKWELL HANDBOOK OF JUDGMENT AND DECISION MAKING, *supra* note 5, at 567, 575–76.

24. *See* POSNER, LAW AND SOCIAL NORMS, *supra* note 8, at 18–20; Richard H. McAdams, *Signaling Discount Rates: Law, Norms, and Economic Methodology*, 110 YALE L.J. 625, 674–75 (2001).

25. For example, an exchange of conciliatory gestures between two warring parties can forge a shared stake in governing institutions. *See* Matthew Hoddie & Caroline Hartzell, *Signals of Reconciliation: Institution-Building and the Resolution of Civil Wars*, 7 INT'L STUD. REV. 21, 38 (2005).

26. John K. Setear, *Responses to Breach of a Treaty and Rationalist International Relations Theory: The Rules of Release and Remediation in the Law of Treaties and the Law of State Responsibility*, 83 VA. L. REV. 1, 94 (1997) (crossed signals can yield spiraling tensions between nations, as when one

country conducts naval exercises which another perceives as an aggressive move; when the second country responds to the perceived aggression, the first country must retaliate).

27. *See* OSIEL, *supra* note 20, at 384–85.

28. *See* David Luban, *The Torture Lawyers of Washington*, *in* LEGAL ETHICS AND HUMAN DIGNITY 162, 176–80, 200–202 (2007); *cf.* ANDREW T. GUZMAN, HOW INTERNATIONAL LAW WORKS: A RATIONAL CHOICE THEORY 212–13 (2008) (discussing holistic signaling about disposition to comply with international law); David H. Moore, *A Signaling Theory of Human Rights Compliance*, 97 NW. U.L. REV. 879 (2003); Christopher J. Borgen, *Hearts and Minds and Laws: Legal Compliance and Diplomatic Persuasion*, 50 S. TEX. L. REV. 769, 774–78 (2009); *see also* Jack Goldsmith & Daryl Levinson, *Law for States: International Law, Constitutional Law, Public Law*, 122 HARV. L. REV. 1791, 1826–28, 1835 (2009) (discussing international and constitutional law as coordinated games in which parties decline short-term benefits to realize gains over time, such as gains flowing from reputation for fair dealing).

29. *See* SOLIS, *supra* note 10, at 254 (distinction entails "duty to take reasonable steps to determine whether or not a person or object is a legitimate target") (citation omitted).

30. *See* Schmitt, et al., Manual, *supra* note 1, at 2.1.1.4 cmt. 5.

31. *See* Additional Protocol I, arts. 52 and 57.

32. *See* GABRIELLA BLUM & PHILIP B. HEYMANNN, LAWS, OUTLAWS & TERRORISTS: LESSONS FROM THE WAR ON TERRORISM 69–91 (2010) (analyzing origins of doctrine, comparative perspective on United States and Israel, and legal concerns about targeted killings); Kenneth Anderson, *Targeted Killing in U.S. Counterterrorism Strategy and Law*, *in* LEGISLATING THE WAR ON TERROR: AN AGENDA FOR REFORM 346 (Benjamin Wittes ed., 2009), *available at* http://ssrn.com/abstract=1415070 (arguing that drone attacks, given appropriate constraints, are consistent with international law of self-defense); William C. Banks & Peter Raven-Hansen, *Targeted Killing and Assassination: The U.S. Legal Framework*, 37 U. RICH. L. REV. 667, 679–81 (2003) (same); Robert M. Chesney, *Who May Be Killed? Anwar Al-Awlaki as a Case Study in the International Legal Regulation of Lethal Force*, 13 Y.B. INT'L HUMANITARIAN L. 3 (2010) (same); Michael N. Schmitt, *Drone Attacks under the* Jus ad Bellum *and* Jus in Bello*: Clearing the "Fog of Law,"* 13 Y.B. INT'L HUMANITARIAN L. 311, 320–21 (2010) (same); *see also* Harold Hongju Koh, *The Obama Administration and International Law*, Annual Meeting of the American Society of International Lawyers, March 25, 2010, *available at* http://www.state.gov/s/l/releases/remarks/139119.htm (reiterating administration's support of IHL); *but see* UN GA, May 28, 2010, Human Rights Council, 14th Sess., Philip Alston, *Report of the Special Rapporteur on extrajudicial, summary or arbitrary executions: Study on targeted killings* (arguing that IHL and international human rights law impose greater restrictions on targeted killing); Mary Ellen O'Connell, *Unlawful Killing with Combat Drones: A Case Study of Pakistan, 2004–2009*, *in* SHOOTING TO KILL: THE LAW GOVERNING LETHAL FORCE IN CONTEXT (Simon Bronitt ed., forthcoming 2012), *available at* http://ssrn.com/abstract=1501144 (same).

33. *See* Richard T. DeGeorge, *A Code of Ethics for Officers*, *in* MILITARY ETHICS: REFLECTIONS ON PRINCIPLES—THE PROFESSION OF ARMS, MILITARY LEADERSHIP, ETHICAL PRACTICES, WAR AND MORALITY, EDUCATING THE CITIZEN-SOLDIER 13, 20 (Malham M. Wakin ed., 1994).

34. *See* Jeffrey K. Walker, *Old Laws and New Wars*, 94 Proceedings of Ann. Meeting of ASIL 312, 313 (April 5–8, 2000) (asserting that "mark of a valorous officer has been an indifference to personal safety, a scorn for injury or death").

35. Noel Gayler, *Nuclear Deterrence—Its Moral and Political Implications*, *in* Military Ethics, *supra* note 33, at 157, 164–65 (arguing that "wars of attrition" are unethical, since they sacrifice soldiers without a reasonable hope of achieving a strategic objective).

36. *See* Ganesh Sitaraman, *Counterinsurgency, the War on Terror, and the Laws of War*, 95 Va. L. Rev. 1745, 1803 (2009) (noting importance of winning over populations in counterterrorism efforts); Gabriella Blum, *On a Differential Law of War*, 52 Harv. Int'l L.J. 163, 187–88 (2011) (arguing that United States should risk more than IHL requires, because of its international position and need to persuade others). On the importance of winning hearts and minds in the detention and trial of suspected terrorists, *see* John B. Bellinger III & Vijay M. Padmanabhan, *Detention Operations in Contemporary Conflicts: Four Challenges for the Geneva Conventions and Other Existing Law*, 105 A.J.I.L. 201 (2011); Vijay M. Padmanabhan, *Norm Internalization Through Trials for Violations of International Law: Four Conditions for Success and Their Application to Trials of Detainees at Guantanamo Bay*, 31 U. Pa. J. Int'l L. 427, 471–90 (2009) (arguing that military commission proceedings that are perceived as illegitimate do not mobilize global communities against terrorism).

37. *See* Margalit & Walzer, *supra* note 3, discussing Asa Kasher & Amos Yadlin, *Assassination and Preventive Killing*, 25(1) SAIS Rev. 45 (Spring 2005).

38. *See* Kasher & Yadlin, *supra* note 37, at 46; *cf.* Noam Neuman, *Applying the Rule of Proportionality: Force Protection and Cumulative Assessment in International Law and Morality*, 7 Y.B. Int'l Humanitarian L. 79, 91 (2004) (noting that the term "military advantage" in proportionality analysis includes force protection, since a commander who fails to protect soldiers has fewer soldiers remaining to achieve the next tactical goal).

39. David Luban, *Was the Gaza Campaign Legal?*, *in* 31 ABA Nat'l Security L. Rep't 2, 5 (January–February 2009) (asserting that in Gaza "the police are the police [and not combatants]"). Luban's more recent work on the duty to risk does not revise this conclusion. *See* Luban, *Risk Taking*, *supra* note 3, at 5–6 (repeating inflated numbers regarding civilian losses in Gaza campaign).

40. *See* Kristen Chick, *In Gaza, Rise of Hamas Military Wing Complicates Reconciliation with Fatah; The Rising Clout of Al Qassam in Gaza Dims Prospects for Mending the Hamas-Fatah Rift. Reconciliation Talks Are Slated for Nov. 9*, Christian Sci. Monitor, Nov. 4, 2010 (quoting veteran member of al Qassam, Hamas's military wing, as stating that "two-thirds of Hamas policemen are police by day and Al Qassam by night"); Ethan Bronner & Jennifer Medina, *Investigator on Gaza Was Guided By His Past*, N.Y. Times, April 20, 2011, at A4 (quoting Hamas spokesperson as acknowledging that most police officers were Hamas fighters); *cf.* Ellen Knickmeyer, *Hamas Asserts Role in Suicide Bombing; Israel Strikes in Gaza a Day after Fatal Attack Within Its Borders; 9 Palestinians Killed*, Wash. Post, Feb. 6, 2008, A14 (reporting claims by Gaza officials that Israeli strike on police station in retaliation for suicide bombing killed police officers and al Qassam members; statement implied that latter were regularly present in significant numbers at site).

41. *See* HCJ 769/02 The Public Committee Against Torture v. The State of Israel, para. 39 [Dec. 13, 2006] (PCAT), *available at* http://elyon1.court.gov.il/files_eng/02/690/007/A34/02007690.a34.pdf (asserting that fighters who make themselves regularly available to terrorist groups for acts of violence are directly participating in hostilities for such time as

they make themselves available; any interlude between acts of violence is merely "preparation" for further violence). In formulating targeting rules in this way, the PCAT Court interpreted broadly interpreted concepts that the International Committee of the Red Cross (ICRC) has interpreted more narrowly. *See* Nils Melzer, Legal Adviser, International Committee of the Red Cross, *Interpretive Guidance on the Notion of Direct Participation in Hostilities under International Humanitarian Law* 54 (May 2009), *available at* http://www.aco.nato.int/resources/20/Legal%20Conference/ICRC_002_0990.pdf (arguing that terrorist bomb maker would be immune from targeting for time not spent making bombs); *see also* Gabor Rona, *US Targeted Killing Policy Unjustified*, Jurist, Feb. 24, 2012, *available at* http://jurist.org/hotline/2012/02/gabor-rona-targeted-killing.php (criticizing United States' targeting criteria as unduly broad); *but see* Michael N. Schmitt, *Deconstructing Direct Participation in Hostilities: The Constitutive Elements*, 42 N.Y.U. J. INT'L L. & POL. 697, 731 (2010) (criticizing narrow view of causation in ICRC Guidance); Kenneth Watkin, *Opportunity Lost: Organized Armed Groups and the ICRC "Direct Participation in Hostilities" Interpretive Guidance*, 42 N.Y.U. J. INT'L L. & POL. 641, 661 (2010) (criticizing ICRC's enabling of "revolving door" mechanism for terrorist groups).

42. *See* United Nations Fact Finding Mission on the Gaza Conflict, *Human Rights in Palestine and Other Occupied Arab Territories*, A/HRC/12/48, September 15, 2009, para. 34, p. 12, *available at* http://www2.ohchr.org/english/bodies/hrcouncil/specialsession/9/docs/UNFFMGC_Report.pdf [hereinafter Goldstone Report]; *but see id.* at para. 34, p. 12 (conceding that "a great number of the Gaza policemen were recruited among Hamas supporters or members of Palestinian armed groups"). For critique of this and similar aspects of the Goldstone Report, *see* Peter Margulies, *Advising Terrorism: Hybrid Scrutiny, Safe Harbors, and Freedom of Speech*, 63 HASTINGS L.J. 455, 490–91 (2012) (discussing evidence that Gaza police were also Hamas fighters, and that this ruse dovetailed with terrorist groups' tendency to game the system of international law); *cf.* Blank, *Finding Facts But Missing the Law*, *supra* note 7, at 292 (Hamas and other terrorist groups use putative civilian sites to store ammunition and conduct hostilities); Samuel Estreicher, *Privileging Asymmetric Warfare? Part I: Defender Duties Under International Law*, 11 CHI. J. INT'L L. 425, 431–37 (2011) (citing evidence that terrorist groups hide operations within civilian populations).

43. *See* Bronner & Medina, *supra* note 40; *cf.* Jenks & Corn, *Siren Song*, *supra* note 7, at 3 (criticizing Goldstone Report for errors of law).

44. *Cf.* PETER BERKOWITZ, ISRAEL AND THE STRUGGLE OVER THE INTERNATIONAL LAWS OF WAR 20–21 (2012) (detailing discrepancies between the Goldstone Report's initial findings and the Israeli Defense Force's comprehensive review).

45. *See* Richard J. Goldstone, *The Role of the International Criminal Court*, *in* MASS ATROCITY CRIMES: PREVENTING FUTURE OUTRAGES 55, 61 (Robert I. Rotberg ed., 2010).

46. *See* Tania Voon, *Pointing the Finger: Civilian Casualties of NATO Bombing in the Kosovo Conflict*, 16 AM. U. INT'L L. REV. 1083, 1098–99 (2001). Consistent with the rationale of the duty to risk, these critics seek to prove their case not with evidence of civilian losses but with inferences based on the lack of NATO casualties. *Id.* at 1099. For a more cautious treatment that acknowledges the concerns behind the duty to risk while still viewing NATO's campaign as justified, *see* Robert D. Sloane, *The Cost of Conflation: Preserving the Dualism of Jus ad Bellum and Jus in Bello in the Contemporary Law of War*, 34 YALE J. INT'L L. 47, 96 (2009) (arguing that NATO deployed aircraft "at heights that reduced the risk to its own combatants to zero at the cost of a substantial increase in the risk to Serbia's civilians"); *but see* Schmitt, *Military*

Necessity and Humanity, *supra* note 10, at 823 (observing that precision-targeting mechanisms require time to accurately pinpoint targets, so that targeting at lower altitudes is actually less accurate).

47. *See* Robert Barnidge, *The Principle of Proportionality Under International Humanitarian Law and Operation Cast Lead*, *in* NEW BATTLEFIELDS, OLD LAWS: CRITICAL DEBATES ON ASYMMETRIC WARFARE 171, 184 (William C. Banks ed., 2011) (rejecting claims that Israeli forces violated proportionality principle in Gaza intervention, while suggesting that principle is not helpful in practice).

48. *See* Luban, *Risk Taking*, *supra* note 3, at 5–6. Luban plans to change the Kosovo casualty estimate in the published version of his paper. *See* David Luban to Peter Margulies, e-mail message, March 2, 2012 (on file with the author).

49. *See* Andreas Laursen, *NATO, the War over Kosovo, and the ICTY Investigation*, 17 AM. U. INT'L L. REV. 765, 767 & n. 8 (2002) (citing Human Rights Watch report suggesting that NATO campaign resulted in approximately five hundred civilian deaths); Bronner & Medina, *supra* note 40 (estimating that approximately seven hundred civilians were killed in the Gaza campaign).

50. *See* Additional Protocol I, art. 57.

51. *See* Schmitt, *Military Necessity and Humanity*, *supra* note 10, at 828–31 (discussing use of warnings in Gaza campaign).

52. *See* Schmitt, *Drone Attacks*, *supra* note 32, at 320 (noting that drones "rely on high resolution imagery . . . transmitted in real time").

53. Schmitt, *Military Necessity and Humanity*, *supra* note 10, at 823.

54. *See* Aaron N. Drake, *Current U.S. Air Force Drone Operations and Their Conduct in Compliance with International Humanitarian Law*, 39 DENV. J. INT'L L. & POL'Y 629, 639 (2011) (discussing role of human intelligence).

55. *See* Rizek Abdel Jawad, *Hamas Executes 2 Convicted Informants*, BOSTON GLOBE, April 16, 2010, at 4.

56. *See* Schmitt, *Military Necessity and Humanity*, *supra* note 10, at 829.

57. *See* YORAM DINSTEIN, THE CONDUCT OF HOSTILITIES UNDER THE LAW OF INTERNATIONAL ARMED CONFLICT 144 (2d ed., 2010) (noting that "desire to achieve surprise may frequently preclude warnings . . . or instigate warnings that are too vague to alert the civilian population to the impending peril").

58. *See* Commentary on the Additional Protocols of June 8, 1977 to the Geneva Conventions of August 12, 1949, at 686 (Yves Sandoz et al. eds., 1987) ("warnings may have general character").

59. *See* DINSTEIN, *supra* note 57, at 136 (noting that "importance of a munitions factory is often so critical that . . . casualties among the labourers may reach very high levels without coming under the rubric of 'excessive'" under the proportionality principle); Michael N. Schmitt, *Human Shields in International Humanitarian Law*, 47 COLUM. J. TRANSNAT'L L. 292, 331 (2009) (citing commentators for view that while at defense plant, workers are "not entitled to the full benefits of civilian status").

60. *See* Barry A. Feinstein, *Proportionality and War Crimes in Gaza under the Laws of Armed Conflict*, 36 RUTGERS L. REC. 224, 242 (2009) (discussing IDF's reliance on warnings); Amos N. Guiora, *Proportionality "Re-Configured,"* 31 ABA NAT'L SECURITY L. REP'T 9, 13 (January–February 2009) (noting that IDF provided at least 250,000 warnings during Gaza campaign).

61. *See* Blank, *supra* note 7, at 296–99.

62. *See* IAN AYRES & JOHN BRAITHWAITE, RESPONSIVE REGULATION: TRANSCENDING THE DEREGULATION DEBATE 4–5, 35–41 (1992); Robert A. Kagan, Dorothy Thornton, & Neil Gunningham, *Explaining Corporate Environmental Performance: How Does Regulation Matter?*, 37 LAW & SOC'Y REV. 51, 73–74 (2003) (tracing lack of efficacy of legal sanctions without buy-in of regulated entities); *cf.* Peter Margulies, *Judging Myopia in Hindsight:* Bivens *Actions, National Security Decisions, and the Rule of Law*, 96 IOWA L. REV. 195, 214–18 (2010) (viewing government officials subject to liability for damages as regulated entities who can either vindicate or frustrate regulation's goals).

63. The law of war has long fashioned norms based on informed predictions that forces would otherwise willfully disobey related rules. For example, consider the crime of perfidy or "false flag," which involves feigning surrender to kill more of the enemy. Perfidy is outlawed because without this prohibition, an attacking force would disregard even genuine efforts at surrender, thereby undermining humanitarian safeguards. *See* SOLIS, *supra* note 10, at 421.

64. *See* Luban, *Risk Taking*, *supra* note 3, at 20 n. 20.

65. *Cf.* Neuman, *Applying the Rule*, *supra* note 38, at 105–109 (arguing morality requires that commander expose soldiers to some greater risks to safeguard civilians, even when this choice is not dictated by principle of proportionality).

66. *Id.* at 28 (asserting that soldiers should on occasion subject themselves to somewhat greater risk than would otherwise be the case to avoid risk to civilians).

67. *Id.*

68. *See* Margalit & Walzer, *supra* note 3.

69. *See* SOLIS, *supra* note 10, at 503–504; Eric Talbot Jensen, *Direct Participation in Hostilities: A Concept Broad Enough for Today's Targeting Decisions*, *in* NEW BATTLEFIELDS, *supra* note 42, at 85, 91–93. One can read Margalit and Walzer's vaguely phrased hypothetical as describing the Lebanese villagers as hostages of Hezbollah. If the villagers were in fact *involuntary* human shields, a state would have to include harm to this group in assessing proportionality.

70. *See* Prosecutor v. Tadić, Case No. IT-94-1-I, Opinion and Judgment, P 568 (Int'l Crim. Trib. for the Former Yugoslavia May 7, 1997) (noting "scope and intensity" as criteria). One commentator has suggested that the IHL/IHRL dichotomy is unduly mechanical, and has argued that the law should require the same pragmatic criteria in each, hinging on the nature of the threat and the tailoring of the state's response. *See* Monica Hakimi, *A Functional Approach to Targeting and Detention*, 110 MICH. L. REV. 1365 (2012); *cf.* Jennifer C. Daskal, *The Geography of the Battlefield: A Framework for Detention and Targeting Outside the "Hot" Conflict Zone*, 161 U. PA. L. REV. (forthcoming 2012), *available at* http://ssrn.com/abstract=2049532 (seeking to distill targeting norms that will bridge IHL-IHRL divide). Another commentator has argued cogently against the functional criteria in *Tadić*, describing the state as a fiduciary with deontologically-derived duties to civilians who are: 1) not directly participating in hostilities, and 2) within its control. *See* Evan J. Criddle, *Proportionality in Counterinsurgency: A Relational Theory*, 87 NOTRE DAME L. REV. 1073 (2012). Criddle does not specifically address whether the acquisitive foreign nationals in Margalit and Walzer's hypothetical should be viewed as under the attacking state's control. Criddle contends that civilians in Gaza should be viewed as under Israel's control, even though Israel has withdrawn from ongoing military supervision over Gaza, because Israel continues to limit aviation and shipping in Gaza and claims the power to intervene militarily to kill or capture suspected terrorists. *Id.* at 1108. This argument parallels the holding of the International Court of Justice in the Israel-Palestinian "Wall" case that terrorist attacks from the West Bank did not

trigger a right to self-defense by Israel. *See* Legal Consequences of the Construction of a Wall in the Occupied Palestinian Territory, Advisory Opinion (Int'l Ct. Justice July 9, 2004), 43 ILM 1009, P 139 (2004). This interpretation is inconsistent with both sound policy and the text of the U.N. Charter. The text of Article 51 refers only to attacks "against" a state, and says nothing about the source of those attacks. *See* YORAM DINSTEIN, WAR, AGGRESSION AND SELF-DEFENCE 204 (4th ed., 2005). Jurists and commentators have vigorously criticized the Court's rationale, arguing that it enshrines artificial distinctions. *See* concurring opinions of Judge Higgins, 43 ILM at 1058, P 16, and Judge Buergenthal, *id.* at 1078, P 3; Sean D. Murphy, *Self-Defense and the Israeli* Wall *Advisory Opinion: An* Ipse Dixit *from the ICJ*, 99 A.J.I.L. 62, 70–72 (2005); Nicholas Rostow, *Wall of Reason: Alan Dershowitz v. the International Court of Justice*, 71 ALB. L. REV. 953, 978–81 (2008); Ruth Wedgwood, *The ICJ Advisory Opinion on the Israeli Security Fence and the Limits of Self-Defense*, 99 A.J.I.L. 52, 57–59 (2005).

71. Prosecutor v. Tadić, Case No. IT-94-1-I, Opinion and Judgment, P 568 (Int'l Crim. Trib. for the Former Yugoslavia May 7, 1997).

72. *See* U.N. Charter, art. 51.

73. *See* Criddle, *supra* note 70.

74. *See* SOLIS, *supra* note 10, at 505.

75. *Id.* at 542–44. For a useful discussion of the appropriate interaction of IHL and IHRL, *see* Jens David Ohlin, *Response to Gabor Rona* (Jan. 17, 2012), *available at* http://www.liebercode.org/2012/01/response-to-gabor-rona.html.

76. *Cf.* Geoffrey S. Corn, *Extraterritorial Law Enforcement or Transnational Counterterrorist Operations: The Stakes of Two Models*, *in* NEW BATTLEFIELDS, *supra* note 70, at 23, 35 (discussing relationship between IHL and law enforcement paradigms).

77. *See* SOLIS, *supra* note 10, at 503–504.

78. App. No. 18984/91, 21 Eur. H.R. Rep. 97 (1995).

79. *Id.* at PP193–212.

80. *Id.* (Ryssdal, J., dissenting), at PP 7–22.

81. IHRL may also inform evaluation of the adequacy of procedures for the detention of suspected terrorists, even when this detention occurs during an armed conflict. A detainee should be able to seek review from an independent decision maker and should be allowed access to representation and relevant evidence. Detailed discussion of detention is beyond the scope of this article; for useful sources, *see* al Maqaleh v. Gates, 605 F.3d 84 (D.C. Cir. 2010) (holding that habeas courts lacked jurisdiction over Afghanistan); Al-Jedda v. United Kingdom, No. 27021/08, Eur. Ct. Hum. Rts. (July 7, 2011) (holding that state violated IHRL by detaining individual in Iraq without judicial review in course of role as part of United Nations–sponsored force); Periodic Review of Individuals Detained at Guantanamo Bay Naval Station Pursuant to the Authorization for Use of Military Force, Exec. Order No. 13,567, 76 Fed. Reg. 13,277 (March 7, 2011) (providing for administrative review of present dangerousness of detainees); Robert M. Chesney, *Who May Be Held? Military Detention Through the Habeas Lens*, 52 B.C. L. REV. 769 (2011) (analyzing continued questions about scope of habeas authority); Matthew C. Waxman, *Administrative Detention of Terrorists: Why Detain, and Detain Whom?*, 3 J. NAT'L SEC. L. & POL'Y 1, 17–23 (2009).

82. *See* OSIEL, *supra* note 20, at 312–14.

83. *See* Blank, *supra* note 7, at 289–93.

84. *See* Additional Protocol I, art. 44(3).

85. Michael A. Newton, *Exceptional Engagement: Protocol I and a World United Against Terrorism*, 45 TEX. INT'L L.J. 323, 344–47 (2009) (describing political agendas that contributed to enactment of Protocol I).

86. *See* Kenneth Watkin, *Warriors Without Rights? Combatants, Unprivileged Belligerents, and the Struggle over Legitimacy* (Harvard University Program on Humanitarian Policy and Conflict Research Occasional Paper Series 2005), *available at* http://www.hpcr.org/pdfs/OccasionalPaper2.pdf (discussing history of provisions). The thorny legal issues associated with combat against violent transborder nonstate actors such as al Qaeda has led some scholars to suggest a third category—transnational armed conflict—that will supplement international and noninternational armed conflict while preserving baseline norms. *See* Geoffrey Corn & Eric Talbot Jensen, *Transnational Armed Conflict: A "Principled" Approach to the Regulation of Counter-Terror Combat Operations*, 42 ISR. L. REV. 46, 53–68 (2009); Eric Talbot Jensen, *Applying a Sovereign Agency Theory of the Law of Armed Conflict*, 12 CHI. J. INT'L L. 685, 707–12 (2012) (outlining related proposal to ensure applicability of baseline norms such as distinction and proportionality).

87. *See* Koh, *The Obama Administration and International Law*, *supra* note 32; Anderson, *supra* note 32.

88. *Cf.* Michael W. Lewis, *Drones and the Boundaries of the Battlefield*, 47 TEX. INT'L L.J. 293 (2012) (suggesting that imposing unduly narrow geographic restrictions on states' ability to target terrorist groups with global operations would grant these groups asymmetric advantage); Michael W. Lewis, *Responses to the Ten Questions*, 37 WM. MITCHELL L. REV. 5021, 5029 (2011) (same); Michael W. Lewis, *Potential Pitfalls in "Strategic Litigation": How the al-Aulaqi Lawsuit Threatened to Undermine International Humanitarian Law*, 9 LOY. U. CHI. INT'L L. REV. 177, 184–84 (same).

89. Lionel K. McPherson, *Is Terrorism Distinctively Wrong?*, 117 ETHICS 524, 541 (2007) (discussing practical checks and balances in nondemocracies).

90. *See* Jon Lee Anderson, *Death of the Tiger: Sri Lanka's Brutal Victory over Its Tamil Insurgents*, NEW YORKER, Jan. 17, 2011, at 41, 46–47 (discussing control exerted by Tamil Tigers).

91. *See* CHARLES TILLY, THE POLITICS OF COLLECTIVE VIOLENCE (2003) (discussing bread riots and other signs of popular opposition in nondemocratic states).

92. *See* Ethan Bronner, *Israelis Feel Tug of Protests, Reviving the Left's Spirits*, N.Y. TIMES, Aug. 1, 2011, at A4 (noting decrease in terrorist attacks); Joshua Mitnick, *Rocket Attacks on Israel: Why Hamas Has an Interest in Keeping Gaza Quiet*, CHRISTIAN SCI. MONITOR, Aug. 2, 2010.

93. *Cf.* Jensen, *supra* note 69, at 94 (noting that narrow definition of direct participation in hostilities gives edge to violent nonstate actors who can shield individuals such as bomb makers); Corri Zoli, *Humanizing Irregular Warfare: Framing Compliance for Nonstate Armed Groups at the Intersection of Security and Legal Analyses*, *in* NEW BATTLEFIELDS, *supra* note 42, at 190, 197 (arguing that provisions of Additional Protocol I to the Geneva Conventions enhance incentives for violent nonstate actors).

94. *See* Luban, *Risk Taking*, *supra* note 3, at 6 n. 9; *but see* Estreicher, *supra* note 42 (discussing nonstate actors' neglect of their duties under IHL).

95. Although Luban fails to address this issue, he claims that imposing a duty to risk will correct for political deficits within states. On this view, if a state has the political will to engage in an intervention, it should persuade the public to tolerate more risk to its own forces. Anything

less, Luban suggests, allows intervention too readily. *See* Luban, *Risk Taking*, *supra* note 3, at 39–40; Kahn, *supra* note 3, at 4. Ken Anderson has criticized this view as trying to reach an illusory level of *jus ad bellum* "efficiency." *See* Kenneth Anderson, *"Efficiency" in Bello and Ad Bellum: Making the Use of Force Too Easy?*, *in* Claire Finklestein, Jens David Ohlen, & Andrew Altman, Targeted Killings: Law and Morality in an Asymmetrical World 374 (Oxford University Press 2012) (describing "techno-war anxiety" that states would use force too readily, because technology makes it so easy and costless; criticizing this view as unduly dismissive of material and institutional constraints that limit warfare).

96. *See* Thomas C. Schelling, *What Purposes Can "International Terrorism" Serve?*, *in* Violence, Terrorism, and Justice 18, 21 (R.G. Frey & Christopher W. Morris eds., 1991) (citing violent "campaign of militant Palestinians against moderate Palestinian leaders"); Anderson, *supra* note 90, at 47 (discussing violence against moderates by Tamil Tigers).

97. *See* Max Abrahms, *What Terrorists Really Want: Terrorist Motives and Counterterrorism Strategy*, 32 Int'l Security 78, 85–86 (2008); *cf.* Christopher Slobogin, *A Jurisprudence of Dangerousness*, 98 Nw. U. L. Rev. 1, 44–46 (2003) (discussing intransigence rooted in ideological commitments of members of terrorist networks).

98. *See* Jordan J. Paust, *Permissible Self-Defense Targeting and the Death of Bin Laden*, 39 Denv. J. Int'l L. & Pol'y 569, 576–77 (2011).

99. *See* Isabel Kershner, *Israel Rebukes 2 in Attack on U.N. Complex*, N.Y. Times, Feb. 2, 2010, at A4.

100. *See* Peter J. Spiro, *Disaggregating U.S. Interests in International Law*, 67 Law & Contemp. Probs. 195 (2004).

101. This, not a disregard for proportionality as the duty to risk champions would have it, is the best reading of Kasher & Yadlin, *supra* note 37.

102. *See* Blum, *supra* note 36; Sitaraman, *supra* note 36; *cf.* Daphne Richemond-Barak, *Nonstate Actors in Armed Conflicts: Issues of Distinction and Reciprocity*, *in* New Battlefields, *supra* note 47, at 106, 120–21 (arguing that compliance with IHL does not hinge on reciprocity, since state owes duty even to parties that fail to observe IHL, while reserving question of whether reciprocity can affect super-IHL obligations).

103. *See* Laurie R. Blank, *Rules of Engagement: Law, Strategy, and Leadership*, *available at* http://ssrn.com/abstract=1872505 (forthcoming Aspects of Leadership: Ethics, Law and Spirituality, Lejeune Leadership Inst. Marine Corps Univ. 2012)).

104. *See generally* Humanitarian Law Project v. Holder, 130 S. Ct. 2705, 2725–26 (2010) (noting that terrorist groups like Hamas assume state-like functions such as provision of social services to cement civilian constituencies' loyalty); Boim v. Holy Land Found. for Relief & Dev., 549 F.3d 685, 698 (7th Cir. 2008) (noting that Hamas's social service programs "mak[e] it more costly . . . to defect" from the group); Eli Berman & David D. Laitin, *Religion, Terrorism, and Public Goods: Testing the Club Model*, 92 J. Pub. Econ. 1942, 1952, 1955 (2008) (same); *see also* Justin Magouirk, *The Nefarious Helping Hand: Anti-Corruption Campaigns, Social Service Provision, and Terrorism*, 20(3) Terrorism & Pol. Violence 356, 358 (2008) (discussing Hamas's provision of social services); Margulies, *Advising Terrorism*, *supra* note 42, at 484–85 (discussing nexus between social services and terrorist recruitment).

105. *See* Watkin, *Combatants and Legitimacy*, *supra* note 86, at 17; III Emmerich de Vattel, The Law of Nations or The Principles of Natural Law, ch. 3, para. 223

(1758) (right to make war "solely belongs to the sovereign power," who is best situated to judge "circumstances of the utmost importance to the state").

106. *See* McPherson, *supra* note 89; *cf.* ERIC A. POSNER & ADRIAN VERMEULE, THE EXECUTIVE UNBOUND: AFTER THE MADISONIAN REPUBLIC 178 (2010) (even in a dictatorship "there will almost always be political forces the dictator(s) must be careful to reward or appease, such as the military or security services, mass public opinion, or an elite 'selectorate' that influences the choice of dictators") (citation omitted).

107. *See* Michael N. Schmitt, *Precision Attack and International Humanitarian Law*, 859 INT'L REV. OF THE RED CROSS 445 (September 2005); *see also* SOLIS, *supra* note 10, at 533–34 (discussing precision weaponry); AP I art. 57(2)(a)(i) (precision may be factor in definition of "feasible" precautions taken to ensure targeting military as opposed to civilian targets and art. 57(2)(a)(ii) in minimizing incidental civilian harm); *Commentary on the Additional Protocols of June 8, 1977 to the Geneva Conventions of August 12, 1949*, Yves Sandoz, Christophe Swinarski & Bruno Zimmerman (eds.), International Committee of the Red Cross, Geneva, 1987, para. 2198 (standard is interpreted according to "common sense and good faith").

108. *See* Final Report to the Prosecutor By the Committee Established to Review the NATO Bombing Campaign Against the Federal Republic of Yugoslavia P73, 75 (ICTY June 13, 2000), *available at* http://www.icty.org/sid/10052 (finding that television station was appropriate target to the extent that NATO forces reasonably believed that station's facilities were used as "radio relay stations and transmitters to support the activities" of the Serbian military).

109. *See* Schmitt, *Precision Attack*, *supra* note 107, at 451–52.

110. *See* SOLIS, *supra* note 10, at 533; *cf.* JACK GOLDSMITH, POWER AND CONSTRAINT: THE ACCOUNTABLE PRESIDENCY AFTER 9/11, at 131 (2012) (noting that involvement of military lawyers in targeting decisions reaps strategic benefits through enhanced perception of legitimacy); Gregory S. McNeal, *The U.S. Practice of Collateral Damage Estimation and Mitigation*, *available at* http://ssrn.com/abstract=1819583 (providing comprehensive account of current U.S. practice). Two scholars have argued that targeted killing of a U.S. citizen abroad might require somewhat more robust procedural safeguards. *See* Richard Murphy & Afsheen John Radsan, *Due Process and Targeted Killing of Terrorists*, 32 CARDOZO L. REV. 404 (2009) (suggesting need for some procedural check); *cf.* Al-Aulaqi v. Obama, 727 F. Supp. 2d 1, 14–35 (D.D.C. 2010) (in challenge to alleged targeting of Anwar al-Awlaki, whom United States regarded as leader of al Qaeda in the Arabian Peninsula, the court held that father lacked standing to bring suit and that subject matter of suit presented political question).

111. *See* Schmitt, *Precision Attack*, *supra* note 107, at 457–58.

112. *Id.* at 450.

113. *See* Goldsmith & Levinson, *supra* note 28, at 1835.

114. *See* Military and Paramilitary Activities In and Against Nicaragua (Nicaragua v. United States), 1986 I.C.J. 14 (June 27) (holding that United States, while it had improperly sponsored the Contras' violent activities, lacked "effective control" over their actions).

115. *See* Legal Consequences of the Construction of a Wall in the Occupied Palestinian Territory, Advisory Opinion (Int'l Ct. Justice July 9, 2004), 43 ILM 1009, P 139 (2004).

116. *See* Harold Hongju Koh, *On American Exceptionalism*, 55 STAN. L. REV. 1479 (2003) (analyzing different meanings, causes, and consequences of American exceptionalism). In arguing that best practices require cooperation with international investigations, I disagree with Peter Berkowitz. *See* BERKOWITZ, *supra* note 44, at 5–6 (arguing that Israeli government acted

prudently in declining to cooperate with Goldstone investigation of Gaza incursion, given ample indications that Goldstone inquiry was flawed from the start). Berkowitz is correct, however, that a state should accord highest priority to its own comprehensive investigation. *Id.*

117. *See* Louis Henkin, Sarah H. Cleveland, Laurence R. Helfer, Gerald L. Neuman, & Diane F. Orentlicher, Human Rights 733–37 (2d ed., 2009).

118. *See* Howard LaFranchi, *US Opposes ICC Bid to Make "Aggression" a Crime under International Law; Obama Administration Has Resisted Efforts by the International Criminal Court to Include "Aggression" as a Crime, Mainly Because It Could Impact US Military Operations Abroad*, Christian Sci. Monitor, June 15, 2010 (reporting view of State Department Legal Adviser Harold Koh that participation in Kampala talks, which produced significant concessions favoring U.S. interests, "reset US relations with the court from hostility to positive engagement"). For an account of the overall importance of accountability through the ICC or other viable tribunals, *see* Diane F. Orentlicher, *Settling Accounts Revisited: Reconciling Global Norms with Local Agency*, 1 Int'l J. Transitional Just. 10 (2007).

119. *See* Henkin et al., *supra* note 117, at 619–22 (discussing United States' ambiguous yet ongoing engagement with IAHCR).

CHAPTER 6

1. An earlier version of this chapter was published in Aspects of Leadership: Ethics, Law and Spirituality (Carroll Connelley & Paolo Tripodi eds., Marine Corps Univ. Press 2012).

2. Assistant professor of law, SMU Dedman School of Law. Professor Jenks previously served in the U.S. Army Judge Advocate General's Corps as the Chief of the International Law Branch, Office of The Judge Advocate General, Washington, D.C.

3. U.S. Dep't of Army, Field Manual 3-24 Counterinsurgency, ¶¶ 7-14, 1-149, 7-23 (Dec. 15, 2006) (emphasis added) [hereinafter Counterinsurgency Manual].

4. By the middle of 2010, Leon Panetta, then head of the Central Intelligence Agency, estimated that there were only fifty to one hundred al Qaeda militants operating in Afghanistan, while the Taliban were "engaged in greater violence" including improvised explosive devices and "going after our troops. There's no question about it." *Fewer Than 100 Al Qaeda in Afghanistan: CIA Chief*, ABC News, June 28, 2010, *available at* http://www.abc.net.au/news/stories/2010/06/28/2938358.htm.

5. International Security Assistance Force, Tactical Directive, Aug. 1 2010, *available at* http://smallwarsjournal.com/documents/isafnewsrelease2.pdf [hereinafter 2010 Tactical Directive].

6. International Security Assistance Force, Tactical Directive, Dec. 2, 2008, *available at* http://www.nato.int/isaf/docu/official_texts/Tactical_Directive_090114.pdf.

7. 2010 Tactical Directive, *supra* note 5.

8. International Security Assistance Force, Tactical Directive, July 2, 2009, *available at* http://www.nato.int/isaf/docu/official_texts/Tactical_Directive_090706.pdf [hereinafter 2009 Tactical Directive].

9. C. J. Chivers, *What Marja Tells Us of Battles Yet to Come*, N.Y. Times, June 10, 2010, at A6.

10. *Id.*

11. Rajiv Chandrasekaran, *Frustration Over War Rules of Engagement Grows Washington Post: Petraeus Will Review Policy to Limit Civilian Deaths That Some Troops Say Hinders Military*

Activity, WASH. POST, July 9, 2010, *available at* http://www.washingtonpost.com/wp-dyn/content/article/2010/07/08/AR2010070806219.html.

12. C. J. Chivers, *General Faces Unease Among His Own Troops, Too*, N.Y. TIMES, June 22, 2010, at A11.

13. This chapter relies on data from the U.N. Assistance Mission in Afghanistan (UNAMA). While the United States compiles its own statistics, certainly for U.S. casualties, for Afghan casualties the United States refers to UNAMA for Afghan civilian casualties and to the Special Inspector General for Afghanistan Reconstruction for Afghan National Army and Police casualties. *See* SUSAN G. CHESSER, CONG. RESEARCH SERVICE, R41084, AFGHANISTAN CASUALTIES: MILITARY FORCES AND CIVILIANS (2012), *available at* http://www.fas.org/sgp/crs/natsec/R41084.pdf. This trend continued in 2011, when 3,021 civilians were killed, an 8 percent increase from 2010. *See* UNITED NATIONS ASSISTANCE MISSION TO AFGHANISTAN HUM. RTS. UNIT, AFGHANISTAN ANNUAL REPORT 2011 ON PROTECTION OF CIVILIANS IN ARMED CONFLICT (2012), *available at* http://unama.unmissions.org/Portals/UNAMA/Documents/UNAMA%20POC%202011%20Report_Final_Feb%202012.pdf [hereinafter 2011 UNAMA Report]. *See also* Chesser, *supra* note 13. Interestingly, at least one news organization, Agence France-Presse, used a Congressional Research Service report which didn't cover of all of 2011, and thus yielded lower civilian death totals, to incorrectly write that "Afghan Civilian Death Toll Ebbs in 2011." *Afghan Civilian Death Toll Ebbs in 2011*, RELIEF WEB, Jan. 24, 2012, *available at* http://reliefweb.int/node/472137. Two months later, with all 2011 data, Agence France-Presse wrote that 2011 was "[t]he Afghan War's Deadliest Year." *The Afghan War's Deadliest Year: Civilian Deaths Climb in 2011*, THE MONTREAL GAZETTE, Feb. 4, 2012.

14. UNITED NATIONS ASSISTANCE MISSION TO AFGHANISTAN HUM. RTS. UNIT, AFGHANISTAN ANNUAL REPORT 2010 ON PROTECTION OF CIVILIANS IN ARMED CONFLICT (2011), at i, *available at* http://unama.unmissions.org/Portals/UNAMA/human%20rights/March%20PoC%20Annual%20Report%20Final.pdf [hereinafter 2010 UNAMA Report].

15. 2010 UNAMA Report, *supra* note 14. According to UNAMA, between 2007 and 2010, 8,832 civilians were killed in Afghanistan. When the range is extended through 2011, that number increases to 11,864. Chesser, *supra* note 13, at 3.

16. 2010 UNAMA Report, *supra* note 14, at ii.

17. *Id.* at i. In 2011, pro-government forces caused the deaths of 410 civilians, a 4 percent decrease from 2010. 2011 UNAMA Report, *supra* note 13, at 2. 410 civilians amount to 14 percent of all conflict related civilian deaths in Afghanistan in 2011. *Id.*

18. 2010 UNAMA Report, *supra* note 14, at i.

19. *Id.* at ii. Civilian deaths from aerial attacks by pro-governmental forces (PGF) also increased in 2011, which is statistically interesting given not only that fatalities attributed to pro-governmental forces as a whole decreased but that there were fewer close air support missions and fewer sorties with weapons release in 2011. 2011 UNAMA Report, *supra* note 13, at 2. Yet the result was increased civilian casualties.

20. 2010 UNAMA Report, *supra* note 14, at i.

21. *Id.* And in 2011, anti-government elements (AGE) caused 2,332 Afghan civilian deaths, a 14 percent increase from 2010. 2011 UNAMA Report, *supra* note 13, at 1.

22. 2010 UNAMA Report, *supra* note 14, at i.

23. UNITED NATIONS ASSISTANCE MISSION TO AFGHANISTAN HUM. RTS. UNIT, AFGHANISTAN MID YEAR BULLETIN ON PROTECTION OF CIVILIANS IN ARMED CONFLICT,

2009, ¶ 4 (2009), *available at* http://unama.unmissions.org/Portals/UNAMA/human%20 rights/09july31-UNAMA-HUMAN-RIGHTS-CIVILIAN-CASUALTIES-Mid-Year-2009-Bulletin.pdf. According to UNAMA, "[i]n the first six months of 2009, 59% of civilians were killed by AGEs and 30.5% by PGF." *Id.* The United Nations characterized this as "a significant shift from 2007 when PGF were responsible for 41% and AGEs for 46% of civilian deaths." *Id.*

24. 2010 UNAMA Report, *supra* note 14, at i. UNAMA claimed that the most alarming trend in 2010 was "the huge number of civilians assassinated" by AGE. The 462 civilians AGE forces assassinated represent "an increase of more than 105% from 2009." The majority of the assassinations occurred in southern Afghanistan, with Helmand province experiencing a 588 percent increase and Kandahar province a 248 percent increase compared to 2009.

25. Brian Montopoli, *Chart: Troop Levels in Afghanistan over the Years*, CBS NEWS, Dec. 1, 2009, http://www.cbsnews.com/8301-503544_162-5855314-503544.html.

26. *Operation Enduring Freedom*, IRAQ CASUALTY COALITION COUNT, icasualties.org [hereinafter Iraq Casualty Count]. In 2011, that figure dropped to 413. Chesser, *supra* note 13, at 1.

27. In 2010, Afghan insurgents emplaced 14,661 improvised explosive devices (IEDs), "a 62 percent increase over 2009 and more than three times as many as the year before." Craig Whitlock, *IED Casualties in Afghanistan Spike Big Increase from 2009 to 2010 Is Result of U.S. Troop Surge*, WASH. POST, Jan. 25, 2011, at A12.

28. *Operation Enduring Freedom: U.S. Wounded Totals*, IRAQ CASUALTY COALITION COUNT, http://icasualties.org/OEF/USCasualtiesByState.aspx. In 2011, that number slightly increased, to 5,183. Chesser, *supra* note 13, at 1.

29. Indeed the number of Afghan civilians killed in 2010 considerably exceeds the sum total of all U.S. fatalities in Afghanistan to date—1476. Iraq Casualty Count, *supra* note 26.

30. These figures seem incongruent with historical trends. In a story on the sixtieth anniversary of the 1949 Geneva Conventions, the British Broadcasting Corporation claimed that "in World War I, the ratio of soldiers to civilians killed was 10 to one. In World War II it became 50–50, and today the figures are almost reversed—up to 10 civilians for every one soldier." Imogen Foulkes, *Geneva Conventions' Struggle for Respect*, BRITISH BROADCASTING CORPORATION NEWS, Aug. 8, 2009, http://news.bbc.co.uk/2/hi/europe/8196166.stm. In Vietnam, the last major counter insurgency operation the U.S. military conducted, the ratio of monthly combat fatalities per 100,000 U.S. service members was 667, or more than fifteen times higher the casualty rate in Afghanistan. Marcus Baram, *Overall, Afghanistan More Lethal for U.S. Soldiers Than Iraq (Chart)*, HUFFINGTON POST, http://www.huffingtonpost.com/2009/10/15/overall-afghanistan-more_n_319194.html.

31. COUNTERINSURGENCY MANUAL, *supra* note 3, ¶¶ 7-14, 1-149. The U.S. military defines an *insurgency* as an organized movement aimed at the overthrow of a constituted government through the use of subversion and armed conflict, and *counterinsurgency* as the military, paramilitary, political, economic, psychological, and civic actions taken by a government to defeat such movements. U.S. DEP'T OF DEFENSE, JOINT PUBLICATION 1-02, DEP'T OF DEFENSE DICTIONARY OF MILITARY AND ASSOCIATED TERMS, 85, 178 (2011).

32. COUNTERINSURGENCY MANUAL, *supra* note 3, ¶ 2-4.

33. *Id.* ¶ 5-65.

34. *Id.* ¶ 7-21.

35. *Id.* ¶ 7-23.

36. *Id.* ¶ 1-149.

37. COUNTERINSURGENCY MANUAL, *supra*, note 3, ¶ 1-150.

38. *Id.* ¶ 1-151.

39. *Id.* ¶ 1-152.

40. U.S. DEP'T OF ARMY, FIELD MANUAL 3-0 OPERATIONS w/change 1, ¶ A-6 (Feb. 22, 2011) (describing the principle of war of "mass" and how "[c]ommanders mass the effects of combat power in time and space to achieve both destructive and constructive results..." in order to overwhelm an opponent or dominate a situation). *Id.*

41. *About ISAF*, AFGHANISTAN INTERNATIONAL SECURITY ASSISTANCE FORCE, http://www.isaf.nato.int/mission.html (last accessed May 21, 2012).

42. 2009 Tactical Directive, *supra* note 8.

43. For example, in September 2011, President Obama awarded former Marine Sergeant Dakota Meyer the Medal of Honor for his heroic actions in Afghanistan in September 2009 in saving the lives of fellow marines and Afghan Army soldiers during a Taliban ambush. The award ceremony retriggered questions of whether the tactical directive and/or risk averse commanders contributed to some of the deaths of U.S. marines in the ambush. *See* Larua Rosen, *Medal of Honor Recipient Highlights Marine's Valor as well as Risks US Troops Faced under Controversial Rules of Engagement*, THE ENVOY, Sept. 14, 2011, http://news.yahoo.com/blogs/envoy/medal-honor-recipient-highlights-marine-valor-well-risks-220928608.html.

44. 2009 Tactical Directive, *supra* note 8 (emphasis added).

45. Chandrasekaran, *supra* note 11.

46. George Will, *An NCO Recognizes a Flawed Afghanistan Strategy*, WASH. POST, June 20, 2010, at A19.

47. Chivers, *supra* note 12.

48. *Id.*

49. William H. McMichael, *Hold Fire, Earn a Medal*, NAVY TIMES, May 11, 2010, http://www.navytimes.com/news/2010/05/military_restraint_medal_051110mar/.

50. *Honoring Courageous Restraint*, AFGHANISTAN INTERNATIONAL SECURITY ASSISTANCE FORCE, http://www.isaf.nato.int/article/caat-anaysis-news/honoring-courageous-restraint.html (last accessed May 21, 2012).

51. McMichael, *supra* note 49.

52. Chivers, *supra* note 9.

53. General McChrystal resigned following the release of a *Rolling Stone* story. Seemingly lost in the controversy over remarks made by members of his staff in the story which prompted the resignation, much of the article dealt with the tactical directive. *See* Michael Hastings, *The Runaway General*, ROLLING STONE, June 22, 2010, *available at* http://www.rollingstone.com/politics/news/the-runaway-general-20100622. The article quoted one U.S. Special Forces operator as claiming, "I would love to kick McChrystal in the nuts. His rules of engagement put soldiers' lives in even greater danger. Every real soldier will tell you the same thing." While the article quotes General McChrystal as acknowledging that "[w]inning hearts and minds in COIN is a coldblooded thing," the general also explains how ISAF can't kill its way out of Afghanistan, that "[t]he Russians killed 1 million Afghans, and that didn't work." *Id.*

54. *Confirmation Hearing: Commander, ISAF/US Forces-Afghanistan: Hearing Before the Senate Armed Services Committee*, 111th Cong. (2010) (opening statement of Gen. David H. Petraeus, Commander, ISAF/U.S. Forces-Afghanistan).

55. ISAF Releases, *General Petraeus Issues Updated Tactical Directive: Emphasizes "Disciplined Use of Force," available at* http://www.isaf.nato.int/article/isaf-releases/general-petraeus-issues-updated-tactical-directive-emphasizes-disciplined-use-of-force.html.

56. David Nakamura, *Afghans Reject Good Guy-Bad Guy Narrative*, WASH. POST, Aug. 14, 2010, at A6.

57. Norine MacDonald, *The Good, the Bad, and the Ugly in Afghanistan*, FOREIGN POLICY, Dec. 16, 2010, *available at* http://afpak.foreignpolicy.com/posts/2010/12/16/the_good_the_bad_and_the_ugly_in_afghanistan. Further supporting an argument that Afghan perceptions of their future in Afghanistan are not positive, beginning in 2008 and continuing through 2010, more Afghans are seeking asylum in foreign countries than at any point since the 2001 U.S. invasion. David Nakamura, *More Afghans Pursue Asylum Disillusioned by War and Instability, Many Have Become Economic Migrants*, WASH. POST, Nov. 28, 2011, at A12. From 2004 through 2007, less than 10,000 Afghans a year sought asylum, while during 2008–2010 the numbers are closer to if not in excess of 20,000. Nonetheless, even this recent spike of asylum applicants is significantly less than the more than 30,000 reported in 2000, the last full year of Taliban rule.

58. Chivers, *supra* note 9.

59. Greg Jaffe, *Combat Generation Elusive Victory Fighting to Get Out of the Way U.S. Troops Battle to Hand Off a Valley Strongly Resistant to Afghan Governance*, WASH. POST, Dec. 27, 2010, at A7.

60. COUNTERINSURGENCY MANUAL, *supra* note 3, ¶ 1-149 (emphasis added).

61. *Id.* ¶ 7-23 (emphasis added).

62. *Id.* ¶¶ 7-24, 7-149.

63. The U.S. Army Chief of Staff established the CAPE at the U.S. Military Academy at West Point, New York, to reinforce the army profession and its ethic. *Cape Home*, CENTER FOR THE ARMY PROFESSION AND ETHIC, http://cape.army.mil/ (last accessed Mar. 21, 2012).

64. Kevin Bell, *How Our Training Fails Us When It Counts*, 61 ARMY 41 (November 2011).

65. *Id.* at 45.

66. *See* Rod Nordland, *War's Risks Shift to Contractors Surrogates for Military Dying in Afghanistan*, WASH. POST, Feb. 22, 2012, at A1; John Tirman, *Do We Care When Civilians Die in War?*, WASH. POST, Jan. 8, 2012, at B1. In a different way, the United States pays condolence money to the families of civilians it kills, but the amount, $2,500, sends a variety of messages about the importance of Afghan civilian lives. *See* John Ryan, *Making Amends for Accidental Deaths*, 18 ARMY TIMES, Jan. 30, 2012.

67. COUNTERINSURGENCY MANUAL, *supra* note 3, ¶¶ 7-14, 1-149, 7-23 (emphasis added).

68. *See Army Values*, U.S. ARMY, http://www.army.mil/values/ (last accessed Mar. 21, 2012).

69. 2008 Tactical Directive, *supra* note 6.

70. 2009 Tactical Directive, *supra* note 8.

71. 2010 Tactical Directive, *supra* note 5.

CHAPTER 7

* Associate Professor Law, Pepperdine University School of Law.

1. Sarah Sewall & Larry Lewis, *Reducing and Mitigating Civilian Casualties: Afghanistan and Beyond*, Joint Civilian Casualty Study: Final Report, 4 (2010) [hereinafter Joint CIVCAS Study].

2. *Id.* at 5.

3. *Id.*

4. *Id.*; *see also* Colin H. Kahl, *How We Fight*, 85 FOREIGN AFFAIRS 83 (2006).

5. Joint CIVCAS Study, *supra* note 1.

6. *Id.*

7. *Id.*

8. *Id.* at 14.

9. Thomas W. Smith, *Protecting Civilians... or Soldiers?: Humanitarian Law and the Economy of Risk in Iraq*, 9 INT'L STUDIES PERSPECTIVES 144, 146 (2008); *see also* Sarah Sewall, *Modernizing U.S. Counterinsurgency Practice: Rethinking Risk and Developing National Strategy*, 86 MILITARY REVIEW 103, 104 (2006).

10. Matthew Waxman, *International Law and the Politics of Urban Air Operations*, RAND Corporation (Project Air Force), July 2000, at 30.

11. Colin H. Kahl, *In the Crossfire or the Crosshairs?: Norms, Civilian Casualties, and U.S. Conduct in Iraq*, 32 INT'L SECURITY 7, 7 (2007).

12. *Id.*

13. U.S. Department of the Army, *Counterinsurgency*, Field Manual (FM) 3-24/Marine Corps Warfighting Publication (MCWP) 3-33.5, Washington D.C.: Department of the Army (Dec. 15, 2006) [hereinafter FM 3-24].

14. *Id.* at 3–67.

15. *Id.* at 7–32.

16. *See* FM 3-24, *supra* note 13; Sewall, *supra* note 9.

17. FM 3-24, *supra* note 13, at 7–36.

18. Sewall, *supra* note 9, at 105.

19. FM 3-24, *supra* note 13, at 7–37.

20. Rebecca J. Wolfe & John M. Darley, *Protracted Asymmetrical Conflict Erodes Standards for Avoiding Civilian Casualties*, 11 PEACE AND CONFLICT: JOURNAL OF PEACE PSYCHOLOGY 55, 56 (2005).

21. *Id.*

22. Luke N. Condra, Joseph H. Felter, Radha K. Iyengar, & Jacob N. Shapiro, *The Effect of Civilian Casualties in Afghanistan and Iraq*, NATIONAL BUREAU OF ECONOMIC RESEARCH, July 2010, at 5–6; *see also* David Kilcullen, *The Accidental Guerilla: Fighting Small Wars in the Midst of a Big One* 85 (2009); Steven Metz, *Psychology of Participation in Insurgency*, SMALL WARS JOURNAL, January 2012, at 7, 9.

23. Condra et al., *supra* note 22, at 5–6.

24. *Id.* at 6–7.

25. *Id.*; *see also* Metz, *supra* note 22, at 7, 10.

26. Condra et al., *supra* note 22, at 6–7.

27. *Id.* at 30.

28. Smith, *supra* note 9.

29. U.S. Department of the Army, *Civilian Casualty [CIVCAS] Mitigation*, Army Tactics, Techniques, and Procedures [ATTP] 3-37.31, Washington D.C.: Headquarters, Department of the Army (December 2011) [hereinafter Army CIVCAS Manual].

30. In the interest of full disclosure I should note that I played a small role as a commenter on an early draft of the manual and hosted the drafting committee at my university for a workshop aimed at vetting and improving an early draft of the manual.

31. Joint CIVCAS Study, *supra* note 1.

32. *Id.* at iii.

33. Joint Forces Command: Joint Center for Operational Analysis, *Adaptive Learning for Afghanistan: Final Recommendations*, 4 (2011) [hereinafter ALA Study].

34. United States Marine Corps, Center for Lessons Learned, *Civilian Casualty Mitigation: Summary of Lessons, Observations and Tactics, Techniques and Procedures from Marine Expeditionary Brigade–Afghanistan (MEB-A)—January-April 2010*, Quick Look Report, 6 (July 29, 2010) [hereinafter USMC CIVCAS Mitigation Report].

35. Army CIVCAS Manual, *supra* note 29, at 2–15.

36. *Id.* at 2–25.

37. *Id.*

38. *Id.* at 2–26.

39. Gregory S. McNeal, *Are Targeted Killings Unlawful? A Case Study in Empirical Claims Without Empirical Proof*, *in* Targeted Killings: Law and Morality in an Asymmetrical World (Claire Finkelstein, Jens David Ohlin, & Andrew Altman, eds., 2012).

40. Joint CIVCAS Study, *supra* note 1, at 14.

41. Army CIVCAS Manual, *supra* note 29, at 2–42.

42. Joint CIVCAS Study, *supra* note 1, at 14.

43. *Id.*

44. Army CIVCAS Manual, *supra* note 29, at 2–70.

45. USMC CIVCAS Mitigation Report, *supra* note 34, at 17.

46. Army CIVCAS Manual, *supra* note 29, at 1–38; *see also*, JP Morgan, *Counter Insurgency Media Applications*, 1 COIN Common Sense 7 (2011).

47. Army CIVCAS Manual, *supra* note 29, at 2–76.

48. USMC CIVCAS Mitigation Report, *supra* note 34, at 50.

49. *Id.* at 71.

50. Army CIVCAS Manual, *supra* note 29, at 2–87.

51. *Id.* at 2–91.

52. *Id.* at 2–96.

53. Joint CIVCAS Study, *supra* note 1, at 77.

54. Army CIVCAS Manual, *supra* note 29, at 2–126; *see also* Joint CIVCAS Study, *supra* note 1, at 5.

55. Joint CIVCAS Study, *supra* note 1, at 10.

56. *Id.*

57. *Id.* at 74.

58. *Id.* at 72.

59. ALA Study, *supra* note 33, at 5.

60. Joint CIVCAS Study, *supra* note 1, at 11.

61. *Id.*

62. *Id.* at 95.

63. ALA Study, *supra* note 33, at 4.

64. *Id.* at 5.

65. *Id.* at 8.

66. *Id.* at 5.

CHAPTER 8

1. Founder & Executive Director of the The International Policy Institute for Counter Terrorism (ICT), and Deputy Dean of the Lauder School of Government at the Interdisciplinary Center, Herzliya, Israel.

2. The laws of war apply in a situation of armed conflict. For the purpose of this paper, "armed conflict" is defined as follows:

1. "International armed conflicts exist whenever there is resort to armed force between two or more States.
2. Non-international armed conflicts are protracted armed confrontations occurring between governmental armed forces and the forces of one or more armed groups, or between such groups arising on the territory of a state [party to the Geneva Conventions]. The armed confrontation must reach a minimum level of intensity and the parties involved in the conflict must show a minimum of organisation" as defined by the International Committee of the Red Cross (ICRC).

Human rights law and other domestic and international laws may govern situations outside conflict settings. *How Is the Term "Armed Conflict" Defined in International Humanitarian Law?* Opinion Paper, INT'L COMMITTEE OF THE RED CROSS (2008).

3. For a discussion on the role, rights, and obligations of nonstate actors in the current legal framework, *see* GARY D. SOLIS, THE LAW OF ARMED CONFLICT: INTERNATIONAL HUMANITARIAN LAW IN WAR (Cambridge Univ. Press 2010); ANDREW CLAPHAM, HUMAN RIGHTS OBLIGATIONS OF NON STATE ACTORS (Oxford Univ. Press 2006); HEATHER A. WILSON, INTERNATIONAL LAW AND THE USE OF FORCE BY NATIONAL LIBERATION MOVEMENTS (Oxford Univ. Press 1988).

4. For the use of lawfare as a weapon, *see* Charles. J. Dunlap, *Lawfare Today—A Perspective*, 3 YALE J. INT'L AFF. 146 (2008).

5. Article 57(2)(iii) of Additional Protocol I to the Geneva Conventions states that "those who plan or decide upon an attack shall refrain from deciding to launch any attack which may be expected to cause incidental loss of civilian life, injury to civilians, damage to civilian objects, or a combination thereof, which would be excessive in relation to the concrete and direct military advantage anticipated." Additional Protocol I to the Geneva Conventions of August 12, 1949, Relating to the Protection of Victims of International Armed Conflicts, art. 57(2)(iii), June 8, 1977, 1125 U.N.T.S. 609, *available at* http://deoxy.org/wc/wc-proto.htm [hereinafter Additional Protocol I].

6. M. N. Schmitt, *Precision Attack and International Humanitarian Law*, 87 INT'L REV. OF THE RED CROSS 445, 456–59 (2005).

7. Article 48 of Additional Protocol I to the Geneva Conventions states that "[i]n order to ensure respect for and protection of the civilian population and civilian objects, the Parties to the conflict shall at all times distinguish between the civilian population and combatants and between civilian objects and military objectives and accordingly shall direct their operations only against military objectives." Additional Protocol I, *supra* note 5, art. 48, *available at* http://www.icrc.org/ihl.nsf/7c4d08d9b287a42141256739003e636b/f6c8b9fee14a77fdc125641e0052b079.

8. *Id.* art. 51(1), (2).

9. *Id.* arts. 51(5)(b) and 57(2)(a). Under Article 51(4), "Indiscriminate attacks" are prohibited. Article 51(5)(b) then states that the following type of attack is indiscriminate and thus unlawful: "an attack which may be expected to cause incidental loss of civilian life, injury to civilians, damage to civilian objects, or a combination thereof, which would be excessive in relation to the concrete and direct military advantage anticipated." *Id.*

10. The proportionality principle has also been widely adopted by military manuals and state practice, and is included in Protocol II and Amended Protocol II of the Convention on Certain Conventional Weapons.

11. Geneva Conventions of August 12, 1949, Relating to the Treatment of Prisoners of War, Oct. 21, 1950, 75 U.N.T.S. 135, *available at* http://www.icrc.org/ihl.nsf/7c4d08d9b287a42141256739003e63bb/6fef854a3517b75ac125641e004a9e68 [hereinafter Geneva Convention III].

12. Among these many rights and obligations, a military force must treat prisoners humanely and protect them from danger, and is obligated to provide them with food, clothing, and medical attention. Geneva Convention III, *supra* note 11, arts. 19 and 20. POWs must also be protected from public curiosity and are entitled to due process guarantees. *Id.* arts. 13 and 11. POWs should not be tortured, the detention camp must meet certain standards, and the work the prisoners are required to perform must not be dangerous, humiliating, or directly connected to wartime operations. When hostilities have ceased, POWs are to be repatriated. *Id.* arts. 13, 16, 17, 22, 23, 51, 52, 56, 84, and 118.

13. *Weapons and International Humanitarian Law*, Int'l Comm. of the Red Cross (2005), *available at* http://www.icrc.org/eng/resources/documents/misc/6jkkbp.htm.

14. Additional Protocol I, *supra* note 5, art. 43.

15. Protocol I states that "[m]embers of the armed forces of a Party to a conflict (other than medical personnel and chaplains covered by Article 33 of the Third Convention) are combatants, that is to say, they have the right to participate directly in hostilities." *Id.*

16. Geneva Convention III, *supra* note 11.

17. Article 43 of Additional Protocol I states: "The armed forces of a Party to a conflict consist of all organized armed forces, groups and units which are under a command responsible to that Party for the conduct of its subordinates, even if that Party is represented by a government or an authority not recognized by an adverse Party. Such armed forces shall be subject to an internal disciplinary system which, inter alia, shall enforce compliance with the rules of international law applicable in armed conflict." Additional Protocol I, *supra* note 5, art. 43.

18. Geneva Convention III, *supra* note 11, art. 4.

19. *Id.* art 4(6). "Inhabitants of a non-occupied territory, who on the approach of the enemy spontaneously take up arms to resist the invading forces, without having had time to form themselves into regular armed units, provided they carry arms openly and respect the laws and customs of war." *Id.*

20. Article 79(2) of Additional Protocol I, read together with Article 51(3), grants civilians all the legal protections conferred by international humanitarian law, unless they participate directly in hostilities and for such time as they so "participate." As stated in Article 79(2), "[t]hey shall be protected as such under the Conventions and this Protocol, provided that they take no action adversely affecting their status as civilians. . . . " Additional Protocol I, *supra* note 5, arts. 51(3) and 79(2). Long-standing dilemmas have surrounded the notion of "direct participation in hostilities," particularly about how the legal concept can be practically applied—especially in light of changing trends in warfare. In seeking to remedy at least some of these dilemmas, in May

2009 the ICRC published an Interpretative Guidance on the concept, providing guidelines on how to interpret international humanitarian law relating to the notion of direct participation in hostilities. The Guidance is informed by discussions held at a number of expert meetings conducted between 2003 and 2008 in The Hague and Geneva, which drew on dozens of legal experts from a variety of disciplines. *See* Nils Melzer, *Interpretive Guidance on the Notion of Direct Participation in Hostilities under International Humanitarian Law,* 90 INT'L REV. OF THE RED CROSS 991 (2009).

21. While there are various interpretations of what constitutes a lawful military objective, Article 52 (2) of Additional Protocol I provides a definition that is considered to reflect customary international law and is largely accepted. It requires that two criteria be met for an object to be considered a military objective:

(1) it has to contribute effectively to the military action of the enemy, through its "nature, location, purpose or use"; and
(2) its destruction, capture, or neutralization has to offer a "definite military advantage" to the attacking side.

Marco Sassòli, *Legitimate Targets of Attacks under International Humanitarian Law—Background Paper prepared for the Informal High-Level Expert Meeting on the Reaffirmation and Development of International Humanitarian Law,* International Humanitarian Law Research Initiative, PROGRAM ON HUMANITARIAN POLICY AND CONFLICT RESEARCH AT HARVARD UNIV. 27–29 (2003).

22. Additional Protocol I, *supra* note 5, arts. 51(5)(b) and 57(2)(a).

23. It has been claimed that the U.S. use of drones in its counterterrorism campaigns is an example of the state's asymmetric advantages in technology. However, this is actually an example of a technological solution to the restraints that the state imposes on itself when exercising the firepower of its air force. The drone attacks are therefore meant to limit the potential collateral damage that might be inflicted by using fighter jets.

24. While definitions of terrorism are widely contested and have yet to be agreed upon, I have proposed defining terrorism as "a form of violent struggle in which violence is deliberately used against civilians in order to achieve political goals (nationalistic, socioeconomic, ideological, religious, etc.)." The "deliberate" targeting of civilians in order to achieve political objectives is what distinguishes a terrorist act from guerrilla warfare, in which military units, rather than civilians, are targeted by nonstate actors.

25. Nonconventional attacks refer to "CBRN terrorism"—that is, chemical, biological, radiological, and nuclear attacks.

26. Boaz Ganor, *"Terrorism Networks: It Takes a Network to Beat a Network," in* THE NETWORK CHALLENGE: STRATEGY, PROFIT, AND RISK IN AN INTERLINKED WORLD 454–56 (Y. Wind Kleindorfer & R. Gunther eds., 2009).

27. A terrorist designated as a "ticking bomb" is involved in the advanced stages of preparation for a concrete attack against civilians.

28. Dinstein states: "Nowadays, customary international law recognizes the principle of proportionality": YORAM DINSTEIN, THE CONDUCT OF HOSTILITIES UNDER THE LAW OF INTERNATIONAL ARMED CONFLICT 120 (Cambridge Univ. Press 2004).

29. *See id.*; Schmitt, *supra* note 6; SOLIS, *supra* note 3.

30. As it has been understood and applied in the Geneva Conventions and in the Statute of the International Criminal Court (ICC), the use of human shields involves "using the presence (or movements) of civilians or other protected persons to render certain points or areas (or military forces) immune from military operations." J. HENCKAERTS & L. DOSWALD-BECK, CUSTOMARY INTERNATIONAL HUMANITARIAN LAW—VOLUME 1: RULES 339 (Cambridge Univ. Press 2004) 339.

31. Dunlap, *supra* note 4.

32. In this regard, the subjectivity of IHL may lead to more civilian casualties, on one hand, and to greater potential for abuse by insurgents or terrorists, on the other hand. The framework proposed in this article is therefore designed to lead to improvements in both areas.

33. It is possible to see from Diagram A that permission of or prohibition against the intentional harming or killing of a combatant or combat-supportive civilian is not influenced by the legitimacy of the actor but rather by his actions and the nature of his involvement in combat.

34. In other words, intentional harm to *nonoperational* combat-supportive civilians is forbidden. For example, intentionally targeting an enemy's food industry, or another civilian industry that supplies nonmilitary resources to enemy combatants, should be forbidden (unlike intentionally targeting an enemy's military industry, operated during wartime by *operational* combat-supportive civilians).

35. An exception to the prohibition of intentional targeting may arise in extreme cases in which terrorists have forced a person to approach an enemy force while carrying an explosive device on his body.

36. Thus, for example, when terrorists hijack an airplane, taking hostage citizens of the country under attack, the state has the right to shoot down the plane, killing all of its passengers, if circumstances warrant (for example, if it is feared that the terrorists plan to crash the plane into a populous area, as was the case on 9/11). Similarly, a state has the right to harm civilians from the terrorist organizations' community of origin, whom the terrorists are holding against their will as "human shields."

37. With former President George W. Bush's signing of the Military Commissions Act of 2006, Title 10 of the United States Code was amended to include a definition of an "unlawful enemy combatant" as a "person who has engaged in hostilities or who has purposefully and materially supported hostilities against the United States or its co-belligerents who is not a lawful enemy combatant (including a person who is part of the Taliban, al Qaeda, or associated forces)." 10 U.S.C. §948A (2010). Under Israel's Incarceration of Unlawful Combatants Law, "'unlawful combatant' is a person who has participated either directly or indirectly in hostile acts against the State of Israel or is a member of a force perpetrating hostile acts against the State of Israel, where the conditions prescribed in Article 4 of the Third Geneva Convention of 12th August 1949 with respect to prisoners-of-war and granting prisoner-of-war status in international humanitarian law, do not apply to him." Israel Incarceration of Unlawful Combatants Law, 5762-2002, *available at* http://www.jewishvirtuallibrary.org/jsource/Politics/IncarcerationLaw.pdf.

38. While President Obama refrained from using the term "unlawful combatants," the principles and substance of his approach have not really changed since the Bush administration. *See* D. Q. Wilber & P. Finn, *U.S. Retires "Enemy Combatants," Keeps Broad Right to Detain*, WASH. POST, Mar. 14, 2009, at A06.

39. Knut Dormann, *The Legal Situation of "Unlawful/Unprivileged Combatants,"* 85 RICR 45, 47 (2003).

40. The value cannot be a negative number or equal to zero.

41. This number actually reflects the level of commitment to proportionality; it may change to reflect the decision(s) of an authorized international forum responding to the changing dynamics of the challenges and circumstances of multidimensional warfare.

42. To a more limited extent, such action is already an obligation under international law, in the form of a precautionary measure codified in Additional Protocol I of the Geneva Conventions, which requires that "effective advance warning shall be given of attacks which may affect the civilian population." As things currently stand, this measure—known as the "obligation of advanced warning"—does not obligate commanders to provide a warning in all cases, if "circumstances do not permit." *See* Additional Protocol I, *supra* note 5, art. 57(2)(c).

CHAPTER 9

* Charles I. Francis, Professor of Law, University of Texas School of Law. This chapter originally appeared, in longer form, as *Who May Be Killed: Anwar Al-Awlaki as a Case Study in the International Legal Regulation of Lethal Force*, 13 Y.B. INT'L HUMANITARIAN L. 3–60 (2011).

1. For a discussion, *see* Robert M. Chesney, *Beyond Conspiracy? Anticipatory Prosecution and the Challenge of Unaffiliated Terrorism*, 80 S. CAL. L. REV. 425, 437–45 (2007) (distinguishing al Qaeda from the larger "global jihad" movement). *See also* Matthew C. Waxman, *The Structure of Terrorism and the Laws of War*, 20 DUKE J. COMP. & INT'L L. 429, 447–51 (2010) (arguing that disagreements about how to understand al Qaeda's structure complicate efforts to apply IHL).

2. *See* Eric Schmitt and Souad Mekhennet, *Qaeda Branch Steps Up Raids in North Africa*, N.Y. TIMES, July 9, 2009, at A1.

3. *See* Craig Whitlock, *Group in Algeria Turned to al-Qaida for Assistance*, WASH. POST, May 30, 2007.

4. *See id.*

5. The account in this paragraph is drawn from THE 9/11 COMMISSION REPORT: FINAL REPORT OF THE NATIONAL COMMISSION ON TERRORIST ATTACKS UPON THE UNITED STATES 152–53 (W.W. Norton & Co. ed., 2004).

6. NATIONAL COMMISSION ON TERRORIST ATTACKS UPON THE UNITED STATES, Chapter 5, http://911research.wtc7.net/mirrors/911commission/report/911Report_Ch5.htm.

7. *See, e.g.*, Alistair Harris, *Yemen: On the Brink—Exploiting Grievances: al-Qaeda in the Arabian Peninsula*, CARNEGIE ENDOWMENT FOR INTERNATIONAL PEACE 3, No. 111 (2010).

8. *Profile: al-Qaeda in the Arabian Peninsula*, BBC NEWS, Oct. 31, 2010, http://www.bbc.co.uk/news/world-middle-east-11483095.

9. *See* Harris, *supra* note 6, at 2.

10. *See, e.g.*, *Understanding Ramadan: The Muslim Month of Fasting with Imam Anwar al-Awlaki, Falls Church Dar Alhijrah Islamic Center*, WASH. POST, Nov. 19, 2001, *available at* http://www.washingtonpost.com/wp-srv/liveonline/01/nation/ramadan_awlaki1119.htm [hereinafter Understanding Ramadan]. *See also* Matthew T. Hall, *Former Local Cleric Seen as "Bin Laden of the Internet,"* U.T-SAN DIEGO, Jan. 10, 2010, *available at* http://www.utsandiego.com/news/2010/jan/10/former-local-cleric-seen-bin-laden-internet/?page=1#article (noting that al-Awlaki had

told *National Geographic* that "[t]here is no way that the people who did this (*i.e.*, the 9/11 attacks) could be Muslim, and if they claim to be Muslim, then they have perverted their religion").

11. *See, e.g.*, Posting of Brian Fishman to Jihadica Blog, http://www.jihadica.com/anwar-al-awlaki-the-infidel/ (Nov. 20, 2009) (discussing al-Awlaki's "personal ideological evolution" with reference to a pre–9/11 episode in which Abdullah al-Faisal, perhaps the most prominent English-language proponent of extremist *jihad*, sharply criticized al-Awlaki's relatively moderate views, as well as a 2004 interview with National Public Radio in which al-Awlaki cited the 2003 invasion of Iraq as having put Western Muslims in a position where they are "torn between solidarity with their religious fellowmen and their fellow citizens"). *See also* Rob Gifford, *U.K. Muslims Struggle with Cleric's Radicalization*, NATIONAL PUBLIC RADIO, Dec. 24, 2009, http://www.npr.org/templates/story/story.php?storyId=121880241 (noting that al-Awlaki was known as a relatively moderate cleric but that his views had grown "increasingly hostile" to "the West" over the years).

12. *See* Understanding Ramadan, *supra* note 9. *See also* Alexander Meleagrou-Hitchens, *Voice of Terror*, FOREIGN POLICY, Jan. 18, 2011, *available at* http://www.foreignpolicy.com/articles/2011/01/18/voice_of_terror (arguing that al-Awlaki has become the most significant English-language propagandist of *jihad* in terms of Western audiences in particular). For a collection of al-Awlaki's videos, some with English subtitles or transcripts, *see* THE MIDDLE EAST MEDIA RESEARCH INSTITUTE TV MONITOR PROJECT, http://www.memritv.org.

13. *See* Opposition to Plaintiff's Motion for Preliminary Injunction and Memorandum in Support of Defendants' Motion to Dismiss, Al-Aulaqi v. Obama, 727 F.Supp.2d 1 (D.D.C. 2010) (No. 10-cv-1469) [hereinafter Government's Brief]; *see also* Unclassified Declaration in Support of Formal Claim of State Secrets Privilege by James R. Clapper, Director of National Intelligence at § 14, Al-Aulaqi v. Obama, *available at* http://www.lawfareblog.com/wp-content/uploads/2010/09/Exhibit-1.pdf [hereinafter Clapper Declaration].

14. *See, e.g.*, Spencer Hsu, *Judge in D.C. Tosses Suit Challenging Placement of Yemeni Cleric on Terror List*, WASH. POST, Dec. 7, 2010 ("Officials say Aulaqi . . . was an operational planner in last year's failed Christmas Day bomb plot against a jetliner over Detroit").

15. *See* Understanding Ramadan, *supra* note 9.

16. *See id.*

17. *See* Scott Shane, *Born in U.S., a Radical Cleric Inspires Terror*, N.Y. TIMES, Nov. 18, 2009, at A1.

18. *See* Carrie Johnson et al., *Obama Vows to Repair Intelligence Gaps behind Detroit Airplane Incident*, WASH. POST, Dec. 30, 2009, *available at* http://www.washingtonpost.com/wp-dyn/content/article/2009/12/29/AR2009122901433.html.

19. Clapper Declaration, *supra* note 12, § 14.

20. Government's Brief, *supra* note 12, at 1 (asserting that "since late 2009, Anwar al-Aulaqi has taken on an increasingly operational role in AQAP, including preparing Umar Farouk Abdulmutallab in his attempt to detonate an explosive device . . . on Christmas Day 2009"); *id.* at 6 ("Since late 2009, Anwar al-Aulaqi has taken on an increasingly operational role in the group, including preparing Umar Farouk Abdulmutallab, who received instructions from Anwar Al–Aulaqi to detonate an explosive device aboard a U.S. airplane over U.S. airspace and thereafter attempted to do so aboard a Northwest Airlines flight from Amsterdam to Detroit on Christmas Day 2009, for his operation."); *id.* at 24 (referring to al-Awlaki as a "senior operational leader"); *id.* at 29 n. 14 (same).

21. *See* Eric Lipton, *U.S.-Born Cleric Justifies the Killing of Civilians*, N.Y. TIMES, May 23, 2010, at A10. For the video, *see* THE MIDDLE EAST MEDIA RESEARCH INSTITUTE TV MONITOR PROJECT, http://www.memritv.org/clip/en/2480.htm.

22. For the translation, *see id.*, http://www.memritv.org/clip_transcript/en/2480.htm.

23. Thomas Hegghammer, *The Case for Chasing al-Awlaki*, FOREIGN POLICY, Nov. 24, 2010, *available at* http://mideast.foreignpolicy.com/posts/2010/11/24/the_case_for_chasing_al_awlaki.

24. *Id.*

25. *Id.*

26. *See Yemeni Air Strike Kills 30, Targets Home of Cleric Linked to Ft. Hood Attack*, CBSNEWS, Dec. 24, 2009, http://www.cbsnews.com/8301-502683_162-6018515-502683.html.

27. *Id.*

28. At a press conference in early January 2010, a reporter asked White House Press Secretary Robert Gibbs whether the government viewed al-Awlaki as merely inspirational or actually an operational figure, and whether the plan was to arrest, capture, or kill him. Gibbs declined to answer, citing intelligence concerns. *See* Briefing by White House Press Secretary Robert Gibbs, Jan. 8, 2010, 2010 WLNR 525020.

29. Dana Priest, *U.S. Military Teams, Intelligence Deeply Involved in Aiding Yemen on Strikes*, WASH. POST, Jan. 27, 2010, *available at* http://www.washingtonpost.com/wp-dyn/content/article/2010/01/26/AR2010012604239_pf.html.

30. *Id.* The *Post* subsequently posted a correction with the article: "The article referred incorrectly to the presence of U.S. citizens on a CIA list of people the agency seeks to kill or capture. After The Post's report was published, a source said that a statement the source made about the CIA list was misunderstood. Additional reporting produced no independent confirmation of the original report, and a CIA spokesman said that The Post's account of the list was incorrect. The military's Joint Special Operations Command maintains a target list that includes several Americans. In recent weeks, U.S. officials have said that the government is prepared to kill U.S. citizens who are believed to be involved in terrorist activities that threaten Americans." *Id.* (posted at the top of the page).

31. *See* Scott Shane, *U.S. Approves Targeted Killing of American Cleric*, N.Y. TIMES, Apr. 6, 2010, at A12. Invoking the principle that an unnamed government official explained that this "was the standard used in adding names to the list of targets." *Id.* Testifying before Congress two months earlier, Director of National Intelligence Dennis Blair had addressed this topic briefly. He explained that when an element of the Intelligence Community intends to take "direct action against terrorists" in circumstances involving a U.S. citizen, the relevant official seek "specific permission" in light of factors including "whether that American is involved in a group that is trying to attack us, [and] whether that American is a threat to other Americans." Eli Lake, *"Permission" Needed to Kill U.S. Terrorists*, WASH. TIMES, Feb. 4, 2010, at A09.

32. *See* Harold Hongju Koh, Legal Advisor, U.S. Dep't of State, Speech to the Annual Meeting of the American Society of International Law, *The Obama Administration and International Law* (Mar. 25, 2010), *available at* http://www.state.gov/s/l/releases/remarks/139119.htm.

33. Posting of Kenneth Anderson to Opinion Juris Blog, http://opiniojuris.org/2010/03/25/bleg-for-harold-kohs-asil-speech/ (Mar. 25, 2010, 20:24 EDT); Posting of Kenneth Anderson to The Volokh Conspiracy Blog, http://volokh.com/2010/03/28/assassination-self-defense-and-the-koh-speech/ (Mar. 28, 2010, 17:30 EDT).

34. *Defending Drones: The Laws of War and the Right to Self-Defense*, WASH. POST, Apr. 13, 2010, at A16, *available at* http://www.washingtonpost.com/wp-dyn/content/article/2010/04/12/AR2010041204086_pf.html.

35. *See, e.g.*, Posting of Kevin Heller to Opinio Juris Blog, http://opiniojuris.org/2010/03/26/when-bad-things-happen-to-good-scholars-koh-and-the-icc-edition/ (Mar. 26, 2010, 21:36 EDT); Keith Johnson, *U.S. Defends Legality of Killing with Drones*, WALL ST. JOURNAL, Apr. 5, 2010, *available at* http://online.wsj.com/article/SB10001424052702303450704575159864237752180.html (citing the ACLU's concern that Koh failed to explain the geographic boundaries of the authority to use force or the criteria for distinguishing legitimate targets from civilians); Posting by Marko Milanovic to EJIL: Talk! Blog, http://www.ejiltalk.org/harold-koh-on-targeted-killings/ (May 27, 2010).

36. Letter from Anthony D. Romero, Executive Director of the American Civil Liberties Union, to Barack Obama, President of the United States (Apr. 28, 2010), at 1, *available at* http://www.aclu.org/files/assets/2010-4-28-ACLULettertoPresidentObama.pdf.

37. *See* U.N. Human Rights Council, Philip Alston, *Report of the Special Rapporteur on Extrajudicial, Summary or Arbitrary Executions*, U.N. Doc. A/HRC/14/24/Add.6 (May 28, 2010).

38. Philip Alston, *Statement of U.N. Special Rapporteur on U.S. Targeted Killings without Due Process*, AMERICAN CIVIL LIBERTIES UNION, Aug. 3, 2010, http://www.aclu.org/national-security/statement-un-special-rapporteur-us-targeted-killings-without-due-process.

39. Complaint at 11, Al-Aulaqi v. Obama, 727 F.Supp.2d 1 (D.D.C. 2010) (No. 10-cv-1469).

40. *See* Al-Aulaqi v. Obama, 727 F.Supp.2d 1 (D.D.C. 2010).

41. U.N. Charter, arts. 1(1), 2(4).

42. *See id.*, arts. 39, 42, 51.

43. *See, e.g.*, U.N. Doc. A/HRC/14/24/Add.6, *supra* note 36, § 37 ("The proposition that a State may consent to the use of force on its territory by another State is not legally controversial."); Michael Byers, *Letting the Exception Prove the Rule*, 17 ETHICS & INT'L AFF. 9 (2003) (asserting, in connection with a 2002 drone strike by the United States in Yemen, that the "right to intervene by invitation is based on the undisputed fact that a state can freely consent to having foreign armed forces on its territory"); Sean D. Murphy, *The International Legality of U.S. Military Cross-Border Operations from Afghanistan to Pakistan*, NAVY WAR COLLEGE, 85 INT'L L. STUD. 109, 118 (2009); YORAM DINSTEIN, WAR, AGGRESSION AND SELF-DEFENCE 112–14 (Cambridge Univ. Press, 4th ed. 2005) (1988).

44. My position on the doctrinal role played by consent differs from that described by Melzer. Melzer refers to a general consensus to the effect that consent is an "exculpatory circumstance" justifying action that infringes Article 2(4). *See* NILS MELZER, TARGETED KILLING IN INTERNATIONAL LAW 41, 75 (Oxford Univ. Press 2008). I argue, in contrast, that where a state consents there is no infringement of Article 2(4) in the first instance and hence no need for exculpation. *See also* Dinstein, *supra* note 42. The result is the same in either case, of course.

45. I do not mean to suggest that demonstration of effective consent—or of the applicability of self-defense under Article 51—suffices to resolve all the international law questions associated with the al-Awlaki scenario. Section II (B) takes up a series of additional concerns sounding in human rights and humanitarian law. Some commentators may object to this sequencing, but I believe it to be the clearest way to proceed. *Cf.* Rise of the Drones II: Examining the Legality of Unmanned Targeting, *Before the U.S. House of Representatives, Committee on Oversight and*

Government Reform, Subcommittee on National Security and Foreign Affairs 50–52 (Apr. 28, 2010) (statement of Mary Ellen O'Connell, Robert and Marion Short professor of law, Notre Dame University), *available at* http://www.fas.org/irp/congress/2010_hr/drones2.pdf [hereinafter Rise of the Drones II] (objecting to arguments involving Yemen's consent to a 2002 drone strike on the ground that states cannot...give consent to a right they do not have); *id.* (arguing that consent to use military force in the form of a drone strike would be *ultra vires* absent the existence of armed conflict permitting the consenting state itself to carry out such an attack).

46. Dana Priest, *U.S. Military Teams, Intelligence Deeply Involved in Aiding Yemen on Strikes*, WASH. POST., Jan. 27, 2010, at A01.

47. *Id.*

48. *Id.*

49. *Id.*

50. *Yemen Warns U.S. on Direct Intervention*, DENVER POST, Jan. 7, 2010, *available at* http://www.denverpost.com/ci_14145679?source=pkg. "The statement underscored the rising concern among Yemen's leadership of a domestic backlash that could politically weaken the government and foment more instability. In recent days, top Yemeni officials have publicly downplayed their growing ties to Washington, fearing they will be perceived by their opponents as weak and beholden to the United States." *Id.*

51. *See* Scott Shane, Mark Mazzetti, & Robert F. Worth, *Secret Assault on Terrorism Widens on Two Continents*, N.Y. TIMES, Aug. 14, 2010, at A1. The article notes that the White House would have preferred to employ drones for these attacks, in light of their capacity to minimize deaths to innocent bystanders, but that the CIA's compliment of armed drones were tied up with operations in Pakistan. *See id.*

52. *Id.*

53. *Id.*

54. The first new story to break this topic observed that drones had been absent from Yemen for years, apparently because of both the demands of combat elsewhere and the since in the mid-2000s that the presence of al Qaeda in Yemen had diminished. Greg Miller, Greg Jaffe, & Karen DeYoung, *U.S. Deploying Drones in Yemen to Hunt for Al-Qaeda, Has Yet to Fire Missiles*, WASH. POST, Nov. 7, 2010, *available at* http://www.washingtonpost.com/wp-dyn/content/article/2010/11/06/AR2010110604454.html?wprss=rss_world/mideast.

55. *See, e.g.*, David Cloud, *Obama Said to Be Weighing Drone Strikes in Yemen*, L.A. TIMES, Nov. 8, 2010; Julian Barnes & Adam Entous, *Yemen Covert Role Pushed: Foiled Bomb Plot Heightens Talk of Putting Elite U.S. Squads in CIA Hands*, WALL ST. JOURNAL, Nov. 1, 2010, *available at* http://online.wsj.com/article/SB10001424052748704477904575586634028056268.html. The *Wall Street Journal* account, notably, added the possibility of placing U.S. special forces units under CIA authority in Yemen expressly in order to establish a capacity to conduct ground operations without disclosure to the Yemeni government. *See id.*

56. *See id. See also* Barnes and Entous 2010, *supra* note 54 (noting the view of a U.S. official to the effect that the Yemeni government "limit[s] us when we are getting too close," and reporting that the Yemenis had "delayed or objected to U.S. operations" in some instances over the past year).

57. *See* Miller, *supra* note 53.

58. Another issue that can arise with consent is fabrication—as when the request for intervention is made by a puppet government acting under the direction of the intervening state. Dinstein suggests that fabricated consent is invalid. *See* Dinstein, *supra* note 42, at 114. There

does not appear to be a basis for treating Yemen's consent as an American fabrication, however. Dinstein also notes that a separate issue arises to the extent that consent is coerced, and he specifically notes that there may be a sense of coercion in the scenario in which the intervening state is attempting to suppress terrorism and makes clear that it will intervene in any event on self-defense grounds if consent is not forthcoming; he does not claim, however, that this scenario would actually amount to coercion to the point of invalidating the consent. *See id.*

59. *See* Murphy, *supra* note 43, at 118–20 for a thorough discussion of this question in the context of Pakistan.

60. *See* Alston, *supra* note 36, at 27 ("If a State commits a targeted killing in the territory of another State, the second State should publicly indicate whether it gave consent, and on what basis.").

61. *See* Murphy, *supra* note 43, at 118–19 (discussing the options for and difficulty of proving private or implicit consent in the context of Pakistan); Mary O'Connell, *Unlawful Killing with Combat Drones: A Case Study of Pakistan, 2004–2009*, *in* SHOOTING TO KILL: THE LAW GOVERNING LETHAL FORCE IN CONTEXT 18 (Simon Bronitt ed., 2010) (Notre Dame Legal Studies Paper No. 09-43), *available at* http://ssrn.com/abstract=1501144 (offering a similar warning).

62. *See* Military and Paramilitary Activities, Jurisdiction and Admissibility, Judgment (Nicar. v. U.S.), 1984 ICJ 392 (Nov. 26); §§ 165–166 (questioning whether the United States had indeed received a request for assistance from El Salvador, Costa Rica, and Honduras), § 199 (concluding that customary law requires an attacked state to actually request assistance from another state where the latter intervenes on the ground of collective self-defense), §§ 232–236 (concluding that El Salvador had not formally requested assistance until some period after the U.S. intervention in Nicaragua began, and though the court conceded that "no strict legal conclusions may be drawn from the date" it nonetheless took this as evidence of whether El Salvador previously believed itself to be the subject of an armed attack from Nicaragua).

63. In the context of Pakistan, Professor O'Connell has raised an important and distinct concern as to the provenance of the host state's consent. Specifically, she has argued that whatever else might be true of consent, it must be the case that consent has been given by the proper domestic authorities rather than, say, some subordinate entity such as a military commander or security service official who may be acting contrary to the preferences of civilian authorities. *See* Posting of Mary O'Connell to IntLawGrrls Blog, http://intlawgrrls.blogspot.com/2009/12/drones-and-law-what-we-know.html (Dec. 7, 2009) ("Pakistani intelligence services or the military have apparently cooperated with the United States on strikes, but under international law, it should be the elected civilian officials who provide a state's consent for foreign military operations"). For present purposes, it suffices to note that the media accounts related above indicate that consent in the Yemeni context flows directly from Yemen's president.

64. At the time this chapter was written, a wave of popular protests against authoritarian rule was sweeping through a number of Arab states, prompting President Saleh to declare in early February 2011 that neither he nor his son would seek the presidency in the 2013 election. *See* Laura Kasinof & Nada Bakri, *Facing Unrest, Yemen's Leader Says He Will Step Down in 2013*, N.Y. TIMES, Feb. 2, 2011, at A13.

65. *See, e.g.*, Sec. Council Res. 1368, U.N. Doc. S/RES/1368 (Sept. 12, 2001) (recognizing, in connection with the 9/11 attacks, "the inherent right of individual or collective self-defence in accordance with the Charter"); Murphy, *supra* note 43, at 129; International Law and the Use

of Drones, Summary of the International Law Discussion Group Meeting Held at Chatham House 5–6 (Oct. 21, 2010) (remarks of Michael Schmitt, Professor of Public International Law, Durham University Law School), *available at* http://www.chathamhouse.org/sites/default/files/public/Research/International%20Law/il211010drones.pdf [hereinafter Chatham House]; Posting of Jordan Paust to EJIL Talk! Blog, http://www.ejiltalk.org/are-us-attacks-in-pakistan-an-armed-attack-on-pakistan-a-response-to-timothy-waters/ (Sept. 30, 2009); Derek Jinks, *State Responsibility for the Acts of Private Armed Groups*, 4 U. Chi. J. Int'l L. 83 (2003); Mary Ellen O'Connell has argued that "[t]errorist attacks are generally treated as criminal attacks and not as the kind of armed attacks that can give rise to the right of self-defense," but noting Israel's situation circa 2006 notes exceptions for circumstances in which the pace and nature of the attacks make them "more than crime" and capable of implicating Article 51. O'Connell, *supra* note 60 at 5. This raises the question whether the 9/11 attacks (or earlier al Qaeda operations), being part of a considerably more episodic pattern, would count as an armed attack on this model. *Cf. id.* at 3 n. 4 (stating that Resolution 1368 "was useful in making a finding that the 9/11 attacks could give rise to a right of self-defense").

66. The ICJ suggested as much in Legal Consequences of the Construction of a Wall in the Occupied Palestinian Territory, Advisory Opinion, 2004 I.C.J. 136, 194 (July 9) ("Article 51 of the Charter thus recognizes the existence of an inherent right of self-defence in the case of armed attack by one State against another State."). *See also* Case Concerning Armed Activities on the Territory of the Congo (Democratic Republic of Congo v. Uganda) 2005 I.C.J. 168, 222–23 (Dec. 19) (treating the question of the DRC's state responsibility as if dispositive, yet also explicitly reserving decision as to whether and when the right of self-defense extends to "irregular forces"). For a thorough debunking based on text, state practice, logic, and policy, *see* Noam Lubell, Extraterritorial Use of Force Against Non-State Actors 30–35 (Oxford Univ. Press ed., 2010). *See also* Elizabeth Wilmshurst, *The Chatham House Principles of International Law on the Use of Force in Self-Defence*, 55 ICLQ 963, 965–71 (2006) ("There is no reason to limit a state's right to protect itself to an attack by another state. The right of self-defence is a right to use force to avert an attack. The source of the attack, whether a state or a non-state actor, is irrelevant to the existence of the right.").

67. An argument might be made for borrowing the standards provided in international law for attribution to a state of a nonstate entity's actions. It is not obvious that it makes sense to transpose such a test to this context, however, and in any event the rules for attribution of state responsibility are not entirely settled themselves. Antonio Cassese, *The Nicaragua and Tadić Tests Revisited in Light of the ICJ Judgment on Genocide in Bosnia*, 18 EJIL 649, 649 (2007) (discussing the contrast between the "effective control" test set forth in *Nicaragua* and the "overall control" test set forth in *Tadić*).

68. That these attempts failed at the last minute should in no way impact their characterization.

69. *See, e.g.*, Chatham House, *supra note* 64, at 3 (remarks of Prof. Mary Ellen O'Connell) ("The ICJ has held on several occasions that the armed attack must be attributable to a state where any counterattack in self-defence occurs."), at 4 ("The ICJ held in *Congo v. Uganda* that Congo's failure or inability to take action against militants carrying out sporadic armed attacks in Uganda did not give rise to any right by Uganda to cross the border and attack the groups themselves."). For additional sources for and against the strict position, *see* Jutta Brunnee & Stephen Toope, Legitimacy and Legality in International Law 295 (Cambridge Univ. Press ed., 2010).

70. Mary O'Connell, *Lawful Self-Defense to Terrorism*, 63 U. PITT. L. REV. 889 (2002).

71. *See, e.g.*, Michael Schmitt, *Responding to Transnational Terrorism under the Jus Ad Bellum: A Normative Framework*, 56 NAVAL. L. REV. 1, 6 (2008) ("It is true that the ICJ, in the *Wall* and the *Congo* cases, appears to have rejected the notion that the right to self-defence arises against an armed attack by a nonstate actor. Yet, those decisions were highly controversial and widely criticized. Indeed, strong dissenting opinions correctly pointed out that not only was the Court ignoring post 9/11 state practice, but that there was nothing in the text of the Article 51 which would indicate that an armed attack cannot be launched by a nonstate actor."). *See also* Lubell, *supra* note 65, at 36–42; Paust, *supra* note 64; Claus Kreß, *Some Reflections on the International Legal Framework Governing Transnational Armed Conflict*, 15 JCSL 245, 248 (2010) (arguing that even prior to 9/11 self-defense extended to attacks from nonstate actors) (citing, inter alia, Claus Kreß, *Gewalverbot and Selbstverteidigungsrecht nach der Satzung der Vereinten Nationen bei staatlicher Verwicklung*, *in* GEWALTAKTE PRIVATER. DUNCKER & HUMBLOT, BERLIN (1995)); Michael Schmitt, *"Change Direction" 2006: Israeli Operations in Lebanon and the International Law of Self-Defense*, 29 U. MICH. L. REV. 127, 145–49 (2008); Dinstein 2005, *supra* note 42.

72. Wilmshurst, *supra* note 65, at 12 (emphasis added). The Chatham House statement expressly rejects a reading of *Democratic Republic of Congo v. Uganda* that would preclude the use of force against a nonstate actor on the territory of the host state absent evidence of the host state's legal responsibility for that nonstate actor. It argues that the decision instead supports no more than the conclusion that absent legal responsibility the defending state's self-defense right does not extend to the host state as such. *See id.* at n. 81. *See also* Schmitt, *Responding to Transnational Terrorism*, *supra* note 70, at 25.

73. Dinstein, *supra* note 42, at 26.

74. *See id. See also* Convention (V) respecting the Rights and Duties of Neutral Powers and Persons in Case of War on Land, The Hague, art. 5, Oct. 18, 1907. *See* Schmitt, *"Change Direction," supra* note 70, at 159–62 (citing, inter alia, Corfu Channel Case (U.K. v. Albania), 1949 I.C.J. 4 (April 9)). *See also* Schmitt, *Responding to Transnational Terrorism*, *supra* note 70, at 20–27.

75. *See* Schmitt, *Responding to Transnational Terrorism*, *supra* note 70, at 27. Note the difficult question of whether the defending state could avoid the obligation of an ultimatum or *demarche* to the host state in circumstances where the defending state suspects the host state will tip off the nonstate actor or even use the warning to enhance its own capacity to repel an attack.

76. The broad state-responsibility position, as described above, could be viewed simply as a very flexible substantive standard for demonstrating state responsibility or, instead, as an argument against requiring state responsibility in the first instance. The difference matters greatly, in that a finding of state responsibility opens the doors to actions directly targeting the state itself, whereas the point of arguing that state responsibility need not be shown is simply to explain why it is justified to attack the nonstate actor within the state's borders, no more and no less. *Cf.* Schmitt, *Responding to Transnational Terrorism*, *supra* note 70, at 27 ("It may not strike any targets of the 'host' government, nor anything else unconnected with the terrorist activity.").

77. *See* O'Connell, *supra* note 69, at 900 (arguing that a state is responsible for a nonstate actor's armed attack if: (1) "agents of that state were involved," (2) the state "sends persons to carry out the attack even if those persons are not the state's officials or agents," and (3) the state "has developed sufficiently close links with the group even if it does not control them" (citing the example of "organizing, coordinating or planning the military actions of the military group, in

addition to financing, training and equipping or providing operational support to that group") (quoting Prosecutor v. Tadić Case No. IT-94-1-T, Opinion and Judgment, ¶ 137 (May 7, 1997)). *Cf.* Derek Jinks, *Self-Defense in an Age of Terrorism*, 97 ASIL 141, 144–46 (2003) (observing that the U.S. government chose to justify its actions against the Taliban in Afghanistan on grounds of state responsibility resting on a "harboring"/"supporting" theory rather than the "overall control" standard of the ICTY in *Tadić* or the "effective control" standard of the ICJ in *Nicaragua*, and raising objections to this approach).

78. *See* O'Connell, *supra* note 69, at 900–901 (citing the examples of Israeli actions against Hezbollah in Lebanese territory and Turkish and Iranian action against Kurdish entities in the Kurdish regions of Iraq during the interwar period).

79. *See id.* at 901.

80. *Cf.* Schmitt, *Responding to Transnational Terrorism*, *supra* note 70, at 1. *See also* Matthew C. Waxman, *The Use of Force against States that Might Have Weapons of Mass Destruction*, 31 MICH. J. INT'L L. 1, 57–77 (2009) (discussing the absence of evidentiary legal standards with respect to use of force, including burdens of proof and their allocation).

81. *Cf.* Jinks, *supra* note 76, at 145–46 (discussing, and critiquing, the arguable post–9/11 shift away from "control" to "harboring" or "supporting" as a standard for state responsibility). *See also* Christian Henderson, *The 2010 United States National Security Strategy and the Obama Doctrine of "Necessary Force,"* 15 JCSL 403 (2010) (contending that the Obama administration has carried forward the harboring standard).

82. O'Connell, *supra* note 69, at 895. O'Connell argues for adoption of a clear-and-convincing evidence standard in this context, as a matter of both law and policy. *See id.* at 895–99.

83. *See, e.g.*, Oil Platforms (Iran v. U.S.) 2003 ICJ 161, 198 (Nov. 6); Legality of the Threat or Use of Nuclear Weapons, Advisory Opinion, 1996 ICJ 226, 245 (July 8); Military and Paramilitary Activities (Nicar. v. U.S.), 1986 I.C.J. 14, 94 (June 27). *See also* Alston, *supra* note 36, at 14; Murphy, *supra* note 43, at 127; Rise of the Drones II, *supra* note 44, at 7–8 (statement of Prof. Kenneth Anderson). Note that I do not discuss arguments for an additional requirement of imminence where there has not yet been an actual armed attack and the "defending" state is acting in an anticipatory mode. That is an important issue, but not one presented here given the attacks that already have been directed at the United States.

84. Murphy, *supra* note 43, at 127 (citing Dinstein, *supra* note 42, at 237).

85. *See, e.g.*, Miller, Jaffe, & DeYoung, *supra* note 53 (discussing conclusion in U.S. government report to the effect that "Pakistan still has not 'fundamentally changed its strategic calculus' regarding insurgent sanctuaries on its territory" and noting that "Pakistan has long resisted U.S. urging to launch all-out attacks against Taliban and al-Qaeda redoubts in the [FATA]").

86. See Murphy, *supra* note 43, at 127 (citing JUDITH GARDAM, NECESSITY, PROPORTIONALITY AND THE USE OF FORCE BY STATES 4–8 (2004)).

87. *See id.* at 129 ("'Proportionality' does not require that the force be a mirror image of the initial attack, or that defensive actions be restricted to the particular geographic location in which the initial attack occurred.").

88. *See* Robert Sloane, *The Cost of Conflation: Preserving the Dualism of Jus Ad Bellum and Jus in Bello in the Contemporary Law of War*, 34 YALE J. INT'L L. 47, 52 (2009) ("ad bellum proportionality asks whether the initial resort to force or particular quantum of force used is proportional to the asserted casus belli").

89. *Id.* at 68.

90. *Id.* (citing ROSALYN HIGGINS, PROBLEMS AND PROCESS: INTERNATIONAL LAW AND HOW WE USE IT (Oxford Univ. Press ed., 1994)). *See also* Murphy, *supra* note 43, at 128. *But see* Kenneth Anderson, *Targeted Killing in U.S. Counterterrorism Strategy and Law*, *in* LEGISLATING THE WAR ON TERROR: AN AGENDA FOR REFORM 5 (2009) (referring to the "customary law standards of necessity and proportionality" in terms of "necessity in determining whom to target, and proportionality in considering collateral damage," adding that "standards in those cases should essentially conform to military standards under the law of war, and in some cases the standard should be still higher").

91. *See* Hamza Hendawi & Ahmed al-Haj, *Yemen Charges US-Born Radical Cleric al-Awlaki*, WASH. POST, Nov. 2, 2010, *available at* http://www.washingtonpost.com/wp-dyn/content/article/2010/11/02/AR2010110200458.html (indicating that "some analysts" believe that Yemen is giving only a "half-hearted effort" to capture al-Awlaki). *Cf. Yemen Sentences Awlaki in Absentia*, AL JAZEERA, Jan. 17, 2011, http://english.aljazeera.net/news/middleeast/2011/01/2011117133558339969.html (noting that Yemen has prosecuted al-Awlaki in absentia, sentencing him to ten years' imprisonment).

92. *See, e.g.*, *Yemeni Forces Kill Suspected al-Qaeda Militant*, CBC NEWS, Jan. 13, 2010, http://www.cbc.ca/news/world/story/2010/01/13/yemen-al-qaeda.html ("The San'a government has little control over Shabwa and large swaths of Yemen.... Powerful, well-armed tribes dominate extensive areas and bitterly resent intrusion by security forces.").

93. *See, e.g.*, Lolita Baldor, *US Officials: Military Wants to Boost Aid to Yemen*, WASH. POST, Nov. 4, 2010 (discussing military and other aid to Yemen and the "need to bolster that country's ability to track and battle militants").

94. *Cf.* Matthew C. Waxman, *Detention as Targeting: Standards of Certainty and Detention of Suspected Terrorists*, 108 COLUM. L. REV. 1365, 1387 n. 76 (2008) (citing *Final Report to the Prosecutor by the Committee Established to Review the NATO Bombing Campaign against the Federal Republic of Yugoslavia*, § 48, PR/P.I.S./510-E (June 13, 2000)).

CHAPTER 10

* Research assistant professor, Institute for National Security and Counterterrorism, Syracuse University.

1. U.S. Dep't of the Army, *Counterinsurgency*, Field Manual (FM) 3-24/Marine Corps Warfighting Publication (MCWP) 3-33.5, Washington, D.C.: Department of the Army, ix (Dec. 15, 2006) [hereinafter FM 3-24].

2. DECLARATION RESPECTING MARITIME LAW, Apr. 16, 1856, reprinted in 1 AJIL Supp. 89 (1907) [hereinafter Paris Declaration].

3. *See* U.N. Security Council, *Security Council Condemns Acts Resulting in Civilian Deaths during Israeli Operation against Gaza-Bound Aid Convoy, Calls for Investigation in Presidential Statement, Also Briefed by United Nations Political Official Who Says Bloodshed Would Have Been Avoided if Israel Had Lifted Unacceptable, Counterproductive Blockade*, SC/9940 (6325th & 6326th Meetings).

4. Turkey was the first Muslim-majority state to recognize Israel on March 28, 1949.

5. Charles Levinson, *Israel's Foes Embrace New Resistance Tactics*, WALL ST. J., July 2. 2010.

6. *Id.*

7. *Id.*

8. Joint doctrine defines insurgency as "an organized, protracted politico-military struggle designed to weaken the control and legitimacy of an established government" and counterinsurgency as the "military, paramilitary, political, economic, psychological, and civic actions taken by a government to defeat insurgency." *See* Joint Publication 3-24, *Counterinsurgency Operations* (Oct. 5, 2009); *see also Department of Defense Dictionary of Military and Associated Terms*, JP 1-02 (Dec. 4, 2001).

9. DEP'T OF DEFENSE, DICTIONARY OF MILITARY AND ASSOCIATED TERMS, JP 1-02 (1994).

10. FM 3-24, *supra* note 1, ¶ 1-2.

11. *See* DEP'T OF DEFENSE, *Quadrennial Defense Review Report*, 32, 47 (February 2010) (noting the use of the concept for air and naval forces to "integrate capabilities across all operational domains," guide "future capabilities needed" to deter and defeat aggression in anti-access environments, and a key element to "rebalance the force"); Bob Poling, *Air-Sea Battle: The Looming Challenges Facing Concept Development*.

12. Sabrina Tavernise & Michael Slackman, *Turkish Funds Helped Group Test Blockade*, N.Y. TIMES, June 1, 2010, at A1.

13. Statements are attributed to Prime Minister Recep Tayyip Erdogan of Turkey, from Isabel Kershner & Neil Macfarquhar, *Pressure Mounts on Israel as Activists Vow to Test Blockade Again*, N.Y. TIMES, June 1, 2010, at A10; French President Nicolas Sarkozy, from Selcan Hacaoglu, *Israeli Commandos Storm Aid Flotilla; 10 Killed*, ASSOCIATED PRESS, May 31, 2010; German Chancellor Angela Merkel called for end of Israel's Gaza blockade and British Prime Minister David Cameron "urged Israel to listen to its critics," from Robert Marquand, *Condemnation across Europe after Israel Raid on Freedom Flotilla*, CHRISTIAN SCIENCE MONITOR, June 1, 2010.

14. Neil MacFarquhar & Alan Cowell, *U.N. Security Council Condemns "Acts" in Israeli Raid*, N.Y. TIMES, June 1, 2010.

15. U.N. Secretary-General Ban Ki-moon said, "I condemn the violence, and Israel must explain," *World Reaction to Israel after Raid on Aid Flotilla Mostly Negative*, CNN, May 31, 2010 [hereinafter World Reaction].

16. *Id.*

17. *LIVE: Israeli Raid on Gaza Flotilla*, BBC NEWS, May 31, 2010; Final Communique of the OIC Expanded Extraordinary Executive Committee Meeting at the Level of Foreign Ministers on the Israeli Aggression against the Freedom Flotilla Shipping Humanitarian Aid to Gaza (Jeddah, June 6, 2010).

18. World Reaction, *supra* note 15.

19. *Id.* Note that by 2010, Israel had been condemned in thirty-two Human Rights Council resolutions—more than any other state for a total of 48 percent of all resolutions—since the Council's inception (2006), and the Council instituted (June 30, 2006) a review of alleged human rights abuses by Israel as a permanent feature of every Council session. *See* Luisa Blanchfield, CRS Report for Congress: *The United Nations Human Rights Council: Issues for Congress*, RL33608 (Nov. 15, 2011), Appendix A & B for human rights abusers as voting members (i.e., Libya, China, Russia, Cuba, Saudi Arabia, Congo, Mauritania, Uganda, etc.,) and disproportionate special sessions devoted to Israel (2006–2011).

20. World Reaction, *supra* note 15.

21. *Id.*

22. According to their website, the "Free Gaza Movement is a human rights group that in August 2008 sent the first international boats to land in the port of Gaza in 41 years. We want to break the siege of Gaza. We want to raise international awareness about the prison-like closure of

the Gaza Strip and pressure the international community to review its sanctions policy and end its support for continued Israeli occupation ... The Free Gaza Movement and our international parteners (*sic*) intend to launch another flotilla at the first available opportunity. After that there will be yet another ... and another ... until this diabolical blockade collapses and steps towards peace are finally taken," *Freegaza—News Section*, The Free Gaza Movement, http://www.freegaza.org/ (last accessed Apr. 16, 2012). *See also* Maev Kennedy, *Free Gaza Movement's Two-year Campaign to Break Blockade*, Guardian, May 31, 2010 (noting "[s]ince 2008, the Free Gaza Movement and a loose coalition of human rights activists and pro-Palestinian groups have been sending boats and landing or attempting to land supplies to break the Gaza blockade, including medical equipment and drugs and building materials").

23. *China Urges "Quick" Security Council Response to Israeli Attack on Aid Convoy*, Xinhuanet, June 1, 2010.

24. Ian Traynor & Dorian Jones, *Gaza Flotilla Attack: World Unites in Condemnation of Israel's Actions*, Guardian, June 1, 2010.

25. Afua Hirsh, *Was the Gaza Flotilla Raid Legal?*, Guardian, June 1, 2010.

26. *OEA Censura Ataque de Israel A Flotilla Humanitaria*, RPP Noticias, May 31, 2010; *Israeli Killings of Gaza Ship Activists Must Be Investigated*, Amnesty International, May 31, 2010.

27. Dan Izenberg, *Israeli Lawyers: Raid Violates International Law, Prosecution Charges that Israel Engaged in Acts of Piracy*, Jerusalem Post, May 31, 2010; Aviad Glickman, *High Court Rejects Flotilla Suits: Soldiers Defended Their Lives*, Ynet, June 3, 2010.

28. George Bisharat, *Gaza Occupation and Siege Are Illegal*, San Francisco Chronicle, June 4, 2010, at A18.

29. Bob Egelko, *Israel's Gaza Blockade Legal, Many Scholars Say*, San Francisco Chronicle, June 5, 2010, at A4.

30. Traynor & Jones, *supra* note 24.

31. On Sept. 12, 2005, the last IDF forces left Gaza, the IDF Commander in Gaza signed the declaration terminating the military administration operating in the territory since the Second or Al Aqsa Intifada (October 2000), and on Sept. 20, 2005, the Minister of Interior established five crossings or "border stations," pursuant to section 7 of the Entry into Israel Law, 5712-1952. *See* Law Implementing Disengagement Plan, 5765-2005, LB 1982 and *Manifest Regarding Termination of Military Rule* (manifest no. 6) (Gaza Region) 5765-2005.

32. *See* Geneva Convention (III) Relative to the Treatment of Prisoners of War art. 2, Aug. 12, 1949, 6 U.S.T. 3316, 75 U.N.T.S. 135 [hereinafter Common Article 2] "... the present Convention shall apply to all cases of declared war or of any other armed conflict which may arise between two or more of the High Contracting Parties, even if the state of war is not recognized by one of them ... The Convention shall also apply to all cases of partial or total occupation of the territory of a High Contracting Party, even if the said occupation meets with no armed resistance."

33. *Israeli Blockade of Gaza at Heart of Crisis*, Associated Press, June 2, 2010.

34. Daniel Henninger, *Beating Up on Israel*, Wall St. J., June 3, 2010.

35. *See* China, *supra* note 23.

36. Traynor & Jones, *supra* note 24.

37. U.N. General Assembly, *Report of the international fact-finding mission to investigate violations of international law, including international humanitarian and human rights law, resulting from the Israeli attacks on the flotilla of ships carrying humanitarian assistance* (chaired

by Judge Hudson Phillips), A/HRC/15/21 [hereinafter *UNHRC Flotilla Report*]; Statement by the representative of the United States of American to the Human Rights Council, Geneva, 9 (June 1, 2010): "we continue to believe the situation in Gaza is unsustainable and is not in the interest of any of those concerned."

38. Statements are attributed respectively to Prime Minister Recep Tayyip Erdogan of Turkey, Kershner & Macfarquhar, *supra* note 13.

39. Shmuel Rosner, *What's to Investigate? We Know What Happened to the Gaza Flotilla. The Debate about the Blockade Is for Another Time*, SLATE, June 1, 2010.

40. Christopher Greenwood, *Scope of Application of Humanitarian Law*, *in* THE HANDBOOK OF INTERNATIONAL HUMANITARIAN LAW 37 (Dieter Fleck ed., 2d ed. 2008).

41. *See* Theodor Meron, *The Humanization of Humanitarian Law*, 94 AM. J. INT'L L. 239, 243–44 (2000) ("The change in direction toward intrastate or mixed conflicts...has drawn humanitarian law in the direction of human rights law"); Dan E. Stigall, Christopher L. Blakesley, & Chris Jenks, *Human Rights and Military Decisions: Counterinsurgency and Trends in the Law of International Armed Conflict*, 30 U. PA. J. INT'L L. 1367 (2008–2009).

42. *See* Federico Sperotto, *Counter-Insurgency, Human Rights, and the Law of Armed Conflict*, 17 HUM. RTS. BRIEF 19 (2009); U.S. ARMY, FIELD MANUAL FM-3.0 OPERATIONS, ¶¶ 2-5, 2-45, 2-46, 2-44 (2008) (defining irregular warfare as "a violent struggle among state and non-state actors for legitimacy and influence over a population" with insurgency, counterinsurgency, and unconventional warfare as "the principal activities").

43. Dan Izenberg, *State: Gaza Blockade Legal, Raid Justified*, JERUSALEM POST, June 2, 2010; Rosner, *supra* note 39; SAN REMO MANUAL ON INTERNATIONAL LAW APPLICABLE TO ARMED CONFLICTS AT SEA 5 (Louise Doswald-Beck ed., 1995) [hereinafter SAN REMO MANUAL].

44. Judy Woodruff, *As Flotilla Inquiry Calls Grow Louder, Legality of Gaza Blockade Examined*, PBS NEWSHOUR, June 2, 2010.

45. SAN REMO MANUAL, *supra* note 43. Article 102 notes, however, that "[t]he declaration or establishment of a blockade is prohibited if: (a) it has the sole purpose of starving the civilian population or denying it other objects essential for its survival; or (b) the damage to the civilian population is, or may be expected to be, excessive in relation to the concrete and direct military advantage anticipated from the blockade." Article 103 further clarifies: "If the civilian population of the blockaded territory is inadequately provided with food and other objects essential for its survival, the blockading party must provide for free passage of such foodstuffs and other essential supplies, subject to: (a) the right to prescribe the technical arrangements, including search, under which such passage is permitted; and (b) the condition that the distribution of such supplies shall be made under the local supervision of a Protecting Power or a humanitarian organization which offers guarantees of impartiality, such as the International Committee of the Red Cross."

46. Yehuda Ben Meir, *A Just Decision Badly Executed*, HAARETZ, June 3, 2010.

47. SAN REMO MANUAL, *supra* note 43, art. 95: "A blockade must be effective. The question whether a blockade is effective is a question of fact."

48. Colum Lunch, *Israel's Flotilla Raid Revives Questions of International Law*, WASH. POST, June 1, 2010.

49. *Id.*

50. Kennedy, *supra* note, 22.

51. *Editorial: Israel and the Blockade*, N.Y. TIMES, June 1, 2010, at A24.

52. Scott Wilson, *Israel Says Free Gaza Movement Poses Threat to Jewish State*, WASH. POST, June 1, 2010.

53. Richard Spencer, *Gaza Flotilla: The Free Gaza Movement and the IHH*, TELEGRAPH, May 31, 2010; Alexander Christie-Miller & James Hider, *Turkish Charity that Sent Aid Convoy to Gaza "has Links to Terrorism,"* TIMES, June 3, 2010 (noting "Jean-Louis Bruguière, a French investigating magistrate and an authority on counter-terrorism, said that in the mid-1990s the group's leader, Bulent Yildirim, made efforts to recruit veteran soldiers in anticipation of the coming holy war. Mr Bruguière testified at the U.S. trial of Ahmed Ressam... that the IHH had played an important role in a failed plot to bomb Los Angeles airport"). *See also* Tavernise & Slackman, *supra* note 12.

54. Matthew Levitt, David Makovsky, & Jeffrey White, *PolicyWatch #1662: Fallout from the Gaza Flotilla Tragedy*, June 1, 2010.

55. *Haniyeh: Gaza Flotilla a Triumph*, JERUSALEM POST, May 29, 2010; Jack Khoury & Yuval Azoulay, *Hamas: Flotilla Shows Whole World Opposes Gaza Siege*, HAARETZ, June 28, 2010.

56. Khoury & Azoulay, *supra* note 55.

57. Anthony Cordesman, *Israel as a Strategic Liability?*, CSIS, June 2, 2010.

58. Mark Landler, *After Flotilla Raid, U.S. Is Torn between Allies*, N.Y. TIMES, June 2, 2010, at A11; Peter Speigel & Joe Lauria, *U.S. Mutes Criticism of Israeli Raid*, WALL ST. J., June 2, 2010.

59. Helene Cooper, *Washington Asks: What to Do about Israel?*, N.Y. TIMES, June 5, 2010.

60. Eyal Benvenisti, *Responsibility for the Protection of Human Rights under the Interim Israeli Palestinians Agreements*, 28 IS. L. REV. 297, 312 (1994) (noting that "[a]fter the 1967 Six-Days War Gaza was administered by the IDF under the laws of belligerent occupation derived from the 1907 Hague Regulations—though Israel contested the formal application of the Fourth Geneva Convention to the occupation of Gaza on the basis of Egypt's lack of sovereignty over the area" and made similar arguments over the legal status of the West Bank captured from Jordan in the 1967 war); Thomas Friedman, *Letter from Istanbul*, N.Y. TIMES, June 15, 2010, at A31 (noting "There is nothing wrong with criticizing Israel's human rights abuses in the territories"... But it is very troubling when Erdogan decries Israelis as killers and, at the same time, warmly receives in Ankara Sudan's president, Omar Hassan al-Bashir, who has been indicted by the International Criminal Court on charges of war crimes and crimes against humanity for his role in the bloodshed in Darfur... Erdogan defended his reception of Bashir by saying: "It's not possible for a Muslim to commit genocide. As one Turkish foreign policy analyst said to me: '"We are not mediating between East and West anymore. We've become spokesmen for the most regressive elements in the East.'").

61. SARI NUSSEIBEH, ONCE UPON A COUNTRY: A PALESTINIAN LIFE 12 (2007) ("Isn't this inability to imagine the lives of the 'other' at the heart of the Israeli-Palestinian conflict?"); JACQUES DERRIDA, ADIEU 51 (Pascale-Anne Brault & Micheal Naas, trans., 1999) ("Is not hospitality an interruption of the self?").

62. *See UNHRC Flotilla Report*, *supra* note 37; Geoffrey Palmer; Alvaro Uribe, J. C. Itzhar, & S. O. Sanberk, *Report of the Secretary-General's Panel of Inquiry on the May 31, 2010 Flotilla Incident* 54 (Sept. 2, 2011) [hereinafter *Palmer Report*] (this report also found "Israel's decision to board the vessels with such substantial force at a great distance from the blockade zone and with no final warning immediately prior to the boarding" to be "excessive and unreasonable," and determined IDF personnel "faced significant, organized and violent resistance from a group

of passengers when they boarded the Mavi Marmara requiring them to use force for their own protection").

63. *Palmer Report*, *supra* note 62, at 4.

64. *See* Justice Emeritus Jacob Türkel et al., *The Public Commission to Examine the Maritime Incident of May 31, 2010* (Jan. 23, 2011) [hereinafter *Turkel Report*]; Turkish National Commission of Inquiry, *Report on the Israeli Attack on the Humanitarian Aid Convoy to Gaza* (February 2011) [hereinafter *Turkish Report*].

65. *Palmer Report*, *supra* note 62, at 9.

66. Gaza must be occupied to claim that Israeli sanctions, including the naval blockade, may constitute collective punishment. *See* Geneva Convention Relative to the Protection of Civilian Persons in Time of War, August 12, 1949, art. 33, 75 U.N.T.S. 287 [hereinafter GC IV] ("Collective penalties and likewise all measures of intimidation or of terrorism are prohibited."). *See also* YORAM DINSTEIN, THE INTERNATIONAL LAW OF BELLIGERENT OCCUPATION (2009).

67. Gabor Rona, *An Appraisal of U.S. Practice Relating to "Enemy Combatants,"* 10 Y.B. OF INT'L HUM. LAW 232, 236 (2009) (noting that while IHL may overlap with human rights law, still "many rules of otherwise applicable human rights law are displaced by conflicting rules of the laws of armed conflict," for example, human rights law prohibits arbitrary detention and presumes the right of habeas corpus to implement such a prohibition in the International Covenant on Civil and Political Rights, art. 9(4) and yet "in international armed conflict... detainees subject to Geneva Convention protections have no right to habeas corpus"). *See also* Kenneth Watkin, *Controlling the Use of Force: A Role for Human Rights Norms in Contemporary Armed Conflict*, 98 AM. J. INT'L L. 1, 22 (2004) (noting that instead of extending "human rights norms to an armed conflict scenario, the appropriate approach is to apply the *lex specialis* of humanitarian law"); Legality of the Threat or Use of Nuclear Weapons, Advisory Opinion, 1996 ICJ REP. 226 (July 8) (noting whether there has been an arbitrary deprivation of the right to life is determined by IHL acting as *lex specialis*).

68. *Palmer Report*, *supra* note 62, at 10 (noting "[w]e observe that the legal views of Israel and Turkey are no more authoritative or definitive than our own... The findings of a Commission of Inquiry bind no one, unlike those of a court. So the legal issues at large in this matter have not been authoritatively determined by the two States involved and neither can they be by the Panel").

69. *See* U.N. Charter, prmbl. If public international law comprises the rules that bind relations between states, the United Nations is that intergovernmental organization instituted to manage those rules and relations via international law to achieve core aims, that is, international peace and security, economic development, social progress, human rights. The U.N. Charter makes these aims explicit: to "save succeeding generations from the scourge of war" and promote "faith in fundamental human rights," but also to "establish conditions under which justice and respect for the obligations arising from treaties and other sources of international law can be maintained." *See also* U.N., Factsheet #5: 2011 Treaty Event, *Towards Universal Participation and Implementation, Role of the United Nations in International Law* ("[e]ncouraging the development of international law as a way to regulate international relations has been a major objective of the United Nations since its very beginning").

70. Jean-Marie Henckaerts & Louise Doswald-Beck (eds.), CUSTOMARY INTERNATIONAL HUMANITARIAN LAW, V I: RULES (2005) (noting it did not distinguish two categories of armed conflict because it was found that states did not make such a distinction in practice).

71. *See* ICRC, *How Is the Term "Armed Conflict" Defined in International Humanitarian Law?*, Opinion Paper, March 2008; Hilaire McCoubrey & Nigel D. White, International Law and Armed Conflict 318 (1992) (noting "the 'grey area' of categorization remains an issue of serious potential concern"); James G. Stewart, *Towards a Single Definition of Armed Conflict in International Humanitarian Law: A Critique of Internationalized Armed Conflict*, 85 Int'l Rev. of the Red Cross 313 (2003); *but see* Lindsay Moir, The Law of Internal Armed Conflict 32 (2002).

72. Gabor Rona, *Interesting Times for International Humanitarian Law: Challenges from the "War on Terror,"* 27 The Fletcher Forum of World Affairs 55, 58 (2003) (noting rules for international armed conflict in the four Geneva Conventions of 1949 (GC I-IV) and their Additional Protocol I (AP I) of 1977, whose scope of application is found in Common Article 2; rules for noninternational armed conflict are found in Common Article 3 and Additional Protocol II (AP II), with their scope of application in both Common Article 3 and in Article 1 of AP II).

73. Gabor Rona, *supra* note 67, at 237 (noting a conflict remains noninternational "even when hostilities spill over or beyond the boundaries of a single state as long as there is no use of force between two or more states," thus making a transnational armed conflict "noninternational so long as no two states are using armed force against each other"). *See also* International Committee of the Red Cross, The Geneva Conventions of August 12, 1949: Commentary 178 (Jean S. Pictet ed., 1958) at 23 [hereinafter ICRC Commentary] (discussing the de facto "armed conflict" standard in Common Article 2). For states party to Additional Protocol I (AP I), Article 1(4) expands the scope of international armed conflict to "include armed conflicts which peoples are fighting against colonial domination and alien occupation and against racist regimes in the exercise of their right of self-determination."

74. *See* Geneva Convention I-IV, *supra* note 72; Common Article 2, *supra* note 32.

75. *See* Prosecutor v. Tadić, Case No. IT–94–1, Opinion and Judgment, ¶ 562 (Int'l Crim. Trib. for the Former Yugoslavia May 7, 1997) (noting that "[t]he test applied by the Appeals Chamber" for "the purposes of the rules contained in Common Article 3 focuses on two aspects of a conflict: the intensity of the conflict and the organization of the parties to the conflict. In an armed conflict of an internal or mixed character, these closely related criteria are used solely for the purpose, as a minimum, of distinguishing an armed conflict from banditry, unorganized and short—lived insurrections, or terrorist activities, which are not subject to international humanitarian law"). One rationale for reduced protections in Common Article 3 conflicts was that states were more likely to use their armed forces against internal threats (even if these did not rise to armed hostilities) because such threats invariably prompted crises of sovereign governance.

76. *See* Watkin, *supra* note 67. *See also* Derek Jinks, *The Applicability of the Geneva Conventions to the "Global War on Terrorism,"* 46 Va. J. Int'l L. 165, 183 (2006) (noting the "purposeful ambiguity" of the drafters in leaving Common Article 3 armed conflict undefined and asserting noninternational conflicts as more challenging because they cut into state sovereignty in ways that international conflict does).

77. Rona, *supra* note 67, at 236.

78. In fact, the examples that have often provoked the most IHL controversy have often been staples of counterinsurgency or counterterrorist strategy, such as detention practices for unprivileged belligerents in Afghanistan and Iraq campaigns, targeted killing by drone strikes, even the "global war on terror" itself.

79. *See* Geoffrey S. Corn, *Multi-National Operations, Unity of Effort, and the Law of Armed Conflict*, Harvard University Program on Humanitarian Policy and Conflict Research (HPCR) *Working Paper Series: International Humanitarian Law and Contemporary Conflicts* (2009) 1, 3; *Commentary*, Convention (1) for the Amelioration of the Condition of the Wounded and Sick in Armed Forces in the Field. Geneva, August 12, 1949, 19–23 (Jean S. Pictet ed., 1960).

80. Jinks, *supra* note 76, at 168.

81. *See* Jelena Pejic, *The Protective Scope of Common Article 3: More than Meets the Eye*, 93 INT'L REV. OF THE RED CROSS 189 (2011). But, the view that noninternational conflict is broad and inclusive of all armed conflicts that do not fit the Common Article 2 definition has been refuted by powerful voices—notably, the U.S. Supreme Court's *Hamdan* opinion—and the history of Common Article 3 itself suggests the term was designed to respond to intrastate armed conflicts. *See* Hamdan v. Rumsfeld, 548 U.S. 557, 629–31 (2006).

82. Prosecutor v. Tadić, (Case No. IT-94-1-AR72), Second Amended Indictment, December 14, 1995, Jurisdiction Decision, ¶ 70 (noting "an armed conflict exists whenever there is a resort to armed force between States or protracted armed violence between governmental authorities and organized armed groups or between such groups within a State"). *See also* Corn, *supra* note 79, at 6 (noting the act of deploying regular troops in sustained operations against internal dissident groups makes it difficult for a "state to credibly disavow the existence of armed conflict").

83. Jinks, *supra* note 76, at 191 (noting that though paradigmatic civil war was on drafters' minds, the legislative history does not support this exclusive view of noninternational conflict, as proposals limiting the application of Common Article 3 to "civil wars" were rejected). *See* II.B FINAL RECORD OF THE DIPLOMATIC CONFERENCE OF GENEVA OF 1949, at 121. This would also create a regulatory gap in the Conventions, covering both international armed conflicts and internal civil wars, but not armed conflicts between a state and foreign (or transnational) armed group or an internal armed conflict that crosses into the territory of another state.

84. *See* Stewart, *supra* note 71 (calling for a single law of armed conflict given the current regime's failure to deal with hybrid conflicts with both international and noninternational dimensions or internationalized armed conflicts). *See, e.g.*, U.S. Dep't of Defense, Dir. 2310.01E, Law of War Program (Feb. 22, 2011) (establishing policy to observe the laws of war in all conflicts and the "principles and spirit of" this law in all other operations).

85. AP II, *supra* note 72, Article 1 restricts the scope of noninternational armed conflict to conflicts of a certain degree of intensity. *See* INTERNATIONAL COMMITTEE OF THE RED CROSS, COMMENTARY ON THE ADDITIONAL PROTOCOLS OF JUNE 8, 1977 TO THE GENEVA CONVENTIONS OF AUGUST 12, 1949, 1348 (Yves Sandoz et al. eds., 1987).

86. *See* MARCO SASSÒLI, ANTOINE BOUVIER, & ANNE QUINTIN, HOW DOES LAW PROTECT IN WAR 347 (3rd ed., ICRC, Geneva, 2011) ("[I]f IHL did not respect the principle of the equality of belligerents before it in non-international armed conflicts, it would have an even smaller chance of being respected").

87. J. Ashley Roach, *The Legal Aspects of Submarine Warfare*, 6 MAX PLANCK YEARBOOK OF UNITED NATIONS LAW 367, 68 (2002) (noting that the Protocols of 1977 Additional "generally ignored the subject").

88. 25th Int'l Conference of the Red Cross, Geneva, October 23–31, 1986, Resolution 7(A)1, *Work on international humanitarian law in armed conflicts at sea and on land*, ICRC, no. 256. The Second Geneva Convention of 1949 deals only with the protection of the wounded, sick and

shipwrecked at sea, with some adjustments in Additional Protocol I of 1977, in particular the extension to shipwrecked civilians of Geneva Convention II protections.

89. J. Ashley Roach, *The Law of Naval Warfare at the Turn of Two Centuries*, 94 AJIL 64, 65 (2000).

90. *Id.*

91. *See* Paris Declaration, *supra* note 2; Helsinki Principles on the Law of Maritime Neutrality, 5.1.2(3) and 5.2.10, adopted by the International Law Association, Taipei, Taiwan, May 30, 1998, Schindler and Toman 2004, Doc. 115, 1425–1430 [hereinafter Helsinki Principles].

92. British, German, Australian, and Canadian governments have also issued similar manuals. *See* Office of the Chief of Naval Operations, U.S. The Commander's Handbook on the Law of Naval Operations (Naval Warfare Pub. NWP 1-14M) (July 2007).

93. Roach, *supra* note 89, at 69.

94. Louise Doswald-Beck, *Background: Development of the San Remo Manual and Its Intended Purpose*, Int'l Rev. of the Red Cross, no. 309, sec. 4, 1995.

95. *Id.*

96. Roach, *supra* note 90, at 69; *see also* The Helsinki Principles (1998).

97. Louise Doswald-Beck, *The San Remo Manual on International Law Applicable to Armed Conflicts at Sea*, 89 AJIL 192 (1995). *See also* Resolution VII of the 25th International Conference of the Red Cross (noting "[i]n view of the extent of uncertainty in the law," experts decided "it was premature to embark on diplomatic negotiations to draft a treaty on the subject." Instead, they "concentrated on finding areas of agreement as to the present content of customary law, far more numerous than initially appeared possible," and as "a second step, discussed controversial issues with a view to reaching an agreed compromise on innovative proposals by way of progressive development." The experts "believed that the drafting of such a document would help clarify the law, thus removing the impression that there was such a degree of disagreement as to render its uniform development in customary law or eventual codification impossible and would encourage the drafting of more national manuals." The Explanation "would also be an indication of which provisions were generally accepted as being declaratory of customary law and which were in the nature of proposals for the progressive development of the law").

98. Unlike AP I, the *San Remo Manual* provides examples of activities that would cause vessels to become military objectives to provide "concrete guidance" (i.e., acting as an auxiliary to an enemy's armed forces by carrying troops or refusing an order to stop or actively resisting visit, search, or capture).

99. *See* Doswald-Beck, *supra* note 94 (noting "it was not always so easy to combine military necessities whilst respecting as far as possible the provisions of that Convention. Controversial areas were rules relating to the protection of the environment, freedom of navigation, and [economic exclusion zones]...Despite this, it may be seen as an important achievement that the Manual specifies that if such zones are created international humanitarian law must be nevertheless respected in its entirety." Moreover, in this way the "should help consolidate contemporary international customary law, promote its coherent development and thereby provide a much firmer foundation for possible future treaty developments than could otherwise have been the case.").

100. *UNHRC Flotilla Report*, *supra* note 37, at 12.

101. *Id.*

102. *See* Chief of Naval Operations, *U.S. Navy Vision for Confronting Irregular Challenges* (January 2010); Molly Dunigan et al., *Characterizing and Exploring the Implications of Maritime Irregular Warfare* (RAND, 2012).

103. Doswald-Beck, *supra* note 94.

104. *See* Tim Benbow, *Maritime Forces and Counter-Insurgency*, 28 CONTEMPORARY SECURITY POLICY 80, 80 (2007); C. E. CALLWELL, SMALL WARS: THEIR PRINCIPLES & PRACTICE (1996).

105. *Id.*

106. *Id.* Defense analysts have recently recommended naval bases as an alternative to controversial bases in conflict zones in the Middle East.

107. *Id.*

108. DEP'T OF DEFENSE, *Sustaining U.S. Global Leadership: Priorities for 21st Century Defense* (Jan. 3, 2012).

Index

J

K

L

www.ingramcontent.com/pod-product-compliance
Ingram Content Group UK Ltd.
Pitfield, Milton Keynes, MK11 3LW, UK
UKHW011024180926
475726UK00004B/68

9 780199 941445